PHASE PORTRAITS OF PLANAR QUADRATIC SYSTEMS

Mathematics and Its Applications

Managing Editor:

M. HAZEWINKEL
Centre for Mathematics and Computer Science, Amsterdam, The Netherlands

Volume 583

PHASE PORTRAITS OF PLANAR QUADRATIC SYSTEMS

By

JOHN REYN
Delft University of Technology, The Netherlands

 Springer

Library of Congress Control Number: 2006927053

ISBN-10: 0-387-30413-4   e-ISBN: 0-387-35215-5

ISBN-13: 978-0-387-30413-7   e-ISBN: 978-0-387-35215-2

Printed on acid-free paper.

AMS Subject Classifications: 34C05, 34C07, 34C15

Printed in the United States of America.

9 8 7 6 5 4 3 2 1

springer.com

To my wife:
Marie Thérèse

Contents

Preface

Solutions of ordinary differential equations usually can not be expressed in terms of well known functions if these equations are nonlinear. As a result numerical and asymptotic methods are developed to obtain approximations to solutions. An alternative approach is furnished by the qualitative theory of differential equations, which seeks to find properties of solutions without actually solving these equations. As such the geometric topological theory of plane autonomous systems $\dot{x} = P(x,y), \dot{y} = Q(x,y)$ due to Poincaré and Bendixson is well known and by now has assumed an almost definite form. A first step within these systems away from the linear systems leads to the case where P and Q are relative prime polynomials of degree at most 2 which are not both linear, and thus to the system

$$\dot{x} = a_{00} + a_{10}x + a_{01}y + a_{20}x^2 + a_{11}xy + a_{02}y^2 , \qquad (0.0.1)$$
$$\dot{y} = b_{00} + b_{10}x + b_{01}y + b_{20}x^2 + b_{11}xy + b_{02}y^2 \qquad (0.0.2)$$

for the functions $x = x(t)$, $y = y(t)$, where a_{ij} ,b_{ij} (i,j $= 0,1,2$) are real (or sometimes complex) constants. In a survey paper on these systems in 1966 by W.A.Coppel these systems were called quadratic systems, by which they have been referred to ever since.

Apart from a variety of problems in various fields of applications that lead to these systems they are also of theoretical interest, being a relatively simple example to study complicated non linear phenomena such as limit cycle behaviour. It then helps to consider the traces of the solutions $x = x(t), y = y(t)$ in the x, y plane, being the orbits that constitute the phase portrait in the x, y (phase) plane.

Although some examples of phase portraits of quadratic systems can already be found in the work of Poincaré, it seems that the first paper dealing exclusively with these systems was published in 1904 by Büchel, although it was mainly a collection of examples. In the remainder of the twentieth century an increasing flow of results produced over a thousand papers; the accumulated number of papers that appeared over those years is illustrated in the accompanying figure.

The present book tries to give a presentation of the advance of our knowledge of phase portraits of quadratic systems as a result of this flow of papers, thereby paying attention to the historical development of the subject. This is also expressed by the chronological structure of the list of references at the end of the book.

Constructing this presentation raises the question of how to order what is known about these portraits. In this book a particular ordering in classes, using the notions of finite and infinite multiplicity and finite and infinite index, is presented and classifications of phase portraits for various classes are given, using the well known methods of phase plane analysis.

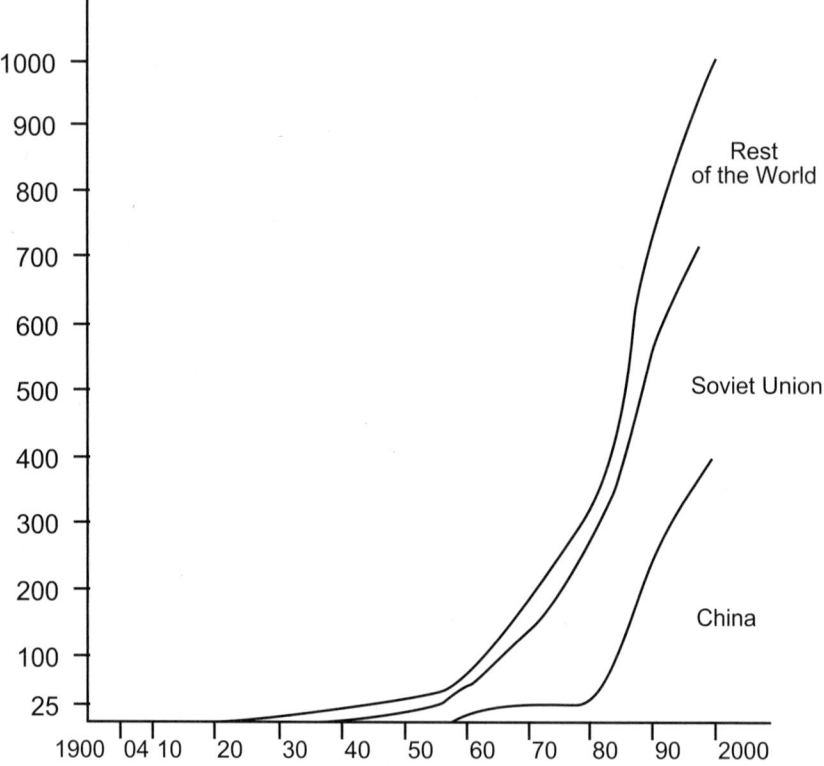

Accumulated number of papers on quadratic systems published in the twentieth century

- In Chapter 1, a review is given of the basic notions of phase portraits, keeping a focus on quadratic systems. It is primarily meant as an

introduction for readers with insufficient knowledge of the Poincaré–
Bendixson theory.

- In Chapter 2, critical points in quadratic systems are discussed, being the first basic elements needed to construct a phase portrait. The
notion of finite multiplicity m_f is introduced,being the sum of the multiplicities of the finite (real or complex) critical points, and it is shown
that in quadratic systems $0 \leq m_f \leq 4$, if the number of critical points
is finite. Finite multiplicity is chosen as the first ordering principle
to define classes of quadratic systems. The partition of the coefficient
space corresponding to the value of m_f is given.Infinite critical points
and infinite multiplicity m_i are defined, where m_i is the sum of the multiplicities of the infinite critical points; if the number of critical points
is finite, then $m_f + m_i = 7$. Further ordering principles are the finite
index i_f $(-2 \leq i_f \leq 2)$ and the infinite index i_i $(-1 \leq i_i \leq 3)$ based on
the Poincaré index definition for a critical point, and it is shown that
$i_f + i_i = 1$. The possible types of finite and infinite critical (real or
complex) points are discussed.

- In Chapter 3, the identification of critical points with base points of
the pencil of conics, being isoclines, leads to an extensive study of these
isoclines, wherein degenerate isoclines and central conics obtain special
attention. A classification of possible degenerate isocline combinations
and character of the central conic leads, within each class characterized
by a given value of m_f $(0 \leq m_f \leq 4)$ and i_f $(-2 \leq i_f \leq 2)$, to
possible combinations of finite and infinite critical points that define a
further subdivision in classes. Following this procedure, 173 classes are
obtained to represent all quadratic systems.

- In Chapter 4, general properties of quadratic systems and results obtained from studying particular classes,characterized by a specific property, are indicated in order to be used in classification of the 173 classes.
The possible types of flow in a quadratic system over a straight line
are classified and the results applied to derive general properties of
quadratic systems.

- In Chapter 5, the phase portraits in class $m_f = 0$ are classified; there
exist 30 possible phase portraits in $m_f = 0$.

- In Chapter 6, quadratic systems with a center point are considered and imbedded in the 173 classes.

- Chapter 7 gives an extensive historic sketch of what has become known about limit cycles in quadratic systems, starting at an elementary level of knowledge about the limit cycle notion.

- In Chapter 8, the phase portraits in the class $m_f = 1$ are classified; there exist 38 possible phase portraits in class $m_f = 1$.

- In Chapter 9, the phase portraits in the class $m_f = 2$ are classified; there exist 230 possible phase portraits in class $m_f = 2$.

- In Chapter 10, the phase portraits in the class $m_f = 3$ are considered. In classes with a degenerate infinite critical point there exist 141 possible phase portraits. For classes with a semi- elementary infinite critical point, 269 possible separatrix structures are given without solving the limit cycle problem.

- In Chapter 11, the phase portraits in class $m_f = 4$ are discussed; the classification in $m_f = 4$ is far from being complete.

Many people have contributed to the development of my interest in quadratic systems and, directly or indirectly, to the motivation to write this book. They are too numerous to mention by name. I hope, in return, that they derive much pleasure in reading it.

I would like to express my thanks to those persons working in Delft University of Technology, Faculty of Technical Mathematics, who helped me to produce the book on the computer. I would like to thank Mr.Wu for installing the Latex program on my home computer and Mr. Kees Lemmens for doing the same, now also introducing the Linux operating system and helping me to start working with the computer to construct an electronic version of the book. My special thanks goes to Mr. Ruud Sommerhalder, who was tireless helping me to overcome the problems that kept coming up during this work.

A lot of work still remains to be done to complete the collection of all phase portraits of quadratic systems. If this book helps in promoting the achievement of this goal, it has served its purpose.

John W. Reyn

Chapter 1

Introduction

This book aims at giving a survey of what has become known so far about the *phase portraits of quadratic systems in the plane*, thereby indicating what still has to be done to complete the picture on the subject.

In the present context, by a *quadratic system* is meant the second order autonomous system of differential equations

$$\dot{x} = a_{00} + a_{10}x + a_{01}y + a_{20}x^2 + a_{11}xy + a_{02}y^2 \equiv P(x, y), \quad (1.0.1)$$
$$\dot{y} = b_{00} + b_{10}x + b_{01}y + b_{20}x^2 + b_{11}xy + b_{02}y^2 \equiv Q(x, y), \quad (1.0.2)$$

for the functions $x = x(t), y = y(t)$, where $\dot{} = d/dt$.

As in the overwhelming majority of papers on the subject, the coefficients $a_{ij}, b_{ij} (i, j = 0, 1, 2)$ are taken to be real, as are the variables t, x and y, although on occasion x and y may also be allowed to take complex values. Degenerate cases are left out of consideration; at least one quadratic term has a coefficient unequal to zero and $P(x, y)$ and $Q(x, y)$ do not have a common factor such that the system is essentially linear.

With regard to the notion of a *phase portrait* of an autonomous system of differential equations in the plane, we use the standard meaning as developed in the Poincaré−Bendixson theory on the qualitative behaviour of differential equations in the plane and as fully explained, for instance, in [ALGM1]. It consists of the (geometrical) picture formed by the solution curves (called orbits, trajectories or integral paths) of (1.0.1), (1.0.2) through the *ordinary points*, which are defined by $P^2(x, y) + Q^2(x, y) \neq 0$ plus the *critical points*, defined by $P^2(x, y) + Q^2(x, y) = 0$. Through every ordinary point (x_0, y_0) in the x, y (or phase)plane there exists a unique orbit, which represents a family of solutions of (1.0.1),(1.0.2) given by $x = x(t; x_0, y_0, t_0), y = y(t; x_0, y_0, t_0)$ satisfying $x(t_0; x_0, y_0, t_0) = x_0, y(t_0; x_0, y_0, t_0) = y_0$, where for a given (x_0, y_0), t_0 may be chosen yielding the same orbit through (x_0, y_0) for an arbitrary value of t_0, t. These solutions may be extended onto the whole interval $-\infty < t < \infty$ unless the orbit escapes to infinity for a finite value of t. The orbit through a critical point (x_0, y_0) coincides with such a point and represents the solutions $x(t; x_0, y_0, t_0) \equiv x_0, y(t; x_0, y_0, t_0) \equiv y_0$ for all values of t, $-\infty < t < \infty$.

In the context of the qualitative theory of real ordinary differential equations the determination of the phase portraits of (1.0.1), (1.0.2) amounts to finding the partition of the space of the coefficients a_{ij} and b_{ij} of (1.0.1),(1.0.2) into regions with *qualitatively equivalent phase portraits*. Two phase portraits are qualitatively equivalent if there exists a homeomorphism between the two phase portraits such that, if two points in one phase portrait lie on the same orbit, their images lie also on the same orbit in the other phase portrait [ALGM1, p. 106].

In order to explore what elements of a phase portrait in a quadratic system seem to be important for its qualitative properties, we will discuss some examples.

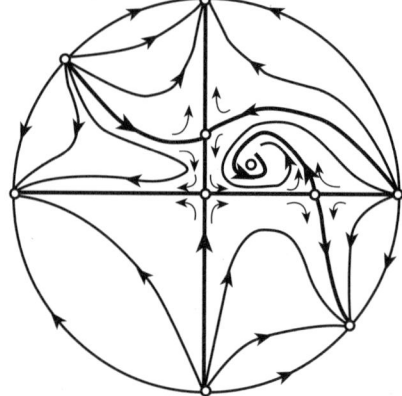

Figure 1.1 Phase portrait of equation (1.0.3),(1.0.4)

An example of an equation occuring frequently in applications is the *Volterra–Lotka equation*

$$\dot{x} = x(a_0 + a_1 x + a_2 y) \equiv P(x, y), \qquad (1.0.3)$$
$$\dot{y} = y(b_0 + b_1 x + b_2 y) \equiv Q(x, y), \qquad (1.0.4)$$

where $a_i, b_i \in \mathbb{R}$. Let $a_0 = 3, a_1 = -1, a_2 < -5, b_0 = -1, b_1 = 1, b_2 \doteq 1$, then Figure 1.1 is obtained.

As usual the phase portrait is given on the unit or Poincaré disk, its interior representing the finite part of the plane and the points on the unit circle "points at infinity". The relation between the x, y plane and the Poincaré disk is visualized in Figure 1.2. Herein the (Poincaré)unit sphere touches the x, y plane at the origin (0,0). A point $M(x, y)$ in the x, y plane is connected with the center C of the sphere through a straight line which intersects the sphere in the points M' ,and M''. Projection of the point M'on the lower hemisphere onto the x, y plane then leads to the point $\bar{M}$ on the Poincaré disk. In particular, if $M(x, y)$ is sent to infinity, $\bar{M}$ approaches the unit circle.

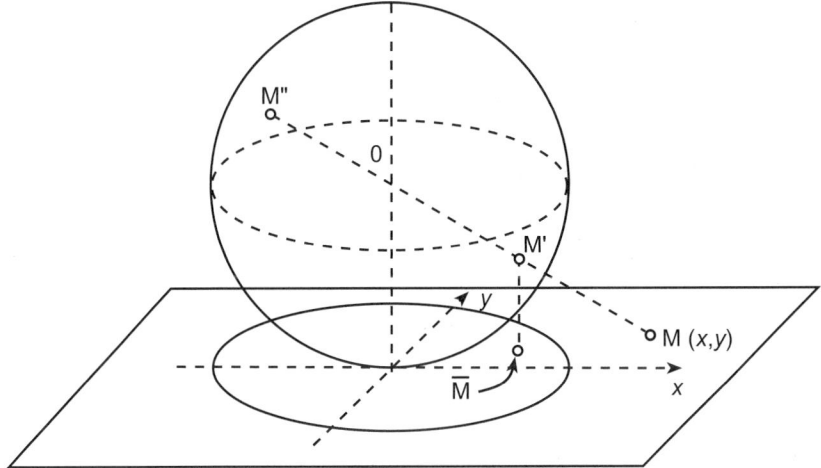

Figure 1.2 Poincaré sphere and disk

In Figure 1.1 a few orbits are sketched. From it, we can see that in regions with ordinary points the orbits are locally nearly parallel, as will also easily follow from a local analysis of the orbits near an ordinary point.Evidently this is not so near a critical point,which for (1.0.3), (1.0.4) are located in $(0,0)$, $(1,0)$, $(0,1)$ and $\frac{(3+a_2,-2)}{1+a_2}$. Moreover, the topological structure of the orbits near the three critical points, situated on the coordinate axes, seems to be the same variance with that near the fourth critical point. An investigation of the *possible critical points in quadratic systems* is thus needed for the determination of all possible phase portraits in quadratic systems. This would include critical points in the finite part of the plane as well as

at infinity. Some orbits through ordinary points of (1.0.3), (1.0.4) appear to
be defined on the whole interval $-\infty < t < \infty$ and extend from one critical
point to another. The critical points are then *limit sets* for these orbits; for
each orbit the critical point corresponding to $t = -\infty$ is then an α *limit point*
and that to $t = \infty$ an ω *limit point*. The occurence of other types of limit
sets in quadratic systems needs to be investigated as well; in fact other types,
and, correspondingly, other types of orbits, will be given in other examples.

As mentioned before, for (1.0.3), (1.0.4) there exist four critical points
in the finite part of the plane, being the intersection points of the conics
$P(x,y)=0$ and $Q(x,y)=0$. For other quadratic systems intersection points
may coincide or become complex. If $P(x,y)=0$ and $Q(x,y)=0$ have a common
asymptotic direction, they intersect in that direction at infinity leaving fewer
finite critical points.

It follows from the presentation on the Poincaré disk that critical points
at infinity are characterized by the condition $\dot{\theta} = 0$, where θ is the polar
angle of the polar coordinates r and θ. If (1.0.1), (1.0.2) are written in these
coordinates there follows

$$\dot{r} = A_1(\theta) + rB_1(\theta) + r^2C_1(\theta), \qquad (1.0.5)$$
$$\dot{\theta} = A_2(\theta) + rB_2(\theta) + r^2C_2(\theta), \qquad (1.0.6)$$

where from $A_i(\theta)$, $B_i(\theta)$, $C_i(\theta)$, $i = (1,2)$ only $C_2(\theta)$ is now explicitly
needed, being the dominant term at large r. As $C_2(\theta) = b_{20}cos^3\theta + (b_{11} - a_{20})sin\theta cos^2\theta + (b_{02} - a_{11})sin^2\theta cos\theta - a_{02}sin^3\theta$, it follows from $C_2(\theta)=0$ that,
in general, there are six pairwise diametrically opposite points with $\dot{\theta} = 0$;
they will be regarded as three *infinite critical points*. For (1.0.3), (1.0.4) they
will be located at $\theta = 0(\pi), \theta = \frac{\pi}{2}(\frac{3\pi}{2})$, and $\theta = \arctan\frac{2}{a_2-1}(\frac{2}{a_2-1} < 0)$.

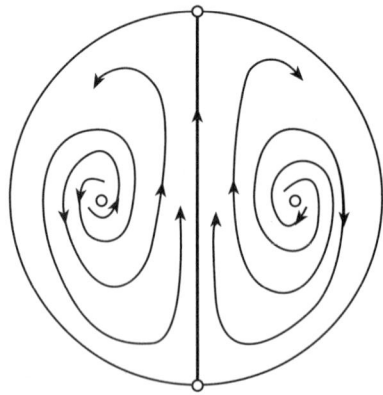

Figure 1.3 Phase portrait of (1.0.7),(1.0.8)

Another example is given by

$$\dot{x} = 2xy \equiv P(x,y), \tag{1.0.7}$$
$$\dot{y} = 1+y-x^2+y^2 \equiv Q(x,y), \tag{1.0.8}$$

of which the phase portrait is given in Figure 1.3. This system has again four finite critical points: two real points: in $(1,0)$ and in $(-1,0)$ and two complex points: in $(0,\frac{1}{2}(-1+i\sqrt{3})$ and $(0,\frac{1}{2}(-1-i\sqrt{3})$. At infinity there exist again three critical points: one real at $\theta = \frac{1}{2}\pi(\frac{3}{2}\pi)$ and two complex: at $\theta = arctan+i$ and $\theta = arctan-i$. Except for the orbit $x \equiv 0$ that runs from the α limit point coinciding with the critical point at $y = -\infty$ to the ω limit point at $y = \infty$, all orbits run from an α limit point that coincides with a finite critical point to an ω limit set that consists of critical points connected by orbits through ordinary points. This shows the second type of limit set that can be encountered in quadratic systems. Figure 1.3 shows another feature(in fact also present in Figure 1.1) important for a qualitative description of a phase portrait. The orbit coinciding with the y axis divides the phase portrait into two regions and thus serves as a *separatrix* between these two regions. The slightest change in $P(x,y)$, however, for instance to $P(x,y) = 2xy+\epsilon$, where ϵ is a small parameter, breaks up this separatrix, leaving the rest of the phase portrait unchanged. As a result of this *bifurcation* a qualitatively different phase portrait is created. This indicates that the *separatrix structure* of phase portraits needs to be considered in the qualitative description of phase portraits and that the study of bifurcation phenomena is needed to find the partition of the coefficient space into regions with qualitatively equivalent phase portraits.

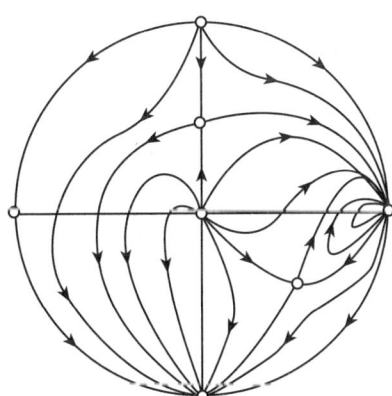

Figure 1.4 Phase portrait of (1.0.9),(1.0.10)

An example that shows that the number of finite critical points is not always equal to four and that critical points may coincide is given by

$$\dot{x} = x(y + \frac{3}{2}), \tag{1.0.9}$$

$$\dot{y} = x + y - 2y^2, \tag{1.0.10}$$

, its phase portrait being given in Figure 1.4. The finite critical points are located in $(0,0)$, $(0, \frac{1}{2})$ and $(6, -\frac{3}{2})$. The fourth intersection point of the degenerate hyperbola $x(y + \frac{3}{2}) = 0$ and the parabola $x + y - 2y^2 = 0$ is "at infinity" in the direction of the common asymptotic direction $y = 0$. The infinite critical point "at the ends of the x axis" also corresponds to a double root, since $C_2(\theta) = -sin^2\theta cos\theta$, and two coinciding infinite critical points. The point can therefore be viewed as the coalescence of three critical points or a critical point with multiplicity 3.

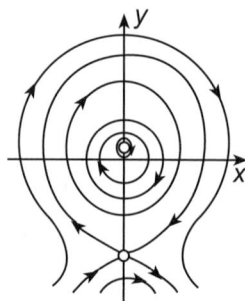

Figure 1.5 Region of closed orbits for system (1.0.13),(1.0.14)

The next example shows that *periodic solutions* can occur in quadratic systems. Periodic solutions are solutions of (1.0.1), (1.0.2) satisfying

$$x(t + T; x_0, y_0, t_0) = x(t; x_0, y_0, t_0), \tag{1.0.11}$$
$$y(t + T; x_0, y_0, t_0) = y(t; x_0, y_0, t_0), \tag{1.0.12}$$

$-\infty < t < \infty$ for some $T > 0$. They are represented in the plane by a *closed orbit*. It is also the limit set of this orbit; all points on the orbit are α limit points as well as ω limit points. The region of closed orbits for the system

$$\dot{x} = -1 + 2y + x^2 + y^2, \tag{1.0.13}$$

$$\dot{y} = -2x, \tag{1.0.14}$$

is given in Figure 1.5. It is bounded by the critical point in $(0, -1 - \sqrt{2})$ and the separatrix $(x^2 + y^2 - 1)e^y = 2(1 + \sqrt{2})e^{-1-\sqrt{2}}$ and completely filled by a nest of periodic orbits.

An important phenomenon in quadratic systems is the occurence of *limit cycles*, being periodic orbits, isolated in the sense that they have an annular neighborhood such that no other periodic orbits exist in that neighborhood.

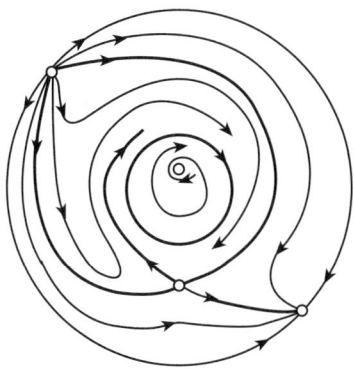

Figure 1.6 Phase portrait of (1.0.15),(1.0.16)

For example the system

$$\dot{x} = -1 + 2y + x^2 + xy + y^2 \equiv P(x, y), \qquad (1.0.15)$$
$$\dot{y} = -2x - x^2 \equiv Q(x, y), \qquad (1.0.16)$$

has a limit cycle coinciding with the unit circle(Figure 1.6). In this example the limit cycle is the ω limit set of all orbits approaching it from the inside as well as from the outside. Limit cycles that are α limit sets or both α and ω limit sets may also occur in quadratic systems.

In summary we can be state that the classification of phase portraits of quadratic systems has to take into account that a phase portrait is topologically characterized by the number,relative location and character of its critical points, both in the finite part of the plane and at infinity, its separatrix structure and the number and location of periodic solutions , if any,in particular of its limit cycles.We start with a discussion of critical points in quadratic systems.

Chapter 2

Critical points in quadratic systems

2.1 Multiplicity of critical points

For the notion of multiplicity of critical points in quadratic systems we follow that developed in the book of Andronov et.al on the bifurcation theory of dynamic systems in the plane [ALGM2].

Definition. A critical point (x_0, y_0) of a quadratic system (1.0.1),(1.0.2) has *multiplicity m* if there exist numbers $\epsilon_0 > 0$ and $\delta_0 > 0$ such that any such system with coefficients $\bar{a}_{ij}, \bar{b}_{ij} (i, j = 0, 1, 2)$ such that $\mid \bar{a}_{ij} - a_{ij} \mid, \mid \bar{b}_{ij} - b_{ij} \mid < \delta < \delta_0$ has at most m critical points in a neighborhood $(x-x_0)^2+(y-y_0)^2 < \epsilon_0$ of (x_0, y_0) , and for any $\epsilon < \epsilon_0$ and $\delta < \delta_0$ there exists a quadratic system which has at least m critical points in $(x - x_0)^2 + (y - y_0)^2 < \epsilon$.

If $m=1$ the critical point is *simple* or *elementary*. Perturbations in the coefficients only lead to the continued presence of a single elementary critical point. At such a point the isoclines $P(x, y) = 0$ and $Q(x, y) = 0$ intersect at an angle unequal to zero ; if the point is located in (x_0, y_0), then $\Lambda(x_0, y_0) = P_x(x_0, y_0)Q_y(x_0, y_0) - P_y(x_0, y_0)Q_x(x_0, y_0) \neq 0$.

Writing these isoclines as

$$P(x, y) = P_x(x_0, y_0)(x - x_0) + P_y(x_0, y_0)(y - y_0) + \tfrac{1}{2}P_{xx}(x_0, y_0)(x - x_0)^2$$
$$+P_{xy}(x_0, y_0)(x - x_0)(y - y_0) + \tfrac{1}{2}P_{yy}(x_0, y_0)(y - y_0)^2 = 0,$$
$$Q(x, y) = Q_x(x_0, y_0)(x - x_0) + Q_y(x_0, y_0)(y - y_0) + \tfrac{1}{2}Q_{xx}(x_0, y_0)(x - x_0)^2+$$
$$Q_{xy}(x_0, y_0)(x - x_0)(y - y_0) + \tfrac{1}{2}Q_{yy}(x_0, y_0)(y - y_0)^2 = 0,$$

yields that the isoclines of (1.0.1),(1.0.2), $\lambda P(x, y) + \mu Q(x, y) = 0, \lambda^2 + \mu^2 \neq$

0, may be written as $(\lambda P_x(x_0, y_0) + \mu Q_x(x_0, y_0))(x - x_0) + (\lambda P_y(x_0, y_0) + \mu Q_y(x_0, y_0))(y - y_0)$+quadratic terms $= \alpha(x - x_0) + \beta(y - y_0)$ + quadratic terms$=0$.

This leads to the system

$$\lambda P_x(x_0, y_0) + \mu Q_x(x_0, y_0) = \alpha,$$
$$\lambda P_y(x_0, y_0) + \mu Q_y(x_0, y_0) = \beta,$$

and shows,with $\Lambda(x_0, y_0) \neq 0$,that to any direction $\alpha(x - x_0) + \beta(y - y_0) = 0$, $\alpha^2 + \beta^2 \neq 0$, there corresponds a unique isocline through an elementary critical point.

In critical points with multiplicity $m > 1$, or *higher order points*, the condition $\Lambda(x_0, y_0) \neq 0$ is not satisfied. Since we do not consider degenerate quadratic systems,the number of finite critical points is finite whereas $m \neq \infty$. Since two conics (conical isoclines) have at most four different intersection points, the multiplicity of a critical point in a quadratic system is at most equal to 4; so $1 \leq m \leq 4$. In relation to the definition of classes in quadratic systems it seems useful to give the following definition.

Definition. The *finite multiplicity* m_f of a quadratic system is the sum of the multiplicities of all finite (real or complex) critical points of such a system.

Example. Consider the system

$$\dot{x} = -\epsilon^2 + x^2 - y^2, \tag{2.1.1}$$
$$\dot{y} = xy, \tag{2.1.2}$$

where ϵ in $\mathbb{R}$. For $\epsilon \neq 0$, the system has two real elementary critical points : in $(\epsilon, 0)$ and in $(-\epsilon, 0)$, and two elementary complex critical points: in $(0, i\epsilon)$ and in $(0, -i\epsilon)$; so $m_f = 4$. Letting $\epsilon \to 0$ not only makes the real but also the complex critical points coincide with the origin and a critical point of multiplicity $m=4$ results. From the system with $\epsilon =0$ four real critical points can be bifurcated by a perturbation leading to the system

$$\dot{x} = x^2 - y^2, \tag{2.1.3}$$
$$\dot{y} = (x - \epsilon_1)(y - \epsilon_2), \epsilon_1 \neq \epsilon_2, \epsilon_i \in R, (i = 1, 2) \tag{2.1.4}$$

for which also $m_f=4$. The finite multiplicity of a quadratic system is invariant under bifurcation of finite critical points.

The partition of the R^{12} coefficient space into regions with equal value of m_f can be found by deriving the equations for the coordinates of the critical points. Then follows [93,R]

Theorem 2.1 *If $a_{02}^2 + b_{02}^2 \neq 0$, the x coordinates of the critical points of (1.1),(1.2) are given by the solutions of*

$$Ax^4 + B_1 x^3 + C_1 x^2 + D_1 x + E_1 = 0. \qquad (2.1.5)$$

To each solution corresponds exactly one finite value of y and thus one (finite) critical point. For $a_{02}^2 + b_{02}^2 = 0$ the equation is not valid and for $a_{02}^2 + b_{02}^2 \neq 0$ the finite multiplicity m_f of (1.1),(1.2) equals

(i) *4 if and only if $A \neq 0$,*

(ii) *3 if and only if $A = 0, B_1 \neq 0$,*

(iii) *2 if and only if $A = B_1 = 0, C_1 \neq 0$,*

(iv) *1 if and only if $A = B_1 = C_1 = 0, D_1 \neq 0$,*

(v) *0 if and only if $A = B_1 = C_1 = D_1 = 0, E_1 \neq 0$,*

(vi) *∞ if and only if $A = B_1 = C_1 = D_1 = E_1 = 0$.*

Theorem 2.2 *If $a_{20}^2 + b_{20}^2 \neq 0$, the y coordinates of the critical points of (1.1),(1.2) are given by the solution of*

$$Ay^4 + B_2 y^3 + C_2 y^2 + D_2 y + E_2 = 0. \qquad (2.1.6)$$

To each solution corresponds exactly one finite value of x and thus one (finite) critical point. For $a_{20}^2 + b_{20}^2 = 0$ the equation is not valid and for $a_{20}^2 + b_{20}^2 \neq 0$ the finite multiplicity m_f of (1.1),(1.2) equals

(i) *4 if and only if $A \neq 0$,*

(ii) *3 if and only if $A = 0, B_2 \neq 0$,*

(iii) *2 if and only if $A = B_2 = 0, C_2 \neq 0$,*

(iv) *1 if and only if $A = B_2 = C_2 = 0, D_2 \neq 0$,*

(v) *0 if and only if $A = B_2 = C_2 = D_2 = 0, E_2 \neq 0$,*

(vi) *∞ if and only if $A = B_2 = C_2 = D_2 = E_2 = 0$.*

Theorem 2.3 *If $a_{20}^2 + a_{02}^2 + b_{20}^2 + b_{02}^2 = 0$, $a_{11}^2 + b_{11}^2 \neq 0$, the finite multiplicity m_f of (1.1),(1.2) equals*

(i) *2 if and only if $c_{25} c_{35} \neq 0$,*

(ii) *1 if and only if either $c_{25} = 0, c_{35} \neq 0, c_{15} + c_{23} \neq 0$,*
    or $c_{25} \neq 0, c_{35} = 0, c_{15} - c_{23} \neq 0$,

(iii) *0 if and only if either $c_{25} = 0, c_{35} \neq 0, c_{15} + c_{23} = 0, c_{13} \neq 0$,*
    or $c_{25} \neq 0, c_{35} = 0, c_{15} - c_{23} = 0, c_{12} \neq 0$,
    or $c_{25} = 0, c_{35} = 0, c_{15} \neq 0$,

(iv) *∞ if and only if either $c_{25} = 0, c_{35} \neq 0, c_{15} + c_{23} = 0, c_{13} = 0$,*
    or $c_{25} \neq 0, c_{35} = 0, c_{15} - c_{23} = 0, c_{12} = 0,$

$$or \ c_{25} = 0, c_{35} = 0, c_{15} = 0.$$

In these theorems the following abbrevations are used.

$A = c_{46}^2 - c_{45}c_{56},$

$B_1 = 2c_{46}(c_{26} - \frac{1}{2}c_{35}) + 2c_{56}(c_{34} - \frac{1}{2}c_{25}),$

$B_2 = -2c_{45}(c_{26} - \frac{1}{2}c_{35}) - 2c_{46}(c_{34} - \frac{1}{2}c_{25}),$

$C_1 = c_{26}(c_{26} - \frac{1}{2}c_{35}) + c_{36}(c_{34} - \frac{1}{2}c_{25}) - \frac{1}{2}c_{56}(3c_{23} + 2c_{15}) + 2c_{16}c_{46},$

$C_2 = c_{24}(c_{26} - \frac{1}{2}c_{35}) + c_{34}(c_{34} - \frac{1}{2}c_{25}) - \frac{1}{2}c_{45}(3c_{23} - 2c_{15}) - 2c_{14}c_{46},$

$D_1 = -c_{23}c_{36} - 2c_{13}c_{56} + 2c_{16}(c_{26} - \frac{1}{2}c_{25}),$

$D_2 = c_{23}c_{24} + 2c_{12}c_{45} + 2c_{14}(c_{34} - \frac{1}{2}c_{25}),$

$E_1 = c_{16}^2 - c_{13}c_{36},$

$E_2 = c_{14}^2 - c_{12}c_{24},$

where,

$c_{12} = a_{00}b_{10} - a_{10}b_{00}, c_{13} = a_{00}b_{01} - a_{01}b_{00}, c_{14} = a_{00}b_{20} - a_{20}b_{00},$

$c_{15} = a_{00}b_{11} - a_{11}b_{00}, c_{16} = a_{00}b_{02} - a_{02}b_{00},$

$c_{23} = a_{10}b_{01} - a_{01}b_{10}, c_{24} = a_{10}b_{20} - a_{20}b_{10}, c_{25} = a_{10}b_{11} - a_{11}b_{10},$

$c_{26} = a_{10}b_{02} - a_{02}b_{10},$

$c_{34} = a_{01}b_{20} - a_{20}b_{01}, c_{35} = a_{01}b_{11} - a_{11}b_{01},$

$c_{45} = a_{20}b_{11} - a_{11}b_{20}, c_{46} = a_{20}b_{02} - a_{02}b_{20},$

$c_{56} = a_{11}b_{02} - a_{02}b_{11}.$

Moreover there exist the relations

$c_{12}c_{34} + c_{14}c_{23} = c_{13}c_{24}, c_{12}c_{35} + c_{15}c_{23} = c_{13}c_{25}, c_{12}c_{16} + c_{16}c_{23} = c_{13}c_{26},$

$c_{12}c_{45} + c_{15}c_{24} = c_{14}c_{25}, c_{12}c_{46} + c_{16}c_{24} = c_{14}c_{26}, c_{12}c_{36} + c_{16}c_{25} = c_{15}c_{26},$

$c_{13}c_{45} + c_{15}c_{34} = c_{14}c_{35}, c_{13}c_{46} + c_{16}c_{34} = c_{14}c_{36}, c_{13}c_{56} + c_{16}c_{35} = c_{15}c_{36},$

$c_{14}c_{56} + c_{16}c_{45} = c_{15}c_{46}, c_{23}c_{45} + c_{25}c_{34} = c_{24}c_{35}, c_{23}c_{46} + c_{26}c_{34} = c_{24}c_{36},$

$c_{23}c_{56} + c_{26}c_{35} = c_{25}c_{36}, c_{24}c_{56} + c_{26}c_{45} = c_{25}c_{46}, c_{34}c_{56} + c_{36}c_{45} = c_{35}c_{46}.$

It may be remarked that under the transformation $x \leftrightarrow y, a_{ij} \leftrightarrow b_{ij}$ it follows that $c_{12} = -c_{13}, c_{14} = -c_{16}, c_{15} = -c_{15}, c_{23} = c_{23}, c_{24} = -c_{36}, c_{25} = -c_{35}, c_{26} = -c_{34}, c_{45} = c_{56}, c_{46} = c_{46}.$ As a result Theorems 1 and 2 may be derived from each other.The conditions on the coefficients that determine m_f may also be written in the form of affine invariants [93,BV]. It should be remarked that (2.1.5) and (2.1.6) were first given by Curtz[80,Cu], without however mentioning the conditions $a_{02}^2 + b_{02}^2 \neq 0, a_{20}^2 + b_{20}^2 \neq 0$, respectively. In fact, if $a_{02}^2 + b_{02}^2 = 0$, then $P(x,y)=0$ and $Q(x,y)=0$ have a common factor y in the quadratic terms, which shows that they have the x axis as common asymptotic direction. As a result, all isoclines $\lambda P(x,y) + \mu Q(x,y) = 0, \lambda^2 + \mu^2 \neq 0$ have this direction in common and the pencil of (conical) isoclines have an *infinite base point* at "the ends of the x axis". A common asymptotic direction of $P(x,y)=0$ and $Q(x,y)=0$ corresponds to a common factor in $P_2(x,y) \equiv a_{20}x^2 + a_{11}xy + a_{02}y^2$ and $Q_2(x,y) \equiv b_{20}x^2 + b_{11}xy + b_{02}y^2$. We can state

Theorem 2.4 *The expressions $P_2(x,y) \equiv a_{20}x^2 + a_{11}xy + a_{02}y^2$ and $Q_2(x,y) \equiv b_{20}x^2 + b_{11}xy + b_{02}y^2$ have a common factor if and only if $A \equiv c_{46}^2 - c_{45}c_{56} = 0$. The common factor is linear if $c_{45}^2 + c_{46}^2 + c_{56}^2 \neq 0$; then there is an infinite base point of the pencil of conics, representing the isoclines in the direction given by $c_{45}x + c_{46}y = 0$ if $c_{45} \neq 0$, and $c_{46}x + c_{56}y = 0$ if $c_{56} \neq 0$. (This direction may also be written as $(c_{45} + c_{46})x + (c_{46} + c_{56})y = 0$.) The common factor is quadratic if $c_{45}^2 + c_{46}^2 + c_{56}^2 = 0$; then there are two (possibly coinciding or complex) directions in which there are infinite base points; these being given by $P_2(x,y) = 0$ and/or $Q_2(x,y) = 0$.*

Proof. We may write
$$P_2(x,y) = (\alpha_1 x + \beta_1 y)(\alpha_2 x + \beta_2 y) = \alpha_1\alpha_2 x^2 + (\alpha_1\beta_2 + \alpha_2\beta_1)xy + \beta_1\beta_2 y^2,$$
$$Q_2(x,y) = (\alpha_3 x + \beta_3 y)(\alpha_4 x + \beta_4 y) = \alpha_3\alpha_4 x^2 + (\alpha_3\beta_4 + \alpha_4\beta_3)xy + \beta_3\beta_4 y^2,$$
with $\alpha_i, \beta_i \in C$, i=1,2,..4. Then follows $c_{45} = \alpha_1\alpha_3(\alpha_2\beta_4 - \alpha_4\beta_2) + \alpha_2\alpha_4(\alpha_1\beta_3 - \alpha_3\beta_1)$, $c_{46} = \alpha_1\alpha_2\beta_3\beta_4 - \alpha_3\alpha_4\beta_1\beta_2$, $c_{56} = \beta_2\beta_4(\alpha_1\beta_3 - \alpha_3\beta_1) + \beta_1\beta_3(\alpha_2\beta_4 - \alpha_4\beta_2)$, so that $c_{46}^2 - c_{45}c_{56} = (\alpha_1\beta_3 - \alpha_3\beta_1)(\alpha_1\beta_4 - \alpha_4\beta_1)(\alpha_2\beta_3 - \alpha_3\beta_2)(\alpha_2\beta_4 - \alpha_4\beta_2)$.

Thus $P_2(x,y)$ and $Q_2(x,y)$ have a common factor (linear or quadratic) if and only if $A \equiv c_{46}^2 - c_{45}c_{56} = 0$. Now let there exist at least one common (linear) factor, then with $\alpha = \alpha_1 = \alpha_3, \beta = \beta_1 = \beta_3$ there follows $c_{45} = \alpha^2(\alpha_2\beta_4 - \alpha_4\beta_2), c_{46} = \alpha\beta(\alpha_2\beta_4 - \alpha_4\beta_2), c_{56} = \beta^2(\alpha_2\beta_4 - \alpha_4\beta_2)$. If there exists only one linear factor in common, then $\alpha_2\beta_4 - \alpha_4\beta_2 \neq 0$, and, since $\alpha^2 + \beta^2 \neq 0, c_{45}^2 + c_{46}^2 + c_{56}^2 \neq 0$. If apart from $\alpha x + \beta y$ there exists another (possibly the same) linear factor in common, then $\alpha_2\beta_4 - \alpha_4\beta_2 = 0$ and $c_{45}^2 + c_{46}^2 + c_{56}^2 = 0$.

If $c_{45} \neq 0$, then $a_{20}^2 + b_{20}^2 \neq 0$ and $a_{11}^2 + b_{11}^2 \neq 0$ and from $P(x,y)=0$, $Q(x,y)=0$ there may be obtained the isoclines
$$c_{14} + c_{24}x + c_{34}y - c_{45}xy - c_{46}y^2 = 0,$$
$$c_{15} + c_{25}x + c_{35}y + c_{45}x^2 - c_{56}y^2 = 0,$$
having a common asymptotic direction given by $c_{45}x + c_{46}y = 0$. Similarly, if $c_{56} \neq 0$ the common asymptotic direction may be written as $c_{46}x + c_{56}y = 0$. $\square$

The question arises whether the equivalence of base points of the pencil of conics formed by the isoclines with the critical points of the vector field

$P(x,y), Q(x,y)$ in (1.0.1),(1.0.2), which applies in the finite part of the plane, also holds at infinity if the presentation of phase portraits on the Poincaré disk is used. As was stated before, from this representation it follows that critical points at infinity are characterized by the condition $\dot{\theta} = 0$, where θ is the polar angle of the polar coordinates r and θ. Infinite base points are then also infinite critical points since all isoclines pass through a base point, including the isocline on which $\dot{\theta} \to 0$ as $r \to \infty$. The reverse statement is not true, however, as already follows from the fact that there exists at most two infinite base points, whereas (1.3) up to three infinite critical points can occur.

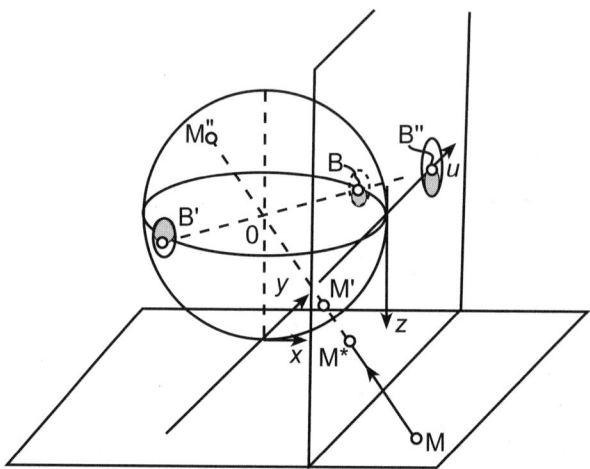

Figure 2.1 Use of the Poincaré sphere to study quadratic systems at infinity

In order to study the behavior of the system (1.0.1),(1.0.2) "near" infinity we will follow Poincaré and use a coordinate system z, u in the plane $x \equiv 1$, thus tangent to the Poincaré(unit) sphere, which is tangent to the x, y plane in $(0,0)$. The origin $(z, u)=(0,0)$ is the point of tangency in $x \equiv 1$, the u axis is parallel to the y axis, pointing in the same direction and the z axis is pointing downwards (see Figure 2.1).The straight line connecting point $M(x, y)$ in the x, y plane with the center of the Poincaré sphere intersects $x \equiv 1$ such that it follows that $z = \frac{1}{x}, u = \frac{y}{x}$ and (1.0.1),(1.0.2) becomes

$$z\dot{z} = -z(a_{20} + a_{10}z + a_{11}u + a_{00}z^2 + a_{01}zu + a_{02}u^2) = -z^3 P(\tfrac{1}{z}, \tfrac{u}{z}), (2.1.7)$$
$$z\dot{u} = b_{20} + b_{10}z + (-a_{20} + b_{11})u + b_{00}z^2 + (-a_{10} + b_{01})zu + (-a_{11} + b_{02})u^2$$
$$- a_{01}zu^2 - a_{00}z^2u - a_{02}u^3 = -z^2uP(\tfrac{1}{z}, \tfrac{u}{z}) + z^2Q(\tfrac{1}{z}, \tfrac{u}{z}). \qquad (2.1.8)$$

Points on $z \equiv 0$ now represent points at infinity in the x, y plane and points on the Poincaré (unit) circle. An ordinary point of (2.1.7), (2.1.8) on $z \equiv 0$ is an *ordinary point at infinity* and a critical point on $z=0$ of these equations, given by

$$f(u) = b_{20} + (-a_{20} + b_{11})u + (-a_{11} + b_{02})u^2 - a_{02}u^3 = 0, z = 0 \qquad (2.1.9)$$

is a *critical point at infinity*. Equation (2.1.9) yields, with $u=\tan\theta$, equation (1.3) found already to characterize an infinite critical point. The representation in the z, u plane near $z \equiv 0$ unites two diametrically opposite critical points on the Poincaré circle, each defined on a half neighborhood ($r \leq 1$)to a single critical point defined in a full neighborhood. As in the finite part of the plane we then define *simple or elementary infinite critical points* and *critical points with higher multiplicity m* , where $1 \leq m \leq 7$. The latter follows since at most three infinite critical points may coincide in an infinite base point of order 4. For a discussion of the points at the ends of the y axis (thus $u = \pm\infty$) the transformation $z = \frac{1}{y}, v = \frac{x}{y}$ can be used. Using (2.1.7),(2.1.8),the eigenvalues of the locally linearized system near a critical point on $z \equiv 0$ can be written as

$$\lambda_z - -a_{20} - a_{11}u - a_{02}u^2, \qquad (2.1.10)$$

$$\lambda_u = f'(u) = b_{11} - a_{20} + 2(b_{20} - a_{11})u - 3a_{02}u^2. \qquad (2.1.11)$$

If $\lambda_z \neq 0$,the critical point is *transversally hyperbolic* with respect to the Poincaré circle (or $z \equiv 0$),whereas for $\lambda_z = 0$ the point will be called *transversally non-hyperbolic*. Similarly,if $\lambda_u \neq 0$ the critical point is called *tangentially hyperbolic* with respect to the Poincaré circle, and if $\lambda_u =0$ *tangentially non-hyperbolic*. For multiple critical points at infinity we will, on occasion,use the notation $M_{p,q}^i$ for a real point and $C_{p,q}^0$ for a complex point. This means that the index of the point equals i, the maximum number of elementary finite critical points that can be bifurcated from $M_{p,q}^i, C_{p,q}^0$ equals p and the maximum number of elementary points that remain after bifurcation on the Poincaré circle equals q. Thus for a transversally non-hyperbolic critical point $1 \leq p \leq 4$ and for a tangentially non-hyperbolic critical point $2 \leq q \leq 3$.

Theorem 2.5 *If the number of critical points at infinity of a quadratic system is finite, then all these points are transversally hyperbolic (p=0) if and only if $A \equiv c_{46}^2 - c_{45}c_{56} \neq 0$, thus if $m_f = 4$.*

Proof. A transversally non-hyperbolic critical point satisfies $f(u) = \lambda_z = 0$; so from (2.1.9),(2.1.10) it follows that

$$a_{20} + a_{11}u + a_{02}u^2 = 0,$$
$$b_{20} + b_{11}u + b_{02}u^2 = 0.$$

As pointed out in Theorem 2.4 these equations contain a common factor if and only if $A \equiv c_{46}^2 - c_{45}c_{56} = 0$.   □

From Theorems 2.4 and 2.5 there follows

Theorem 2.6 *A transversally non-hyperbolic critical point at infinity coincides with an infinite base point of the pencil of (conical) isoclines and occurs only if $A=0$. If $c_{45}^2 + c_{46}^2 + c_{56}^2 \neq 0$ there exists only one such point; it is located in the direction given by $(c_{45}+c_{46})x+(c_{46}+c_{56})y = 0$. If $c_{45}^2 + c_{46}^2 + c_{56}^2 = 0$ there are two (possibly coinciding or complex)such points; they are located in the directions determined by $P_2(x,y) = 0$ and /or $Q_2(x,y) = 0$.*

Complementary to the finite multiplicity m_f of a quadratic system we define the *infinite multiplicity* m_i of a quadratic system as the *sum of the multiplicities of the infinite critical points*; it may also be finite or infinite. Moreover we define the *total multiplicity* $m_t = m_f + m_i$, being the sum of the multiplicities of all critical points.

Theorem 2.7 *If m_t is finite,then $m_t = 7$.*

Proof. Assume that there exist no critical points on the y axis; if not rotate and/or shift the y axis a small amount. Eqs.(2.1.7),(2.1.8) may then be written as

$$z\dot{z} = -z\bar{P}(z,u),$$
$$z\dot{u} = \bar{Q}(z,u) - u\bar{P}(z,u),$$

where

$$\bar{P}(z,u) = a_{20} + a_{10}z + a_{11}u + a_{00}z^2 + a_{01}zu + a_{01}u^2,$$
$$\bar{Q}(z,u) = b_{20} + b_{10}z + b_{11}u + b_{00}z^2 + b_{01}zu + b_{02}u^2,$$

and only critical points in the finite part of the (z,u) plane need be considered. Counting multiplicity on $z = \bar{Q}(z,u) - u\bar{P}(z,u) = 0$ yields a contribution of three, whereas $\bar{P}(z,u) = \bar{Q}(z,u) - u\bar{P}(z,u) = 0$ yields a contribution of four to the total multiplicity $m_t = 7$.   □

2.2   Poincaré index of critical points

Apart from its multiplicity, the Poincaré index is also an interesting property to characterize a critical point. The theory of the Poincaré index can be found in many books [ALGM1],[G],[CL],[JS],[BP]; we only state briefly what we need in the present context.

Definition. Let C be a closed, non-selfintersecting curve in the phase plane not passing through any of the critical points of $(P(x,y), Q(x,y))$ in (1.0.1),(1.0.2) and let φ be the angle of (P,Q) with the positive x axis (measured positive in the anti-clockwise direction). Then the *index I(C) of the closed curve C with respect to the vector field P(x,y),Q(x,y)on C* is $\frac{\Delta\varphi}{2\pi}$,where $\Delta\varphi$ is the total change in φ as C is traversed once in the positive (anti-clockwise)direction.

Clearly, I(C) is an integer. This observation makes it easy to see that if the vector field is not only defined on C but also in its interior, wherein also a single critical point exists, a gradual contraction of C does not change I(C) as long as the critical point remains in its interior, so that I(C), where C contracts upon the critical point, may be defined as the *Poincaré index i of a critical point* of the vector field $(P(x,y), Q(x,y))$. In analogy with the finite multiplicity m_f we may also define the *finite index i_f of a quadratic system* , it being the *sum of the indices of the finite critical points*. For a non-degenerate quadratic system, i_f is finite.(It will be shown later that $-2 \leq i_f \leq 2$.) A useful method to determine the index of a critical point,in particular if the point is of higher order, is to make use of Bendixson's formula $i=1+\frac{1}{2}(e-h)$, where e is the number of elliptic sectors and h that of the hyperbolic sectors adjacent to the critical point.

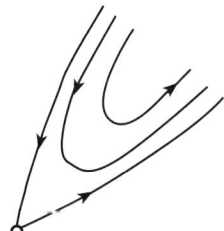

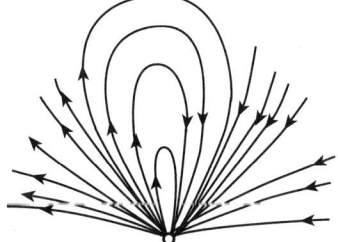

Figure 2.2 Hyperbolic sector   Figure 2.3 Elliptic sector

A *hyperbolic sector* is sketched in Figure 2.2. It is characterized by the existence of two orbits each approaching the critical point in a definite direction, such that there exists a neighborhood of the critical point bounded by these orbits wherein only orbits exist leaving this neighborhood for $t \rightarrow \infty$ and $t \rightarrow \infty$. An *elliptic sector* is sketched in Figure 2.3. It is characterized by the existence of a neighborhood of the critical point containing a loop, being an orbit, defined on $-\infty < t < \infty$ and approaching the critical point for $t \to -\infty$ and $t \to \infty$, such that the interior of the loop only contains orbits which are the same type of loops as well.

The *Poincaré index of critical points* , on the Poincaré circle enclosing the Poincaré disk, and representing the behavior of the system *at infinity*, may be simply taken to be equal to the index of the corresponding critical points in the (z, u) plane $((z, v)$plane).This can be motivated by the observation that the number and character of the sectors adjacent to the critical point are invariant under the projections of the Poincaré sphere onto the Poincaré disk and on the plane $x \equiv 1(y \equiv 1)$.The neighborhood within the Poincaré disk of two diametrically opposite points on the Poincaré(unit)circle then are mapped to a full neighborhood of the corresponding critical point in the (z, u)plane $((z, v)$plane).

We define the *infinite index* i_i of a quadratic system as the *sum of the indices of the infinite critical points*. It is only defined if the number of infinite critical points is finite. Moreover we define the *total index* i_t, where $i_t = i_f + i_i$.

Theorem 2.8 *If the number of finite and infinite critical points is finite,then $i_t = 1$.*

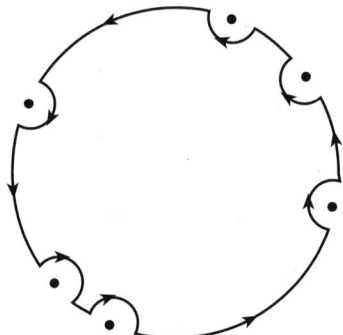

Figure 2.4 The curve C

Proof. Consider, as is sketched in Figure 2.4, a curve C consisting of the Poincaré circle minus the part cut out by the circle with radius ϵ around each critical point on the circle plus that part of these circles located within the Poincaré disk. Determine the rotation of the vector field when the curve C is traversed anti-clockwise. Since the part of C coinciding with the Poincaré circle consists of orbits,the rotation of the vector field on this part tends to 2π as $\epsilon \to 0$.The contribution of the rotation on the circles around the infinite critical points tends to $-2\pi_i$ since they are traversed in the clockwise direction. The result follows from $2\pi - 2\pi_i = 2\pi i_f$.   □

2.3   Types of critical points

Since the determination of the behavior of orbits near a critical point is a local problem, i.e. the notion of a limit may be used, methods are available to investigate what kind of critical points are possible in quadratic systems. In fact, application of these methods has led to a virtually definite body of knowledge on the critical points in quadratic systems. A distinction will be made between finite and infinite critical points.

2.3.1   Finite critical points

Let (x_0, y_0) be a critical point of (1.0.1),(1.0.2), so $P(x_0, y_0) = Q(x_0, y_0) = 0$, then a first division in classes of critical points may be obtained by considering the locally linearized system

$$\dot{x} = P_x(x_0, y_0)(x-x_0)+P_y(x_0, y_0)(y-y_0),   \qquad (2.3.1)$$
$$\dot{y} = Q_x(x_0, y_0)(x-x_0)+Q_y(x_0, y_0)(y-y_0).   \qquad (2.3.2)$$

It is well known that the product $\lambda_1\lambda_2$ of the eigenvalues of the coefficient matrix equals $\Lambda(x_0, y_0) = P_x(x_0, y_0)Q_y(x_0, y_0) - P_y(x_0, y_0)Q_x(x_0, y_0)$ and its sum $\lambda_1 + \lambda_2$ equals $d(x_0, y_0) = P_x(x_0, y_0) + Q_y(x_0, y_0)$, which also is the expression for the divergence of the vector field (P,Q) in (x_0, y_0). The following cases can be distinguished:

(i). Elementary critical points, then $\Lambda(x_0, y_0) \neq 0$;
(ii). Semi-elementary critical points, then $\Lambda(x_0, y_0) = 0, d(x_0, y_0) \neq 0$;
(iii). Degenerate critical points, then $\Lambda(x_0, y_0) = d(x_0, y_0) = 0$.

The latter can further be divided into a) Nilpotent critical points, then $P_x^2 + P_y^2 + Q_x^2 + Q_y^2 \neq 0$ in (x_0, y_0) and b) Critical points in essentially homogeneous systems, then $P_x^2 + P_y^2 + Q_x^2 + Q_y^2 = 0$ in (x_0, y_0).

Elementary critical points

A discussion on elementary critical points can be found in several books (see section 2.2); we recall some basic facts. The qualitative behavior of the orbits near elementary critical points occuring in quadratic systems are sketched in Figure 2.5. If $\Lambda(x_0, y_0) < 0$ the point is a sadddle (Figure 2.5a); it has four hyperbolic sectors and the separatrices approach the point at different angles; the Poincaré index equals -1. If $\Lambda(x_0, y_0) > 0$ the Poincaré index equals $+1$; for this reason it is often called an antisaddle. Further distinction gives: if $\Lambda(x_0, y_0) < \frac{1}{4}d^2(x_0, y_0)$ the point is a node with two approaching directions for the orbits (Figure 2.5 b),if $\Lambda(x_0, y_0) = \frac{1}{4}d^2(x_0, y_0)$ the point is a node with either one or an infinite number of approaching directions for the orbits (Figure 2.5 c,d), and if $\Lambda(x_0, y(_0)) > \frac{1}{4}d^2(x_0, y_0)$ the point is a (weak) focus or a center point (Figure 2.5 e,f), depending on the quadratic terms. In fact, investigation of the conditions on the quadratic terms for which the critical

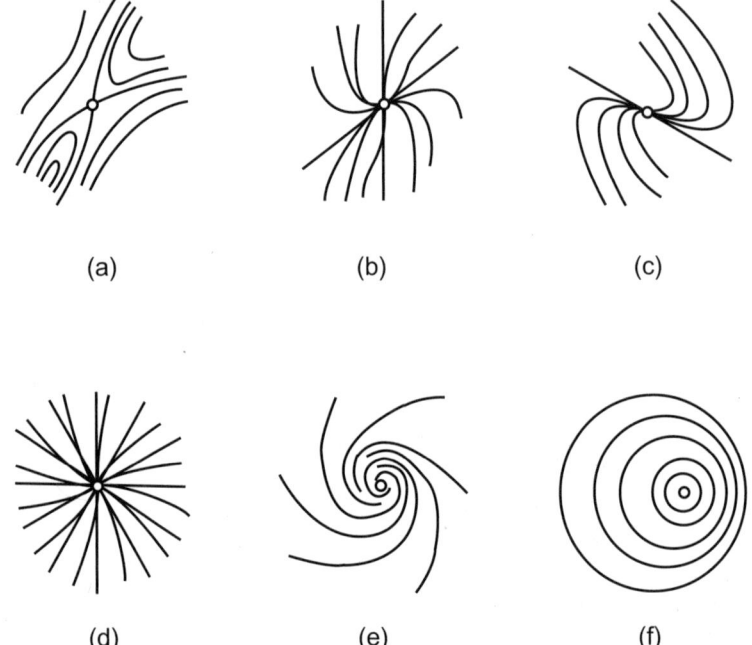

(a)   (b)   (c)

(d)   (e)   (f)

Figure 2.5 Qualitative behavior of orbits near elementary critical points in quadratic systems

point is a center had already begun at the beginning of the twentieth century and has been given ample attention since. In Figure 2.5 the arrows point in the direction of increasing t, yielding the cases of stable node and focus; the unstable case corresponds to reversal of the arrows. For the saddle point and the center point the (in)stability is not affected by reversal of the arrows. For a saddle point we will use the notation e^{-1}, for an antisaddle e^1 and for a complex critical point c_1^0. For higher order points we will use the notation m_q^i, if the point is real and c_q^0 if complex; i then means the value of its Poincaré index and q its multiplicity.

Semi-elementary critical points

If a quadratic system contains a semi-elementary critical point, by a shift it can be transformed to the origin and the system can be written as

$$\dot{x} = a_{20}x^2 + a_{11}xy + a_{02}y^2, \tag{2.3.3}$$
$$\dot{y} = y + b_{20}x^2 + b_{11}xy + b_{02}y^2, \tag{2.3.4}$$

[ALGM1], as a result of which the center manifold(s) are in (0,0) tangent to the x axis. This also applies to all orbits except for two orbits approaching along the positive and negative y axis, respectively. System (2.3.3),(2.3.4) may be realized by using theorem 65 in [ALGM1]. Apart from, possibly, the ∞ isocline, all isoclines are also tangent in (0,0) to the center manifold(s) and can be represented by $y + (ka_{20} + b_{20})x^2 + (ka_{11} + b_{11})xy + (ka_{02} + b_{02})y^2 = 0$, where $-\infty < k < \infty$. The function $\Lambda(x, y) = 0$ can be written as $\Lambda(x,y) = 2a_{20}x + a_{11}y + 2c_{45}x^2 + 4c_{46}xy + 2c_{56}y^2 = 0$.

There exist four types of semi-elementary critical points:
$(i) m_2^0$: a second order saddle node; if $a_{20} \neq 0$,
$(ii) m_3^{-1}$: a third order saddle; if $a_{20} = 0, a_{11}b_{20} > 0$,
$(iii) m_3^1$: a third order node; if $a_{20} = 0, a_{11}b_{20} < 0$,
$(iv) m_4^0$: a fourth order node; if $a_{20} = a_{11} = 0, a_{02}b_{20} \neq 0$.

A sketch of the second order saddle node $m_2^0(a_{20} > 0)$ and of the fourth order saddle node $m_4^0(a_{20} > 0)$ is given in Figure 2.6. The saddle part is then on the negative side of the x axis and the nodal part on the positive side. For $a_{20} < 0, a_{02} < 0$ these parts are interchanged and the flow is reversed. For the third order saddle m_3^{-1} and node m_3^1 the topological structure is as in Figure 2.5(a) and 2.5(b),respectively.

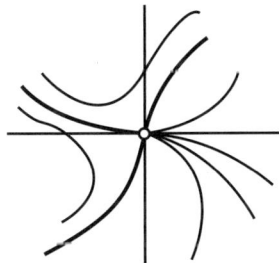

Figure 2.6 A second or fourth order saddle node

In m_2^0 the ∞ isocline intersects (0,0) in two (different or coinciding, real or complex)directions, whereas the other isoclines are all in (0,0) tangent to the center manifold(s), all having a different curvature there.The curve $\Lambda(x,y) = 0$ intersects the center manifold(s) transversally.

In m_3^{-1} and m_3^1 the ∞ isocline consists of one branch passing through (0,0) transversal to the center manifold(s) and one branch doing so tangentially. All other isoclines are in (0,0) tangent to the center manifold and have there an equal non-zero value of the curvature. The curve $\Lambda(x,y) = 0$ is tangent to the x axis on the other side and has the opposite curvature.

In m_4^0 all isoclines are in (0,0) tangent to the center manifold(s) and, apart from the ∞ isocline,having there the same non-zero curvature.The curve $\Lambda(x,y)$=0 consists of two branches: one tangent to the center manifold(s) and one transversally to it.

Nilpotent critical points

If a quadratic system contains a nilpotent critical point (x_0, y_0), by a shift it can be transferred to the origin and the system can be written as

$$\dot{x} = y + a_{20}x^2 + a_{11}xy + a_{02}y^2 , \qquad (2.3.5)$$
$$\dot{y} = b_{20}x^2 + b_{11}xy + b_{02}y^2, \qquad (2.3.6)$$

[ALGM1], as a result of which the eigenvector of the coefficient matrix corresponding to the linear part is directed along the x axis. This also applies to all orbits approaching the critical point. System (2.3.5),(2.3.6) may be analyzed using theorems 66 and 67 in [ALGM1]. Apart from, possibly, the o isocline, all isoclines are in (0,0) also directed along the x axis and can be represented by $y + (a_{20} + kb_{20})x^2 + (a_{11} + kb_{11})xy + (a_{02} + kb_{02})y^2 = 0$, where$-\infty < k < \infty$. The function $\Lambda(x,y) = 0$ can be written as $-2b_{20}x - b_{11}y + 2c_{45}x^2 + 4c_{46}xy + 2c_{56}y^2 = 0$.

There exist four types of nilpotent critical points in quadratic systems:

(i) m_2^0: a second order cusp;if $b_{20} \neq 0$,

(ii) m_3^{-1}: a third order saddle;if $b_{20} = 0, a_{20}b_{11} < 0$,

(iii) m_3^1: a third order point with an elliptic and a hyperbolic sector; if $b_{20} = 0, a_{20}b_{11} > 0$,

(iv)m_4^0: a fourth order saddle node;if $b_{20} = b_{11} = 0, a_{20}b_{02} \neq 0$.

A sketch of these critical points is given in Figure 2.7. For 2.7(a), $b_{20} > 0$, 2.7(b), $a_{20} < 0$, 2.7(c), $a_{20} > 0$, 2.7(d), $a_{20} > 0$. The behavior of the isoclines and the curve $\Lambda(x, y){=}0$ in $(0,0)$ is similar to that near a semi-elementary critical point.

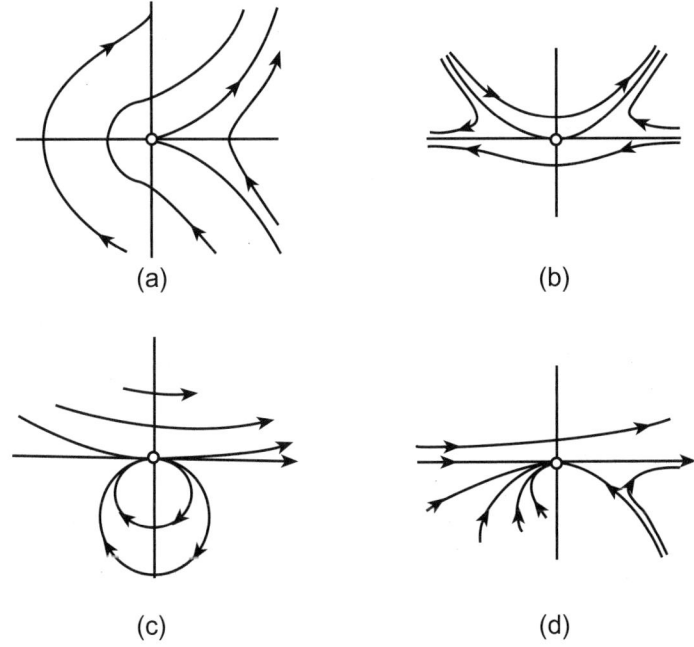

(a)   (b)

(c)   (d)

Figure 2.7 Nilpotent critical points in quadratic systems

In the cusp m_2^0 the 0 isocline intersects $(0,0)$ in two (different or coinciding, real or complex) directions, whereas the other isoclines are all in $(0,0)$ tangent to the x axis, all having a different curvature there.The curve $\Lambda(x, y) = 0$ intersects the x axis in $(0,0)$ transversally.

In the third order points m_3^{-1} and m_3^1 the 0 isocline consists of one branch passing through $(0,0)$ transversal to the x axis and one branch coinciding with it. All other isoclines are in $(0,0)$ tangent to the x axis and have there an equal non-zero value of the curvature. The curve $\Lambda(x, y) = 0$ is tangent to the x axis on the other side and has the opposite curvature.

In m_4^0 all isoclines are in $(0,0)$ tangent to the x axis and, apart from the 0 isocline,having there the same non-zero curvature. The curve $\Lambda(x, y) = 0$ consists of two branches: one coinciding with the x axis and one transversal to it.

Critical points in essentially homogeneous quadratic systems.

If a quadratic system has a critical point (x_0, y_0) such that $\Lambda(x_0, y_0) = d(x_0, y_0) = 0$ and moreover in (x_0, y_0) there is $P_x^2 + P_y^2 + Q_x^2 + Q_y^2 = 0$, then by a shift the point can be transferred to the origin so that the homogeneous system

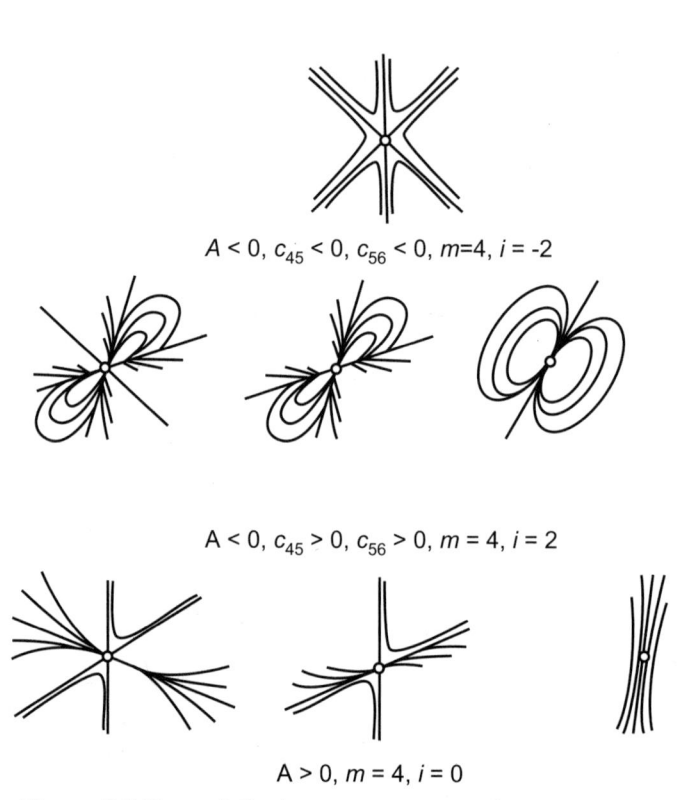

$A < 0, c_{45} < 0, c_{56} < 0, m=4, i = -2$

$A < 0, c_{45} > 0, c_{56} > 0, m = 4, i = 2$

$A > 0, m = 4, i = 0$

Figure 2.8 Essentially homogeneous quadratic systems

$$\dot{x} = a_{20}x^2 + a_{11}xy + a_{02}y^2, \qquad (2.3.7)$$
$$\dot{y} = b_{20}x^2 + b_{11}xy + b_{02}y^2, \qquad (2.3.8)$$

results. We recall that $c_{45} = a_{20}b_{11} - a_{11}b_{20}$, $c_{46} = a_{20}b_{02} - a_{02}b_{20}$, $c_{56} = a_{11}b_{02} - a_{02}b_{11}$, $A = c_{46}^2 - c_{45}c_{56}$, and $A \neq 0$ if $(0,0)$ is the only critical point. Then we may distinguish the following cases for the topological structure near the critical point. See Figure 2.8.

(i)$A < 0, c_{45} + c_{46} < 0$. Then $C_2(\theta) = 0$ has three different real solutions each representing a straight line solution through $(0,0)$ and yielding six hyperbolic sectors. The Poincaré index equals -2, so the point is an m_4^{-2} point.

(ii)$A > 0$. Then $C_2(\theta) = 0$ has either one, two or three real solutions and possibly two complex solutions. There exist two hyperbolic sectors either with an opening angle π or less. In the latter case there also exist two parabolic sectors in between the hyperbolic sectors. The Poincaré index equals 0, so the point is an m_4^0 point.

(iii)$A < 0, c_{45} + c_{46} > 0$. Then $C_2(\theta) = 0$ also has either one, two or three solutions and possibly two complex solutions. There exist two elliptic sectors, either with opening angle $\leq \pi$. In the latter case there also exist two parabolic sectors in between the elliptic sectors. The Poincaré index equals 2, so the point is an m_4^2 point.

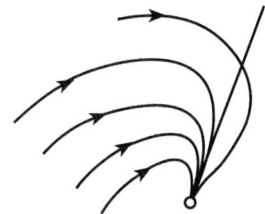

Figure 2.9 Parabolic sector

A *parabolic sector* is sketched in Figure 2.9. It is characterized by the existence of two orbits each approaching the critical point in the same direction, such that there exists a neighborhood of the critical point bounded by these orbits wherein only orbits exist that also approach the critical point in this direction. These orbits either all approach the critical point for $t \to \infty$ or all for $t \to -\infty$.

The statements given above will be proved in a later discussion of homogeneous quadratic systems.

2.3.2 Infinite critical points

In [95,RK1] an answer is given to the question of what types of infinite critical points are possible and what their relation is to the coefficients a_{ij} and b_{ij} in (1.0.1),(1.0.2).For the analysis (2.1.7),(2.1.8) may be used ,which is obtained from (1.0.1),(1.0.2) by the transformation $z = \frac{1}{x}, u = \frac{y}{x}$:

$$z\dot{z} = -z(a_{20} + a_{10}z + a_{11}u + a_{00}z^2 + a_{01}zu + a_{02}u^2) \equiv \hat{P}(z, u), \qquad (2.3.9)$$
$$z\dot{u} = b_{20} + b_{10}z + (-a_{20} + b_{11})u + b_{00}z^2 + (-a_{10} + b_{01})zu + (-a_{11} + b_{02})u^2$$

$$-a_{01}zu^2 - a_{00}z^2 u - a_{02}u^3 \equiv \hat{Q}(z,u), \tag{2.3.10}$$

or the system obtained by the transformation $z = \frac{1}{y}, v = \frac{x}{y}$:

$$z\dot{z} = -z(b_{02} + b_{01}z + b_{11}v + b_{00}z^2 + b_{10}zv + b_{20}v^2) \equiv \tilde{P}(z,v), \tag{2.3.11}$$
$$z\dot{v} = a_{02} + a_{01}z + (-b_{02} + a_{11})v + a_{00}z^2 + (a_{10} - b_{01})zv + (a_{20} - b_{11})v^2$$
$$-b_{10}zv^2 - b_{00}z^2 v - b_{20}v^3 \equiv \tilde{Q}(z,v). \tag{2.3.12}$$

The critical points at infinity are given by $(0,u)$ or/and $(0,v)$, where u and v are the solutions of

$$f(u) = b_{20} + (-a_{20} + b_{11})u + (-a_{11} + b_{02})u^2 - a_{02}u^3 = 0, \tag{2.3.13}$$
$$g(v) = b_{20}v^3 + (-a_{20} + b_{11})v^2 + (-a_{11} + b_{02})v - a_{02} = 0, \tag{2.3.14}$$

respectively.

Obviously, if $b_{20} = -a_{20} + b_{11} = -a_{11} + b_{02} = a_{02} = 0$, all points at infinity are critical points. The possible character of these points will not be discussed here but will follow from the analysis of the phase portraits of quadratic systems in classes having finite multiplicity.

Isolated infinite critical points are connected by orbits on the Poincaré circle ($z \equiv 0$). So all real infinite critical points have at least one orbit approaching the point in a definite direction (modulo π).This means that not all types of finite critical points are possible at infinity ; yet new types become evident as well.

As for finite critical points a first division into classes of infinite critical points may be obtained by considering the system linearized in the critical point.

We recall that for the eigenvalues we can find

$$\lambda_z = \hat{P}_z(0,u) = -a_{20} - a_{11}u - a_{02}u^2, \tag{2.3.15}$$
$$\lambda_u = \hat{Q}_u(0,u) = f'(u) = b_{11} - a_{20} + 2(b_{02} - a_{11})u - 3a_{02}u^2, \tag{2.3.16}$$

corresponding to (2.3.9),(2.3.10), whereas corresponding to (2.3.13),(2.3.14)we find

$$\lambda_z = \tilde{P}_z(0,v) = -b_{02} - b_{11}v - b_{20}v^2, \tag{2.3.17}$$
$$\lambda_v = \tilde{Q}_v(0,v) = g'(v) = -a_{11} + b_{02} + 2(-a_{20} + b_{11})v + 3b_{20}v^2, \tag{2.3.18}$$

and u and v are the solutions of (2.3.13), and (2.3.14),respectively.

If Λ_c is the product of the eigenvalues of the system, linearized in the critical point, and d_c their sum,the following cases can be distinguished:

(i)Elementary critical points,then $\Lambda_c \neq 0$,

(ii)Semi-elementary critical points,then $\Lambda_c = 0, d_c \neq 0$,

(iii)Degenerate critical points,then $\Lambda_c = d_c = 0$,

They can further be divided into

a. Nilpotent critical points,then $\hat{P}_z^2 + \hat{P}_u^2 + \hat{Q}_z^2 + \hat{Q}_u^2 \neq 0 (\tilde{P}_z^2 + \tilde{P}_v^2 + \tilde{Q}_z^2 + \tilde{Q}_v^2 \neq 0)$ in the critical point.

b. Critical points of homogeneous quadratic systems, then $\hat{P}_z^2 + \hat{P}_u^2 + \hat{Q}_z^2 + \hat{Q}_u^2 = 0 (\tilde{P}_z^2 + \tilde{P}_v^2 + \tilde{Q}_z^2 + \tilde{Q}_v^2 = 0)$ and the quadratic terms have no common factor.

c. Critical points of cubic systems that cannot occur in quadratic systems,then $\hat{P}_z^2 + \hat{P}_u^2 + \hat{Q}_z^2 + \hat{Q}_u^2 = 0 (\tilde{P}_z^2 + \tilde{P}_v^2 + \tilde{Q}_z^2 + \tilde{Q}_v^2 = 0)$ and the quadratic terms have (a) common factor(s).

Elementary critical points

As was already remarked, all real infinite critical points have at least one orbit approaching the point in a definite direction (modulo π).This rules out the center point and the focus and leaves the saddle for $\Lambda_c < 0$ and the nodes for $\Lambda_c > 0$. Analytically, this can be seen as follows. Rotate the axes such that the critical point under consideration is at the end of the x axis so that $\Lambda_c = a_{20}(a_{20} - b_{11}), d_c = -2a_{20} + b_{11}$. Then $\Lambda_c - \frac{1}{4}d_c^2 - \frac{1}{4}(d + 2a_{20})^2 \leq 0$. All three types of nodes are possible. For a saddle point we will use the notation E^{-1} and for a node E^1. Elementary complex points will be indicated by $C_{0,1}^0$.

Semi-elementary critical points

As for finite semi-elementary critical points there exist three topological types: saddles ,nodes and saddle nodes. However, a distinction should be made as to whether the center manifold(s) is (are) tangent to the Poincaré circle or transversal to it. Moreover, the possible multiplicity of the critical points have to be reconsidered, since now we must also take into account that finite and infinite critical points may coincide.

If in the critical point the *center manifold(s) is (are) tangent to the Poincaré circle,* the point is transversally hyperbolic and tangentially at most of order 3 since up to three infinite critical points can coincide.That leaves the points $M_{0,2}^i$ and $M_{0,3}^i$. In fact the analysis in [95,RK1] shows that there can exist a second order saddle node $M_{0,2}^0$, a third order saddle $M_{0,3}^{-1}$ or a third order node $M_{0,3}^1$; the latter two evidently can only exist if $m_f = 4$.

The conditions on the coefficients in (1.0.1),(1.0.2) if the critical point is at the ends of the x axis are also given in [95,RK].

If in the critical point the *center manifold(s) is (are) transversal to the Poincaré circle,* the point is tangentially hyperbolic and transversally at most of order 4 since up till 4 finite critical points can coincide with an infinite critical point. This leaves the possibilities $M_{1,1}^i, M_{2,1}^i, M_{3,1}^i$ and $M_{4,1}^i$. In fact the analysis in [95,RK] shows that there can exist a second order or a fourth order saddle node $M_{1,1}^0$ or $M_{3,1}^0$, a third order or a fifth order saddle $M_{2,1}^{-1}$ or $M_{4,1}^{-1}$ or a third order or a fifth order node $M_{2,1}^1$ or $M_{4,1}^1$; the points $M_{4,1}^i$ evidently can only exist if $m_f=0$.The conditions on (1.0.1),(1.0.2) if the critical point is at the ends of the x axis are also given in [95,RK1].

Nilpotent critical points

Since a nilpotent critical point at infinity $M_{p,q}$ is non-hyperbolic in both tangential and transversal directions with respect to the Poincaré circle, there should be $1 \leq p \leq 4, 2 \leq q \leq 3$. This rules out the second order cusp as encountered in the finite plane. Saddles,saddle nodes and points with an elliptic and a hyperbolic sector may, however, exist.The impossibility of a cusp also follows from the fact that $z \equiv 0$ is an orbit.It follows that there are possible a third order nilpotent saddle $M_{1,2}^{-1}$ (only for $m_f = 3$) and a fifth order saddle $M_{3,2}^{-1}$ (only for $m_f = 1$);a third order point with an elliptic and a hyperbolic sector $M_{1,2}^1$ and a fifth order point of that type $M_{3,2}^1$ (only for $m_f = 1$); fourth order nilpotent saddle nodes $M_{2,2}^0$ (only for $m_f = 2$) and $M_{1,3}^0$ (only for $m_f = 3$) and a sixth order nilpotent saddle node $M_{4,2}^0$ (only for $m_f = 0$). Moreover there also exists a fifth order node $M_{2,3}^1$ (only for m_f =2) having only one approaching direction to the critical point (the Poincaré circle) [95,RK].

Highly degenerate critical points

For such a critical point there is $f(u) = \Lambda(z, u) = d(z, u) = \hat{P}_z^2 + \hat{P}_u^2 + \hat{Q}_z^2 + \hat{Q}_u^2 = 0$ in $(0,u)$; the coordinates of the point. Let, furthermore, the highly degenerate critical point be located in $(0,0)$, then (2.3.9),(2.3.10) can be written as

$$z\dot{z} = -a_{10}z^2 - a_{11}zu - a_{00}z^3 - a_{01}zu - a_{02}zu^2, \tag{2.3.19}$$
$$z\dot{u} = b_{00}z^2 + (-a_{10}+b_{01})zu + (-a_{11}+b_{02})u^2 - a_{00}z^2u - a_{01}zu^2 - a_{02}u^3. \tag{2.3.20}$$

The quadratic terms have no common factor if $\gamma \equiv (a_{11} - b_{02})(a_{10}^2 b_{02} - a_{10}a_{11}b_{01} + a_{11}^2 b_{00}) \neq 0$. It appears from the analysis in [95,RK] that, depending on the coefficients, all types of critical points possible in a homogeneous quadratic system can be realized for $\gamma \neq 0$. They are of fourth order and of types $M_{2,2}^{-2}$, $M_{2,2}^0$ and $M_{2,2}^2$ and can occur only for $m_f = 0$, 1 or 2 .If $\gamma = 0$ the quadratic terms are no longer dominant and the critical point is only possible in a cubic (or of higher order). They are of fifth, sixth or seventh order and occur only for $m_f = 0, 1$ or 2. The analysis in [95,RK] shows that points of fifth order are possible for the cases $M_{3,2}^{-1}$(one type and for $m_f=1$), $M_{3,2}^1$(three types and for $m_f=0$ and 1),$M_{2,3}^{-1}$(one type and for $m_f=2$)and $M_{2,3}^1$(three types and for $m_f=2$). For points of sixth order there are the cases:$M_{4,2}^0$(six types and $m_f=0$), $M_{4,2}^2$(one type and $m_f=0$) and $M_{3,3}^0$(four types and $m_f=1$). For points of the seventh order there are six types of points $M_{4,3}^1$ and they only occur for $m_f=0$. A discussion of these higher order critical points will be given later in relation to the classification of phase portraits for $m_f=0,1$ and 2.

For complex critical points there exist the possibilities $C_{0,1}^0, C_{1,1}^0$ and $C_{2,1}^0$.

Chapter 3

Isoclines, critical points and classes of quadratic systems

3.1 Introduction

As was remarked before, a phase portrait is characterized by its critical points, separatrix structure and periodic solutions. In practice the analysis of a phase portrait usually starts with determination of the number, relative location and character of the critical points, both in the finite part of the plane and at infinity. This is so because, in itself, it contains vital information about the phase portrait, whereas it is also important for determination of the separatrix structure and the periodic solutions of the system at hand. As a result it seems natural, if classes of quadratic systems are to be defined, to base this on properties of the critical points, being the zero points of the vector field $(P(x,y), Q(x,y))$ in (1.0.1),(1.0.2).

Now, since orbits of (1.0.1),(1.0.2) are obtained by integration of the vector field $(P(x,y), Q(x,y))$, making a drawing that shows the direction of this field in a sufficient number of points can already give valuable information on the properties of critical points and orbits of the system under investigation. Rather than considering the direction of $(P(x,y), Q(x,y))$ in a number of discrete points for a specific system, the direction field may also be considered to be given by the isoclines, being curves of equal direction of $(P(x,y), Q(x,y))$, thus enabling an analytical and geometrical approach to the direction field for classes of systems. In this book we make a systematic study of these isoclines and investigate how knowledge of their properties helps to obtain information about critical points. The geometrical properties of the isoclines then lead in a natural way to a division into classes of quadratic systems, as defined by the possible combinations of finite and infinite critical points in these systems. A next step in the classification would then be to give for each of these classes a more detailed description of its critical points, its separatrix structure and the number and location of its periodic solutions.

The isoclines of $(1.0.1),(1.0.2)$ are given by

$$\lambda P(x,y) + \mu Q(x,y) = \lambda a_{00} + \mu b_{00} + (\lambda a_{10} + \mu b_{10})x + (\lambda a_{01} + \mu b_{01})y$$
$$+(\lambda a_{20} + \mu b_{20})x^2 + (\lambda a_{11} + \mu b_{11})xy + (\lambda a_{02} + \mu b_{02})y^2 = 0, \qquad (3.1.1)$$

where $\lambda, \mu \in \mathbb{R}$, $\lambda^2 + \mu^2 \neq 0$, which is a pencil of conics. (On occasion λ and μ may also be allowed to take complex values.) For real values of λ and μ, $(3.1.1)$ is of elliptic, parabolic or hyperbolic type according as $D_1 < 0, = 0$ or > 0, respectively, where the discriminant D_1 is given by

$$D_1 = (\lambda a_{11} + \mu b_{11})^2 - 4(\lambda a_{02} + \mu b_{02})(\lambda a_{20} + \mu b_{20}) =$$
$$(a_{11}^2 - 4a_{20}a_{02})\lambda^2 - 2(2a_{20}b_{02} - a_{11}b_{11} + 2a_{02}b_{20})\lambda\mu + (b_{11}^2 - 4b_{20}b_{02})\mu^2. (3.1.2)$$

Considered as a function of λ and μ, $(3.1.2)$ is of elliptic, parabolic or hyperbolic type according as $D_2 < 0, = 0$, or > 0, respectively, where the discriminant D_2 is given by

$$D_2 = 4(2a_{20}b_{02} - a_{11}b_{11} + 2a_{02}b_{20})^2 - 4(a_{11}^2 - 4a_{20}a_{02})(b_{11}^2 - 4b_{20}b_{02})$$
$$= 16(c_{46}^2 - c_{45}c_{56}) = 16A, \qquad (3.1.3)$$

and where $c_{45} = a_{20}b_{11} - a_{11}b_{20}, c_{46} = a_{20}b_{02} - a_{02}b_{20}, c_{56} = a_{11}b_{02} - a_{02}b_{11}$. Now if $A < 0$, so $D_2 < 0$, D_1 is for all values of $(\lambda, \mu) \neq (0,0)$ either positive or negative. All isoclines are hyperbolic if $D_1 > 0$ and ellipses if $D_1 < 0$. Later analysis will show, however, that $D_1 < 0$ cannot occur, so that only hyperbolic isoclines are possible if $A < 0$. If $A = 0$, and $(3.1.2)$ is not degenerate, in addition to hyperbolic isoclines there exist two coinciding real isoclines of parabolic type, whereas for $A = 0$ and if $(3.1.2)$ is degenerate, i.e. $a_{11}^2 - 4a_{20}a_{02} = 2a_{20}b_{02} - a_{11}b_{11} + 2a_{20}b_{20} = b_{11}^2 - 4b_{20}b_{02} = 0$, all real isoclines are of parabolic type. If $A > 0$ there exist real isoclines of both elliptic and hyperbolic type as well as two real different isoclines of parabolic type.

All members of the pencil of conics, being the isoclines $(3.1.1)$ of $(1.0.1)$, $(1.0.2)$ go through the base points of the pencil; they coincide with the critical points of the vector field $(P(x,y), Q(x,y))$. Interesting elements of a pencil of conics are its degenerate conics and the central conic. A degenerate conic consists of two lines, intersecting, parallel or coinciding, real or complex, in the finite part of the plane or at infinity. The central conic is the geometrical locus of the midpoints of the conics in the pencil.

Our goal, then , is to find the relation of the various types of degenerate isoclines and the character of the central conic on one hand with the properties of the critical points , such as number, relative location and character, on the other hand.

3.2 Pencil of isoclines, central conic and degenerate isoclines

The *degenerate conics of the pencil of conics formed by the isoclines* (3.1.1) are determined by the vanishing of the Hessian determinant H of $\lambda P(x, y) + \mu Q(x, y) = 0$, where H is given by

$$\begin{vmatrix} \lambda a_{00} + \mu b_{00} & \frac{1}{2}(\lambda a_{10} + \mu b_{10}) & \frac{1}{2}(\lambda a_{01} + \mu b_{01}) \\ \frac{1}{2}(\lambda a_{10} + \mu b_{10}) & \lambda a_{20} + \mu b_{20} & \frac{1}{2}(\lambda a_{11} + \mu b_{11}) \\ \frac{1}{2}(\lambda a_{01} + \mu b_{01}) & \frac{1}{2}(\lambda a_{11} + \mu b_{11}) & \lambda a_{02} + \mu b_{02} \end{vmatrix}$$

So

$$\alpha \lambda^3 + \beta \lambda^2 \mu + \gamma \lambda \mu^2 + \delta \mu^3 = 0, \tag{3.2.1}$$

where

$\alpha = a_{00} a_{20} a_{02} - \frac{1}{4} a_{00} a_{11}^2 - \frac{1}{4} a_{10}^2 a_{02} + \frac{1}{4} a_{10} a_{01} a_{11} - \frac{1}{4} a_{01}^2 a_{20},$

$\beta = a_{00} a_{20} b_{20} + a_{00} a_{20} b_{02} + a_{20} a_{02} b_{00} - \frac{1}{2} a_{00} a_{11} b_{11} - \frac{1}{4} a_{11}^2 b_{00} - \frac{1}{2} a_{10} a_{02} b_{10}$
$- \frac{1}{4} a_{10}^2 b_{02} + \frac{1}{4} a_{10} a_{11} b_{01} + \frac{1}{4} a_{10} a_{01} b_{11} + \frac{1}{4} a_{01} a_{11} b_{10} - \frac{1}{2} a_{01} a_{20} b_{01} - \frac{1}{4} a_{01}^2 b_{20},$

$\gamma = a_{00} b_{20} b_{02} + a_{02} b_{00} b_{20} + a_{20} b_{00} b_{02} - \frac{1}{4} a_{00} b_{11}^2 - \frac{1}{2} a_{11} b_{00} b_{11} - \frac{1}{4} a_{02} b_{10}^2$
$- \frac{1}{2} a_{10} b_{10} b_{02} + \frac{1}{4} a_{10} b_{01} b_{11} + \frac{1}{4} a_{11} b_{10} b_{01} + \frac{1}{4} a_{01} b_{10} b_{11} - \frac{1}{4} a_{20} b_{01}^2 - \frac{1}{2} a_{01} b_{01} b_{20},$

$\delta = -\frac{1}{4} b_{00} b_{11}^2 + b_{00} b_{20} b_{02} - \frac{1}{4} b_{10}^2 b_{02} + \frac{1}{4} b_{10} b_{01} b_{11} - \frac{1}{4} b_{01}^2 b_{20}.$

If $\alpha \neq 0, \mu = 0$ is not a solution of (3.2.1) and we write

$$f(\frac{\lambda}{\mu}) \equiv (\frac{\lambda}{\mu})^3 + \frac{\beta}{\alpha}(\frac{\lambda}{\mu})^2 + \frac{\gamma}{\alpha}\frac{\lambda}{\mu} + \frac{\delta}{\alpha} = 0, \tag{3.2.2}$$

then

$$f'(\frac{\lambda}{\mu}) = 3(\frac{\lambda}{\mu})^2 + 2\frac{\beta}{\alpha}\frac{\lambda}{\mu} + \frac{\gamma}{\alpha}, \tag{3.2.3}$$

and (3.2.2) has a double or triple root if $f(\frac{\lambda}{\mu}) = f'(\frac{\lambda}{\mu}) = 0$, which leads to the requirement D=0,where

$$D = \frac{1}{(3\alpha)^4}[(-9\alpha\delta + \beta\gamma)^2 - 4(3\alpha\gamma - \beta^2)(3\beta\delta - \gamma^2)]. \qquad (3.2.4)$$

From the additional condition $f''(\frac{\lambda}{\mu}) = 0$ it may be seen that a triple root prevails if $3\alpha\gamma - \beta^2 = 0$ and a double root (plus a simple root) if $3\alpha\gamma - \beta^2 \neq 0$. The triple root is given by $\lambda/\mu = -\beta/3\alpha$ and this occurs for $3\alpha/\beta = \beta/\gamma = \gamma/3\delta$. The double root is given by $\lambda/\mu = (9\alpha\delta - \beta\gamma)/2(\beta^2 - 3\alpha\gamma)$ and occurs for D=0, $3\alpha\gamma - \beta^2 \neq 0$. It can further be shown that (3.2.2) has three simple real roots if D<0 and one simple real and two complex conjugate roots if D>0.

If $\alpha = 0, \mu = 0$ is a solution of (3.2.1) and the equation may be solved as a quadratic equation if $\beta \neq 0$, as a linear equation if $\beta = 0, \gamma \neq 0$, and has a triple solution $\mu = 0$ if $\beta = \gamma = 0, \delta \neq 0$. Finally if $\alpha = \beta = \gamma = \delta = 0$ the equation is degenerate.

The three solution pairs (λ, μ) of (3.2.1) each determine a degenerate conic which is either of hyperbolic type, then consisting of two intersecting real lines, of elliptic type, then consisting of two complex lines intersecting at a real point, or of parabolic type, then consisting of two parallel lines, real different, coinciding or complex; none, one or both of these lines may be located at infinity. There exist thus *six isoclinic lines*. If (3.2.1) is degenerate, all pairs (λ, μ) are solutions and there exist *infinitely many degenerate isoclines*.

The geometrical locus of the midpoints of the conics in the pencil $\lambda P(x, y) + \mu Q(x, y) = 0$ will be indicated as the *central conic of the pencil* and is given by

$$\Lambda(x, y) \equiv P_x(x, y)Q_y(x, y) - P_y(x, y)Q_x(x, y) =$$

$$c_{23} - 2(c_{34} - \frac{1}{2}c_{25})x + 2(c_{26} - \frac{1}{2}c_{35})y + 2c_{45}x^2 + 4c_{46}xy + 2c_{56}y^2 = 0. (3.2.5)$$

Obviously the type of the central conic is determined by the sign of the determinant $A \equiv c_{46}^2 - c_{45}c_{56}$. If $A < 0$ it is of elliptic type, if A=0 of parabolic type and if A>0 of hyperbolic type. The central conic is degenerate if H=0, where H is the Hessian determinant

$$H = -4c_{23}(c_{46}^2 - c_{45}c_{56})$$

$$-2[c_{45}(c_{26} - \frac{1}{2}c_{35})^2 + 2c_{46}(c_{26} - \frac{1}{2}c_{35})(c_{34} - \frac{1}{2}c_{25}) + c_{56}(c_{34} - \frac{1}{2}c_{25})^2](3.2.6)$$

The center of the central conic is located in $(x_c, y_c) = -1/4A(B_1, B_2)$, whereas $\Lambda(x_c, y_c) = -H/4A$(A statement about our notation can be found following Theorem 2.3). Both A and H are sign invariant under translation and rotation of the coordinate axes. By a translation such that the center of the central conic is in the origin and a rotation such that the axes are along the coordinate axes, for $A \neq 0, H \neq 0$, (3.2.5) may be written as

$$\frac{4AS_1}{H}x^2 + \frac{4AS_2}{H}y^2 = 1, \qquad (3.2.7)$$

where $S_i(i=1,2)$ are the roots

$$S_{1,2} = c_{45} + c_{56} \pm \sqrt{(c_{45} + c_{56})^2 + 4(c_{46}^2 - c_{45}c_{56})} \qquad (3.2.8)$$

of

$$S^2 - 2(c_{45} + c_{56})S - 4(c_{46}^2 - c_{45}c_{56}) = 0 \qquad (3.2.9)$$

showing the third sign invariant $c_{45} + c_{56}$, invariant under translation and rotation.

3.3 Central conic and finite multiplicity of a quadratic system

In Theorems 2.1-2.3 the finite multiplicity of a quadratic system was characterized by conditions in parameter space $a_{ij}, b_{ij}(i, j = 0, 1, 2)$. It is also possible to a certain extent to characterize m_f geometrically.

Theorem 3.1 *The finite multiplicity m_f of a quadratic system is equal to 4 if and only if the central conic is of elliptic or hyperbolic type and equal to 3 if and only if the central conic is a non-degenerate parabola. If the central conic is a degenerate parabola m_f equals 0,1,2 or ∞.*

Proof. From Theorems 2.1-2.3 it follows that $m_f = 4$ if and only if $A \neq 0$, thus if and only if the central conic is of elliptic or hyperbolic type. It also follows from these theorems that $m_f = 3$ if and only if $A = 0, B \equiv B_1^2 + B_2^2 \neq 0$. The central conic should therefore be of parabolic type. Furthermore

$$B \equiv B_1^2 + B_2^2 = 4(c_{45}^2 + c_{46}^2)(c_{26} - \frac{1}{2}c_{35})^2$$

$$+8c_{46}(c_{45} + c_{56})(c_{26} - \frac{1}{2}c_{35})(c_{34} - \frac{1}{2}c_{25}) + 4(c_{46}^2 + c_{56}^2)(c_{34} - \frac{1}{2}c_{25})^2,$$

or, with $A = c_{46}^2 - c_{45}c_{56} = 0$, also

$$B = 4(c_{45} + c_{56})[c_{45}(c_{26} - \frac{1}{2}c_{35})^2 + 2c_{46}(c_{26} - \frac{1}{2}c_{35})(c_{34} - \frac{1}{2}c_{25})$$

$$+c_{56}(c_{34} - \frac{1}{2}c_{25})^2] = -2(c_{45} + c_{56})H, \qquad (3.3.1)$$

thus $A = 0$, $B \neq 0$ implies $H \neq 0$ and the parabola is non-degenerate. Also, if the central conic is a non-degenerate parabola, then $A = 0$, $H \neq 0$ so $B \neq 0$ since $c_{45} + c_{56} \neq 0$ as $c_{45} + c_{56} = 0$ implies $c_{45} = c_{46} = c_{56} = 0$ and H=0. So $m_f = 3$. □

Remark. Note that $m_f = 3(A = 0, B \neq 0)$ also implies $c_{45}^2 + c_{46}^2 + c_{56}^2 \neq 0$ so that there exists only one transversally non-hyperbolic infinite critical point. The axis of the parabolic central conic points in the direction of this point, as may be shown using Theorem 2.4.

The properties of the central conic are less conclusive if a quadratic system has a finite multiplicity $0 \leq m_f \leq 2$. A distinction should be made whether there exist one or two transversally non-hyperbolic infinite critical points.

Theorem 3.2 *A quadratic system with a finite multiplicity m_f has one transversally non-hyperbolic infinite critical point and $0 \leq m_f \leq 2$ if and only if the central conic is a parabola , degenerated into two real parallel or coinciding lines.*

Proof. From Theorems 2.1-2.4,2.6 it follows that necessary conditions for a quadratic system with one transversally non-hyperbolic infinite critical point to be of finite multiplicity $0 \leq m_f \leq 2$ are that $A = B = 0$, $c_{45}^2 + c_{46}^2 + c_{56}^2 \neq 0$. Now $c_{45} + c_{56} \neq 0$ since $c_{45} + c_{56} = 0$ implies that $c_{45} = c_{46} = c_{56} = 0$ if A=0. Thus B=0 implies, with (3.3.1), that $c_{45}(c_{26} - \frac{1}{2}c_{35})^2 + 2c_{46}(c_{26} - \frac{1}{2}c_{35})(c_{34} - \frac{1}{2}c_{25}) + c_{56}(c_{34} - \frac{1}{2}c_{25})^2 = 0$ from which ,with A=0 , from (3.3.1) also follows that H=0, and the central conic is a degenerate parabola. Without loss of generality it may be assumed that the transversally non-hyperbolic critical point is at the ends of the x axis, then $a_{20} = b_{20} = 0$, $c_{56} \neq 0$ and $B_1 = 0$ yields $c_{25} = 0$.The central conic can then be written as

$$c_{23} + 2(c_{26} - \frac{1}{2}c_{35})y + 2c_{56}y^2 = 0,$$

which, since $c_{23}c_{56} + c_{26}c_{35} = c_{25}c_{36} = 0$, may be written as

$$(y - \frac{c_{35}}{2c_{56}})(y + \frac{c_{26}}{c_{56}}) = 0,$$

which shows that the degenerate parabola consists of two parallel lines, co-inciding if $c_{26} + \frac{1}{2}c_{35} = 0$. Conversely, if the central conic is a parabola de-generated into two parallel or coinciding lines they may, by a rotation of the axes, be directed along the x axis, thus (3.2.5) yields $c_{45} = c_{46} = 0, c_{56} \neq 0$, thus $a_{20} = b_{20} = 0$ and y is a common factor of the quadratic terms in (1.0.1),(1.0.2). Thus there exists a transversally non-hyperbolic infinite criti-cal point at the ends of the x axis, which is the only one since $c_{45}^2 + c_{46}^2 + c_{56}^2 \neq 0$. From (3.2.5) it follows that $c_{25} = 0$, thus $B_1 = 0$ and $m_f \leq 2$ is finite so $0 \leq m_f \leq 2$.   □

Remark. As in the previous remark it follows that the axis of the (de-generate)parabola points in the direction of the transversally non-hyperbolic infinite critical point.

If there exist two transversally non-hyperbolic infinite critical points there follows $c_{45}^2 + c_{46}^2 + c_{56}^2 = 0$ and the expression for the central conic may be-come meaningless. Further statements are only possible after a more detailed analysis of the various situations with regard to the critical points.

We will now proceed by investigating each class of quadratic systems with a given value of m_f and study the relation between degenerate isoclines, central conic and critical points in order to give a further specification of classes.

3.4 Quadratic systems with $m_f = 4$

In this section the various possible combinations of degenerate isoclines, cen-tral conic and critical points are studied, resulting in the summary sketched in Figure 3.4. This leads to the definition of 65 classes of quadratic systems in $m_f = 4$, being 65 types of combinations of critical points. They are given in Table 1 in subsection 3.4.5.

3.4.1 Systems with four finite critical points

Systems with four real finite critical points

It was already remarked in 1959 by Tung Chin Chu (Dong Jinzhu) [59,D] that the line through two critical points in a quadratic system is an isoclinic line. This can readily be seen by rotating the axes such that there exist critical points in $(x_1, 0)$ and $(x_2, 0)$. Then it follows, that on the line $y \equiv$

$0, \dot{x} = a_{20}(x - x_1)(x - x_2), \dot{y} = b_{20}(x - x_1)(x - x_2)$ and the result follows. If there exist four real critical points there are six isoclinic lines, which will now be identified as three degenerate conics in the pencil of conical isoclines.

Theorem 3.3 *If a quadratic system has four real finite critical points, then there exist six different real isoclinic lines forming three degenerate isoclines.*

(i)If the central conic is a hyperbola, thus if $A>0$, then the critical points form a convex quadrangle; there are two saddles and two antisaddles , and the sum of the indices of the critical points equals zero; so $i_f = 0$. Both elliptic and hyperbolic isoclines exist and there are two parabolic isoclines. The three degenerate isoclines are hyperbolae unless the central conic is degenerate; then one or two degenerate isoclines are parabolae, corresponding to two or four sides of the quadrangle being parallel.

(ii)If the central conic is of elliptic type, thus if $A<0$, it is a non-degenerate real ellipse and the critical points form a non-convex quadrangle. Either $c_{45} + c_{56} >0$, thus $c_{45}, c_{56} > 0$, and there exists one saddle and three antisaddles, the sum of their indices being equal to 2, so $i_f = 2$ or $c_{45} + c_{46} < 0$, thus $c_{45}, c_{56} < 0$, and there are three saddles and one antisaddle, the sum of the indices being equal to -2, and $i_f = -2$. All isoclines, including the degenerate isoclines, are hyperbolae.

Proof. Use of linear transformations allows us to locate the critical points in A:(0,0), B:(1,0), C:(0,1), D:(α, β),where α, β are real parameters , obeying $\alpha\beta(\alpha + \beta - 1) \neq 0$. System (1.0.1),(1.0.2)then becomes

$$\dot{x} = a_1 x(1 - x) + b_1 y(1 - y) + (\frac{\alpha - 1}{\beta}a_1 + \frac{\beta - 1}{\alpha}b_1)xy, \qquad (3.4.1)$$

$$\dot{y} = b_2 x(1 - x) + a_2 y(1 - y) + (\frac{\beta - 1}{\alpha}a_2 + \frac{\alpha - 1}{\beta}b_2)xy, \qquad (3.4.2)$$

where $a_1 a_2 - b_1 b_2 \neq 0, a_i, b_i \in \mathbb{R}, (i = 1, 2)$.

The central conic of the pencil of conical isoclines of (3.4.1),(3.4.2) is given by

$$\alpha\beta + \beta(-1 - 2\alpha + \beta)x + \alpha(-1 + \alpha - 2\beta)y - 2\beta(\beta - 1)x^2 + 4\alpha\beta xy - 2\alpha(\alpha - 1)y^2 = 0,$$

the type of which is determined by the sign of

$$A \equiv c_{46}^2 - c_{45}c_{56} = \frac{\alpha + \beta - 1}{\alpha\beta}(a_1 a_2 - b_1 b_2)^2.$$

In Figure 3.1 we illustrate how the sign of A depends on the location of the critical point (α, β). It follows that if A>0, and the central conic is then hyperbolic, the four critical points form a convex quadrangle . If A < 0-and the central conic is then of elliptic type-,they form a non-convex quadrangle. The Hessian determinant of the central conic reads

$$H = \frac{1}{2}\alpha\beta(\alpha + \beta - 1)(\alpha - 1)(\beta - 1)(\alpha + \beta).$$

The central conic is thus degenerate if and only if at least two sides of the quadrangle of critical points are parallel; (α, β) is then in the region A>0 and the central conic is a degenerate hyperbola, consisting of two real intersecting lines.

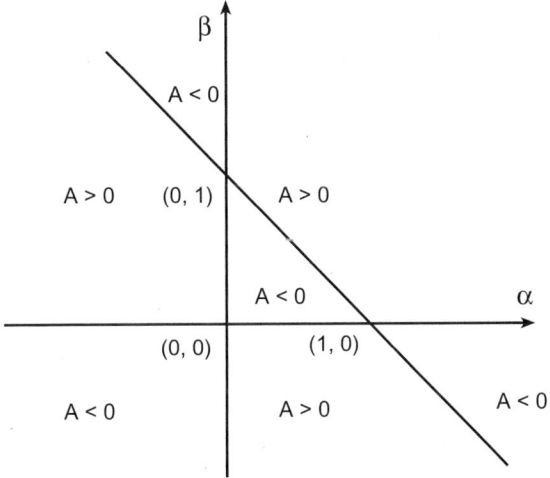

Figure 3.1 Sign of A in dependence on the location of the critical point (α, β)

The degenerate isoclines are given by

$$(a_1\lambda + b_2\mu)(b_1\lambda + a_2\mu)[(\alpha a_1 + \beta b_1)\lambda + (\alpha b_2 + \beta a_2)\mu] = 0,$$

the three real different (since $a_1 a_2 - b_1 b_2 \neq 0$) solution pairs (λ, μ) of which determine degenerate hyperbolae,unless $\alpha = 1$ and/or $\beta = 1$ and/or $\alpha + \beta = 0$, in which cases one or two degenerate parabolae,degenerated into two parallel lines occur. See Figure 3.2. As stated in Chapter 1, for A>0 there

exist both elliptic and hyperbolic isoclines as well as two parabolae. If A<0
all isoclines are of the same type , thus are hyperbolae since the degenerate
isoclines are of that type.

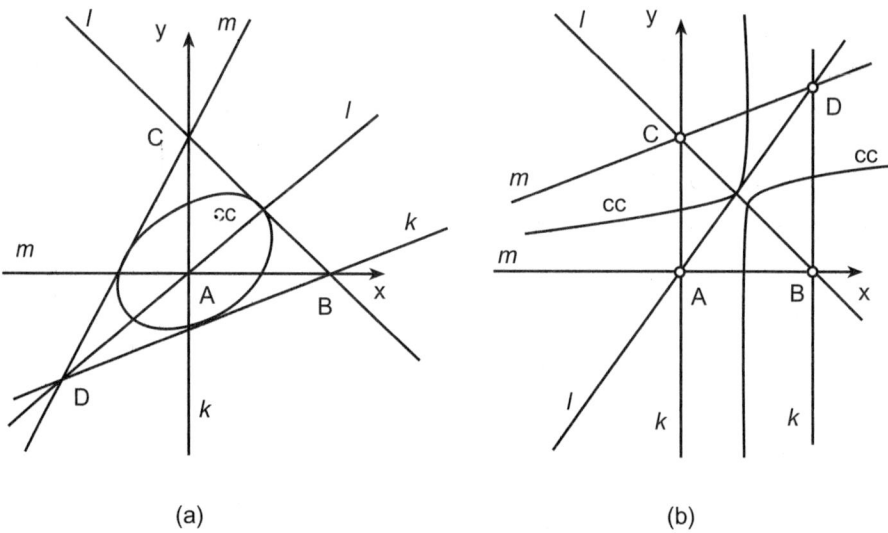

(a)   (b)

Figure 3.2 Degenerate isoclines and central conic
a)A<0 $(\alpha, \beta <0)$,b)A>0$(\alpha, \beta >0)$

By calculating $\Lambda(x,y)$ in the critical points, Berlinski's theorem may be
obtained [60,B] which states that if the quadrangle formed by the critical
points is convex, two opposite critical points are saddles and the other two
antisaddles. However, if the quadrangle is not convex, then either the three
exterior vertices are saddles and the interior vertex is an antisaddle or the
exterior vertices are antisaddles and the interior vertex is a saddle. In fact
there is $\Lambda(0,0) = a_1 a_2 - b_1 b_2$, $\Lambda(1,0) = -(\alpha + \beta - 1)/\alpha\Lambda(0,0)$, $\Lambda(0,1) =$
$-(\alpha + \beta - 1)/\beta\Lambda(0,0)$, $\Lambda(\alpha, \beta) = (\alpha + \beta - 1)\Lambda(0,0)$. So, if, for instance,
$\alpha > 0, \beta > 0, \alpha + \beta - 1 > 0$, then the quadrangle formed by the four critical
points is convex. Then it follows, since $\Lambda(\alpha, \beta)\Lambda(0,0) > 0$, that (0,0) and
(α, β) are both saddles or both antisaddles, whereas from $\Lambda(1,0)\Lambda(0,0) < 0$
and $\Lambda(0, 1)\Lambda(0,0) < 0$ it follows that (1,0) and (0,1) have the opposite index

from $(0,0)$ and (α, β). If $\alpha > 0, \beta > 0, \alpha + \beta - 1 < 0$ the quadrangle is not convex and the result follows from $\Lambda(\alpha, \beta)\Lambda(0,0) < 0, \Lambda(1,0)\Lambda(0,0) > 0, \Lambda(0,1)\Lambda(0,0) > 0$. The argument may be repeated to cover all the cases for α, β.

As a result, it now also follows that the sum of the indices of the four real finite critical points is equal to zero, so $i_f = 0$ if the quadrangle is convex and the central conic of hyperbolic type ($A>0$), and equal to ± 2, so $i_f = \pm 2$, if the quadrangle is not convex and the central conic is of elliptic type ($A<0$) and, in fact, a real non-degenerate ellipse. In order to distinguish the latter case it should be observed that the critical point in the interior of the (elliptic) central conic, as given by $\Lambda(x,y)=0$, the sign of Λ equals that of Λ_c in the center of the central conic. Now $\Lambda_c = \Lambda(x_c, y_c) = -H/4A$, so if $H<0$ the interior point is a saddle and $i_f = 2$, whereas for $H>0$ this point is an antisaddle and $i_f = -2$. These conditions may be formulated using the third invariant $c_{45} + c_{56}$. Since the central conic is a real (non-degenerate) ellipse, (3.2.7) shows that $AS_iH > 0$ (i=1,2) so $S_iH < 0$ (i=1,2). If $A<0$ it follows furthermore, using (3.2.8), that $S_i > 0(< 0)$ as $c_{45} + c_{56} > 0(< 0)$, so that $H< 0(> 0)$ corresponds to $c_{45} + c_{56} > 0(< 0)$.   □

Finite index of a quadratic system with finite multiplicity 4

The finite index of a quadratic system can be obtained by calculating the rotation of the vector field in (1.0.1),(1.0.2) along a circle large enough to encircle all finite critical points. Since variation of the coefficients of a quadratic system with finite multiplicity 4 does not lead to finite critical points going to infinity if the system remains within the class $m_f = 4$, this circle can be taken large enough to encircle all finite critical points for all changes of the coefficients such that the system stays in the class $m_f=4$. For each of the three classes in the previous section, as given by $A>0$; $A<0, c_{45} + c_{56} < 0$ and $A<0, c_{45} + c_{56} >0$, all changes of the coefficients within each class thus leave the finite index invariant, yet generate all systems in $m_f = 4$. So the results obtained for the case with four real critical points may be used for all quadratic systems in $m_f=4$.

Theorem 3.4 *The finite index i_f of a quadratic system of finite multiplicity 4 equals 0 if $A>0$, equals -2 if $A<0$, $c_{45} + c_{56} <0$, thus $c_{45}, c_{56} < 0$ and equals 2 if $A<0$, $c_{45} + c_{56} >0$, thus $c_{45}, c_{56} >0$. In short: for $m_f=4$ there is $i_f = [1 - sgn(c_{46}^2 - c_{45}c_{56})]sgn(c_{45} + c_{56})$.*

Proof. The index of a curve C with respect to $(P(x,y), Q(x,y))$ in

(1.0.1),(1.0.2) may be written as

$$I_C = \frac{1}{2\pi} \oint \frac{PdQ - QdP}{P^2 + Q^2} \ ,$$

where C is the circle around the origin with radius R, to be traversed in anticlockwise direction. Using polar coordinates we write $P = a_0(\theta) + a_1(\theta)R + a_2(\theta)R^2$, $Q = b_0(\theta) + b_1(\theta)R + b_2(\theta)R^2$, where $a_i(\theta), b_i(\theta)$ (i=0,1,2) only contain coefficients of terms of degree i in (1.0.1),(1.0.2). Then,with $a_i' = \frac{da_i(\theta)}{d\theta}, b_i' = \frac{db_i(\theta)}{d\theta}$, we may write, with R → ∞,

$$I_C = \frac{1}{2\pi} \oint \frac{a_2(\theta)b_2'(\theta) - b_2(\theta)a_2'(\theta)}{a_2^2(\theta) + b_2^2(\theta)} d\theta \ ,$$

which is a convergent integral, as from $A \neq 0$ it follows that $a_2(\theta)$ and $b_2(\theta)$ have no common zeros, and which only depends on the quadratic terms in (1.0.1),(1.0.2).   □

Systems with complex elementary critical points

Theorem 3.5 *If a quadratic system has four elementary critical points among which two or four complex critical points there exist six different isoclinic lines, forming two complex degenerate isoclines and one (real) degenerate hyperbola or parabola,consisting of two real parallel or coinciding lines.*
   There exist two possibilities:
   (i) The two complex degenerate isoclines each are formed by two different conjugate complex lines ; then there are four complex critical points. These complex degenerate isoclines are parabolae or ellipses both with real coefficients. The central conic is a hyperbola ; degenerate or non-degenerate.
   (ii) The two complex degenerate isoclines each are formed by two non-conjugate complex lines; then there exist two complex and two real critical points. If the central conic is of hyperbolic type (A>0) it is a non-degenerate hyperbola and the two real critical points are a saddle and an antisaddle. If the central conic is of elliptic type (A<0) it is a real ellipse or a degenerate ellipse consisting of two complex conjugated lines. The two real critical points are both saddles if $c_{45} + c_{56} < 0$ and both antisaddles if $c_{45} + c_{56} > 0$.
   Proof. Quadratic systems having critical points in $(\alpha + i\beta, \gamma + i\delta)$ and $(\alpha - i\beta, \gamma - i\delta)$ may be linearly transformed so that the critical points are in A(i,i) and B(−i,−i). They can be represented by

$$
\begin{aligned}
\dot{x} &= a_{20} + a_{11} + a_{02} + a_{10}x - a_{10}y + a_{20}x^2 + a_{11}xy + a_{02}y^2, &(3.4.3)\\
\dot{y} &= b_{20} + b_{11} + b_{02} + b_{10}x - b_{10}y + b_{20}x^2 + b_{11}xy + b_{02}y^2, &(3.4.4)
\end{aligned}
$$

where $a_{ij}, b_{ij} \in \mathbb{R}$ (i,j=0,1,2).

The degenerate isoclines of (3.4.3),(3.4.4) are determined by

$$[(a_{20} + a_{11} + a_{02})\lambda + (b_{20} + b_{11} + b_{02})\mu][(-\frac{1}{4}a_{10}^2 + a_{20}a_{02} - \frac{1}{4}a_{11}^2)\lambda^2$$

$$+(-\frac{1}{2}a_{10}b_{10} - \frac{1}{2}a_{11}b_{11} + a_{20}b_{02} + a_{02}b_{20})\lambda\mu$$

$$+(-\frac{1}{4}b_{10}^2 + b_{10}b_{02} - \frac{1}{4}b_{11}^2)\mu^2] = 0. \qquad (3.4.5)$$

The solution of (3.4.5)corresponding to the first factor,linear in λ and μ, yields the degenerate hyperbola or parabola

$$l : (x - y)[c_{24} + c_{25} + c_{26} + (c_{45} + c_{46})x + (c_{46} + c_{56})y] = 0. \qquad (3.4.6)$$

The branch $x - y = 0$ of l contains the critical points A and B, and is thus "a real line containing complex critical points". The other branch is also a real line. Together they form a degenerate hyperbola if $c_{45} + 2c_{46} + c_{56} \neq 0$,a degenerate parabola if $c_{45} + 2c_{46} + c_{56} = 0$, consisting of two parallel real lines if $c_{24} + c_{25} + c_{26} \neq 0$ and of two coinciding lines $(x-y)^2 = 0$ if $c_{24} + c_{25} + c_{26} = 0$. The isocline l is only a parabola if A >0 since from $c_{46} = -\frac{1}{2}(c_{45} + c_{56})$ it follows that $A = \frac{1}{4}(c_{45} - c_{56})^2 \geq 0$, whereas $c_{45} - c_{56} \neq 0$ since otherwise A=0 and $m_f \neq 4$.

The square factor in (3.4.5) has the discriminant D, where

$$D = c_{24}c_{26} - \frac{1}{4}c_{25}^2 + c_{46}^2 - c_{45}c_{56}. \qquad (3.4.7)$$

Since $a_{20}^2 + b_{20}^2 \neq 0$, as otherwise $A = c_{46}^2 - c_{45}c_{56} = 0$, equation (2.1.5) for the x coordinate of the critical points holds and since $x = \pm i$ should satisfy (2.1.5) we have

$$(x^2 + 1)(Ax^2 + B_1x + E_1) = 0, \qquad (3.4.8)$$

where

$$B_1 = 2c_{46}(c_{26} + \frac{1}{2}c_{25}) - 2c_{56}(c_{24} + \frac{1}{2}c_{25}),$$
$$E_1 = c_{26}(c_{24} + c_{25} + c_{26}) + (c_{46} + c_{56})^2.$$

The square factor in (3.4.8) has real or complex roots depending on the sign of $B_1^2 - 4AE_1 = 4D(c_{46} + c_{50})^2$. Since $a_{20}^2 + b_{20}^2 \neq 0$ as otherwise A=0, equation (2.1.6) for the y coordinate of the critical points holds and since $y=\pm i$ should satisfy (2.1.6) there is

$$(y^2 + 1)(Ay^2 + B_2 y + E_2) = 0, \qquad (3.4.9)$$

where

$$B_2 = -2c_{45}(c_{26} + \tfrac{1}{2}c_{25}) + 2c_{46}(c_{24} + \tfrac{1}{2}c_{25}),$$
$$E_2 = c_{24}(c_{24} + c_{25} + c_{26}) + (c_{45} + c_{46})^2.$$

The square factor in (3.4.9) has real or complex roots depending on the sign of $B_2^2 - 4AE_2 = -4D(c_{45} + c_{46})^2$. Since $c_{45} + c_{46}$ and $c_{46} + c_{56}$ are not simultaneously equal to zero as, otherwise A=0, there follows that if D<0, thus if the (λ, μ) equation has two complex roots, there are real critical points in addition to the two complex critical points A and B; if D=0 there exists a real double point apart from A and B; and if D>0, thus if the (λ, μ) equation has three real roots, all four critical points are complex. From (3.4.8), (3.4.9) it follows that apart from A and B there exist the critical points C and D:

$$C: 1/2A[(-B_1 + 2(c_{46} + c_{56})\sqrt{-D}), (-B_2 - 2(c_{45} + c_{46})\sqrt{-D})],$$
$$D: 1/2A[(-B_1 - 2(c_{46} + c_{56})\sqrt{-D}), (-B_2 + 2(c_{45} + c_{46})\sqrt{-D})].$$

The critical points C and D are located on the branch $c_{24} + c_{25} + c_{26} + (c_{45} + c_{46})x + (c_{46} + c_{56})y = 0$ of the isocline l. Apart from the possibility that C and D may coincide if D=0, also C may coincide with A(or B) and D with B(or A) if $B_1 = B_2 = 0$, yielding $c_{26} + \tfrac{1}{2}c_{25} = c_{24} + \tfrac{1}{2}c_{25} = 0$, thus D=A>0, and if $c_{45} + c_{46} = -(c_{46} + c_{56}) = -\sqrt{A}(\text{or}\sqrt{A})$. This gives $c_{45} + 2c_{46} + c_{56} = c_{24} + c_{25} + c_{26} = 0$, so the two branches l coincide onto the line $(y - x)^2 = 0$. Conversely, if the two branches of l coincide on $(y - x)^2 = 0$, for the x coordinate of a critical point it follows from P(x,x)=Q(x,x)=0 that $(a_{20} + a_{11} + a_{02})(1 + x^2) = (b_{20} + b_{11} + b_{02})(1 + x^2) = 0$ and since $(a_{20} + a_{11} + a_{02})^2 + (b_{20} + b_{11} + b_{02})^2 \neq 0$ as $A \neq 0$, there is x=$\pm i$ and correspondingly y= $\pm i$,thus A and B are double points c_2^0 and there exist four critical points on l.

Apart from the real degenerate isocline l there exist two degenerate complex isoclines k and m.

(i) If D>0, all critical points are complex. There exist the isocline k, consisting of the lines AC and BD and the isocline m, consisting of the lines

AD and BC. So the isocline k is given by the complex conjugate isoclinic lines;

$$AC : -\frac{1}{2}(B_1 - B_2) + i(c_{45} + 2c_{46} + c_{56})\sqrt{D}$$

$$+[-A - (c_{45} + c_{46})\sqrt{D} + \frac{1}{2}iB_2]x$$

$$+[A - (c_{46} + c_{56})\sqrt{D} - \frac{1}{2}iB_1]y = 0, \qquad (3.4.10)$$

$$BD : -\frac{1}{2}(B_1 - B_2) - i(c_{45} + 2c_{46} + c_{56})\sqrt{D}$$

$$+[-A - (c_{45} + c_{56})\sqrt{D} - \frac{1}{2}iB_2]x$$

$$+[A - (c_{46} + c_{56})\sqrt{D} + \frac{1}{2}iB_1]y = 0. \qquad (3.4.11)$$

They form the conic with real coefficients given by

$$[-\frac{1}{2}(B_1 - B_2) + [-A - (c_{45} + c_{46})\sqrt{D}]x + [A - (c_{46} + c_{56})\sqrt{D}]y]^2$$

$$+[(c_{45} + 2c_{46} + c_{56})\sqrt{D} + \frac{1}{2}B_2 x - \frac{1}{2}B_1 y]^2 = 0, \qquad (3.4.12)$$

which is of elliptic type if $\kappa_1 \neq 0$ and of parabolic type if $\kappa_1 = 0$, where κ_1 is given by

$$\kappa_1 = [\frac{1}{2}(B_1 - B_2) + (c_{24} + c_{25} + c_{26})\sqrt{D}]A. \qquad (3.4.13)$$

For $\kappa_1 \neq 0$, then, (3.4.12) is an ellipse degenerated into two complex conjugated lines, intersecting in the only real point on the conic. If $\kappa_1 = 0$ the degenerate parabola consists of two complex lines which may coincide into one real line if and only if $B_1 = B_2 = c_{45} + 2c_{46} + c_{56} = 0$, from which follows $c_{26} + \frac{1}{2}c_{25} = c_{24} + \frac{1}{2}c_{25} = 0$. Then k coincides with l: $(y - x)^2 = 0$ and there exist two complex double critical points c_2^0.

The isocline m is given by the complex conjugated lines

$$AD : -\frac{1}{2}(B_1 - B_2) - i(c_{45} + 2c_{46} + c_{56})\sqrt{D}$$

$$+[-A + (c_{45} + c_{46})\sqrt{D} + \frac{1}{2}iB_2]x$$

$$+[A + (c_{46} + c_{56})\sqrt{D} - \frac{1}{2}iB_1]y = 0, \qquad (3.4.14)$$

$$BC : -\frac{1}{2}(B_1 - B_2) + i(c_{45} + 2c_{46} + c_{56})\sqrt{D}$$

$$+[-A + (c_{45} + c_{46})\sqrt{D} - \frac{1}{2}iB_2]x$$

$$+[A + (c_{46} + c_{56})\sqrt{D} + \frac{1}{2}iB_1]y = 0. \tag{3.4.15}$$

They form the conic with real coefficients given by

$$[-\frac{1}{2}(B_1 - B_2) + [-A + (c_{45} + c_{46})\sqrt{D}]x + [A + (c_{46} + c_{56})\sqrt{D}]y]^2$$

$$+[-(c_{45} + 2c_{46} + c_{56})\sqrt{D} + \frac{1}{2}B_2x - \frac{1}{2}B_1y]^2 = 0, \tag{3.4.16}$$

which is of elliptic type if $\kappa_2 \neq 0$ and of parabolic type if $\kappa_2 = 0$, where κ_2 is given by

$$\kappa_2 = [\frac{1}{2}(B_1 - B_2) - (c_{24} + c_{25} + c_{26})\sqrt{D}]A. \tag{3.4.17}$$

For $\kappa_2 \neq 0$, (3.4.16) is an ellipse degenerated into two complex conjugated lines intersecting in the only real point on the conic. If $\kappa_2 = 0$ the degenerate parabola consists of two complex lines that may coincide into one real line if and only if $B_1 = B_2 = c_{45} + 2c_{46} + c_{56} = 0$, from which follows $c_{26} + \frac{1}{2}c_{25} = c_{24} + \frac{1}{2}c_{25} = 0$. Then m coincides with l: $(y - x)^2 = 0$ and there exist two complex double critical points c_2^0. In fact k,l and m coincide and there are six coinciding isoclinic lines.

(ii) If $D<0$, apart from the complex critical points A:(i,i) and (B,B):$(-i,-i)$ there exist the real critical points C and D. The complex degenerate isocline k is now given by the non-conjugate complex lines

$$AC : -\frac{1}{2}(B_1 - B_2) + (c_{45} + 2c_{46} + c_{56})\sqrt{-D}$$

$$+[-A + i(\frac{1}{2}B_2 + (c_{45} + c_{46})\sqrt{-D})]x$$

$$+[A + i(-\frac{1}{2}B_1 + (c_{46} + c_{56})\sqrt{-D})]y = 0, \tag{3.4.18}$$

$$BD : -\frac{1}{2}(B_1 - B_2) - (c_{45} + 2c_{46} + c_{56})\sqrt{-D}$$

$$+[-A + i(-\frac{1}{2}B_2 + (c_{45} + c_{46})\sqrt{-D})]x$$

$$+[A + i(\frac{1}{2}B_1 + (c_{46} + c_{56})\sqrt{-D})]y = 0. \tag{3.4.19}$$

Similarly, the complex degenerate isocline m is now given by the non-conjugate complex isoclinic lines

$$AD : -\frac{1}{2}(B_1 - B_2) - (c_{45} + 2c_{46} + c_{56})\sqrt{-D}$$

$$+[-A - i(-\frac{1}{2}B_2 + (c_{45} + c_{46})\sqrt{-D})]x$$

$$+[A - i(\frac{1}{2}B_1 + (c_{46} + c_{56})\sqrt{-D})]y = 0, \tag{3.4.20}$$

$$BC : -\frac{1}{2}(B_1 - B_2) + (c_{45} + 2c_{46} + c_{56})\sqrt{-D}$$

$$+[-A - i(\frac{1}{2}B_2 + (c_{45} + c_{56})\sqrt{-D})]x$$

$$+[A - i(-\frac{1}{2}B_1 + (c_{46} + c_{56})\sqrt{-D})]y = 0. \tag{3.4.21}$$

Note that k and m cannot coincide with l, since the conditions $c_{24} + \frac{1}{2}c_{25} = c_{26} + \frac{1}{2}c_{25} = c_{45} + 2c_{46} + c_{56} = 0$ cannot be satisfied as then $D = A = \frac{1}{4}(c_{45} - c_{56})^2 \geq 0$.

When considering the central conic, again D<0 and D>0 will be taken apart.

(i)If D>0 all critical points are complex.

The central conic is given by

$$(c_{24} + \frac{1}{2}c_{25})x + (\frac{1}{2}c_{25} + c_{26})y + c_{45}x^2 + 2c_{46}xy + c_{56}y^2 = 0, \tag{3.4.22}$$

and its Hessian determinant by

$$H = -2[c_{45}(\frac{1}{2}c_{25} + c_{26})^2 - 2c_{46}(\frac{1}{2}c_{25} + c_{26})(\frac{1}{2}c_{25} + c_{24}) + c_{56}(\frac{1}{2}c_{25} + c_{24})^2]. \tag{3.4.23}$$

The central conic is a hyperbola(degenerate or not) since A>0 as $i_f = 0$ and Theorem 3.4 applies.

(ii)If D<0 there are two complex and two real critical points. Locating the real critical points in $(-1,0)$ and $(1,0)$ brings (1.0.1),(1.0.2) into the form

$$\dot{x} = a_{01}y + a_{20}(x^2 - 1) + a_{11}xy + a_{02}y^2, \tag{3.4.24}$$

$$\dot{y} = b_{01}y + b_{20}(x^2 - 1) + b_{11}xy + b_{02}y^2, \tag{3.4.25}$$

yielding for the y coordinates of the critical points

$$y^2[Ay^2 + (c_{35}c_{45} - 2c_{34}c_{46})y + (c_{34}^2 - c_{45}^2)] = 0,$$

so that there exist two complex critical points if

$$(c_{35}c_{45} - 2c_{34}c_{46})^2 - 4(c_{34}^2 - c_{45}^2)A < 0, \qquad (3.4.26)$$

so $(c_{34}^2 - c_{45}^2)A > 0$. As $\Lambda(-1,0)\Lambda(1,0) = -4(c_{34}^2 - c_{45}^2)$ this means that $A\Lambda(-1,0)\Lambda(1,0) < 0$ should be satisfied.

The central conic for (3.4.24),(3.4.25) reads

$$-2c_{34}x - c_{35}y + 2c_{45}x^2 + 4c_{46}xy + 2c_{56}y^2 = 0, \qquad (3.4.27)$$

and its Hessian determinant

$$H = 2(c_{34}c_{36} - \frac{1}{4}c_{35}^2)c_{45}. \qquad (3.4.28)$$

If A>0, the central conic is of hyperbolic type and is a non-degenerate hyperbola. In fact, (3.4.26) may be written as $-2c_{45}H + 4c_{45}^2 A < 0$, which cannot be satisfied for H=0,A>0. Since $A\Lambda(-1,0)\Lambda(1,0) < 0$ there exist a saddle and an antisaddle, so $i_f = 0$, in accordance with Theorem 3.4.

If A<0, the central conic is of elliptic type. Now $\Lambda(-1,0)\Lambda(1,0) > 0$ and there exist either two saddles or two antisaddles. Also $\Lambda(-1,0) = 2(c_{34} + c_{45})$ and $\Lambda(1,0) = 2(-c_{34} + c_{45})$. So if there exist two saddles, then $c_{45} < 0$ and $i_f = -2$ in accordance with Theorem 3.4. Similarly, if there are two antisaddles, then $c_{45} > 0$, and $i_f = 2$. The real degenerate isocline $y(c_{34} + c_{45}x + c_{46}y) = 0$ intersects itself in its midpoint $(-c_{34}/c_{45}, 0)$, which is a real point on the central conic as is $(0,0)$. So the central conic is a real ellipse if $c_{34} \neq 0$ or if $c_{34} = 0, c_{35} \neq 0$. If $c_{34} = c_{35} = 0$, then H=0 and the central conic is a degenerate ellipse, consisting of two complex lines, intersecting at the origin. (Note that $c_{45} \neq 0$ since A<0).   □

3.4.2   Systems with three finite critical points

Theorem 3.6 *If a quadratic system with $m_f = 4$ has three finite critical points, then there exist two (real or complex) elementary critical points and one second order critical point m_2^0 (a cusp or a semi-elementary saddle node). There exist two coinciding degenerate isoclines, each consisting of two real different, coinciding (but then there exist two critical points) or complex conjugate lines; if they intersect, they do so at m_2^0. The remaining degenerate isocline is a degenerate hyperbola or parabola of which one branch contains the second order point m_2^0 and is tangent there to all non-degenerate isoclines. On the* *other branch the two elementary (real or complex) critical points are located. For A>0, the central conic is a degenerate or non-degenerate hyperbola; then the elementary critical points are a saddle and an antisaddle or two complex critical points; whereas a second order m_2^0 point is also possible. For A<0, the central conic is a real ellipse; then the two elementary critical points are both saddles if $c_{45} + c_{56} < 0$ or both antisaddles if $c_{45} + c_{56} > 0$. The second order point m_2^0 is on the central conic.*

Proof. If the second order point m_2^0 is *semi- elementary* it is a saddle node and we can use (2.3.3),(2.3.4)

$$\dot{x} = a_{20}x^2 + a_{11}xy + a_{02}y^2 \equiv P(x,y), \tag{3.4.29}$$
$$\dot{y} = y + b_{20}x^2 + b_{11}xy + b_{02}y^2 \equiv Q(x,y), \tag{3.4.30}$$

where $a_{20} \neq 0$. The degenerate isoclines of (3.4.29),(3.4.30) are given by

$$(a_{20}\lambda + b_{20}\mu)\mu^2 = 0, \tag{3.4.31}$$

and the central conic by

$$2a_{20}x + a_{11}y + 2c_{45}x^2 + 4c_{46}xy + 2c_{56}y^2 = 0, \tag{3.4.32}$$

which has the Hessian determinant

$$H = -\frac{1}{2}(a_{11}^2 - 4a_{20}a_{02})c_{45}. \tag{3.4.33}$$

In m_2^0 all non-degenerate isoclines are tangent to each other and to the center manifold(s) in that point. To $\mu^2 = 0$ correspond the two coinciding degenerate isoclines k,m: $P(x,y)=0$, each consisting of two isoclinic lines, real different or coinciding or complex, forming a degenerate hyperbola,parabola or ellipse,respectively. They are not tangent to the center manifold(s) in m_2^0. To $a_{20}\lambda + b_{20}\mu = 0$ corresponds the degenerate isocline l:$(a_{20}+c_{45}x+c_{46}y)y = 0$, the branch $y \equiv 0$ of which contains m_2^0 and is tangent there to the center manifold(s) and the non-degenerate isoclines. The other branch contains the elementary critical points which are real, coinciding or complex according as $a_{11}^2 - 4a_{20}a_{02} >0,=0,\text{or}<0$, respectively.

If the central conic is a hyperbola, then A>0, $i_f=0$, and there exist a saddle and an antisaddle if $a_{11}^2 - 4a_{20}a_{02} >0$, another second order point m_2^0 if $a_{11}^2 - 4a_{20}a_{02}=0$, then the central conic is degenerate, and two complex critical points if $a_{11}^2 - 4a_{20}a_{02} <0$. The central conic is also degenerate if

$c_{45}=0$; then l consists of two real parallel lines thus forming a non-degenerate parabola.

If the central conic is of elliptic type, then A<0 and $i_f = -2$ if $c_{45}+c_{56} <0$ and $i_f=2$ if $c_{45}+c_{56} >0$, yielding either two saddles on the branch $a_{20}+c_{45}x+ c_{56}y=0$ of l if $i_f = -2$ or two antisaddles if $i_f=2$. So $a_{11}^2 - 4a_{20}a_{02} >0$ and since $c_{45} \neq 0$ it follows that $H \neq 0$. Moreover for $c_{45} + c_{56} <0(>0)$, there is $S_i <0(>0)$, (i=1,2) and H>0(<0) so that from $AS_i/H >0(i=1,2)$ it follows that the central conic is a real ellipse.

If the second order point m_2^0 is *nilpotent* it is a cusp and we can use (2.3.5),(2.3.6)

$$\dot{x} = y + a_{20}x^2 + a_{11}xy + a_{02}y^2 \equiv P(x,y), \qquad (3.4.34)$$
$$\dot{y} = b_{20}x^2 + b_{11}xy + b_{02}y^2 \equiv Q(x,y), \qquad (3.4.35)$$

where $b_{20} \neq 0$. The degenerate isoclines of (3.4.34),(3.4.35) are given by

$$(a_{20}\lambda + b_{20}\mu)\lambda^2 = 0, \qquad (3.4.36)$$

and the central conic by

$$-2b_{20}x - b_{11}y + 2c_{45}x^2 + 4c_{46}xy + 2c_{56}y^2 = 0, \qquad (3.4.37)$$

which has the Hessian determinant

$$H = -\frac{1}{2}(b_{11}^2 - 4b_{20}b_{02})c_{45}. \qquad (3.4.38)$$

In m_2^0 all non-degenerate isoclines are tangent to each other and to the eigenvector of the linearized system in m_2^0. To $\lambda^2=0$ correspond the two coinciding degenerate isoclines k,m: Q(x,y)=0, each consisting of two isoclinic lines, real different or coinciding or complex, forming a degenerate hyperbola, parabola or ellipse, respectively. They are not tangent to the non-degenerate isoclines in m_2^0. To $a_{20}\lambda + b_{20}\mu =0$ corresponds the degenerate isocline l:$(-b_{20} + c_{45}x + c_{46}y)y=0$, the branch $y\equiv0$ of which contains m_2^0 and is tangent there to all non-degenerate isoclines. The other branch contains the elementary critical points which are real, coinciding or complex according as $b_{11}^2 - 4b_{20}b_{02} >0,=0$ or <0, respectively.

If the central conic is hyperbolic,than A>0, $i_f=0$, and there exist a saddle and an antisaddle if $b_{11}^2 - 4b_{20}b_{02} >0$, another second order point m_2^0 if $b_{11}^2 - 4b_{20}b_{02}=0$, then the central conic is degenerate, and two complex critical

points if $b_{11}^2 - 4b_{20}b_{02} < 0$. The central conic is also degenerate if $c_{45}=0$; then
l consists of two real parallel lines forming a degenerate parabola.

If the central conic is of elliptic type, then A<0 and $i_f = -2$ if $c_{45}+c_{56} <0$
and $i_f=2$ if $c_{45}+c_{56} >0$, yielding two saddles or two antisaddles on the branch
$-b_{20} + c_{45}x + c_{46}y=0$ of the isocline l, respectively. So $b_{11}^2 - 4b_{20}b_{02} >0$ and
since $c_{45} \neq 0$ it follows that H $\neq 0$. Moreover for $c_{45} + c_{56} <0(>0)$ there is
$S_i <0(>0)$, (i=1,2) and H$>0(<0)$ so that from A$S_i/H >0$(i=1,2) it follows
that the central conic is a real ellipse.   □

3.4.3   Systems with two finite critical points

Theorem 3.7 *If a quadratic system with $m_f =4$ has two finite critical points,
there exist two possibilities;*

*(i)There exist two second order critical points; they are both real, then
there exist the combination $m_2^0 m_2^0$ or both are complex and then $c_2^0 c_2^0$. Then
A$>0, i_f=0$.*

*If they are both real, each of them is either a saddle-node or a cusp. There
exist two degenerate isoclines, both parabolae, each formed by two coinciding
lines, thus yielding four isoclinic lines coinciding with the line connecting the
second order points. The third degenerate isocline is a degenerate hyperbola
or parabola, which in the points m_2^0 is tangent to all non-degenerate isoclines
through these points. The central conic is a degenerate hyperbola, one branch
of it connecting the points m_2^0.*

*If they are both complex there is $c_2^0 c_2^0$. All degenerate isoclines coincide,
each of them consisting of two coinciding (real) isoclinic lines, thus forming
six coinciding isoclinic lines (or three degenerate parabolae). The central
conic is a degenerate hyperbola, one branch of the two lines coinciding with
the six coinciding isoclinic lines.*

*(ii)There exist an elementary and a third order critical point, which is
a semi-elementary or nilpotent saddle m_3^{-1}, a semi-elementary node m_3^1 or
a nilpotent point with an elliptic and a hyperbolic sector m_3^1. There exist
three coinciding degenerate isoclines; they are hyperbolae, one set of coinciding branches is tangent to the center manifold(s) in m_3 or in the direction
of the eigenvector of the linearized system in m_3 as well as to the non- degenerate isoclines. The other set of coinciding branches is transversal to this
direction and contains the elementary critical point. The central conic is
a non-degenerate hyperbola or real ellipse, in m_3 also tangent to the center
manifold or eigenvector and having a curvature opposite to that of the non-*

degenerate isoclines. If the central conic is a hyperbola, so A>0, i_f=0, the elementary critical point is a saddle if m_3 is an m_3^1 point, and an antisaddle if m_3 is an m_3^{-1} point(saddle). If the central conic is an ellipse the elementary critical point is also a saddle if m_3 is a saddle m_3^{-1} and an antisaddle if m_3 is an m_3^1 point.

Proof (i) The situation with two real second order critical points $m_2^0 m_2^0$ was already met as a limiting case in Theorem 3.6. We may work with the system

$$\dot{x} = a_{01}y + a_{20}(x^2 - 1) + a_{11}xy + a_{02}y^2, \qquad (3.4.39)$$
$$\dot{y} = b_{01}y + b_{20}(x^2 - 1) + b_{11}xy + b_{02}y^2, \qquad (3.4.40)$$

which has finite critical points in $(-1,1)$ and $(1,0)$, and since $\Lambda(-1,0) = c_{34} + c_{45}, \Lambda(1,0) = c_{34} - c_{45}$ they are both second order if $c_{34} = c_{45}=0$. Since i_f=0, then A>0 so $c_{46} \neq 0$, whereas from $c_{34}c_{56} + c_{36}c_{45} = c_{35}c_{46}$ then it follows that c_{35}=0.

The degenerate isoclines are given by

$$(a_{20}\lambda + b_{20}\mu)^2[b_{20}(-a_{20}a_{02} + \frac{1}{4}a_{11}^2 - \frac{1}{4}a_{01}^2)\lambda + a_{20}(-b_{20}b_{02} + \frac{1}{4}b_{11}^2 - \frac{1}{4}b_{01}^2)\mu] = 0.$$
$$(3.4.41)$$

To $a_{20}\lambda + b_{20}\mu$=0 corresponds the degenerate parabola y^2=0, so the second order critical points are connected by four coinciding isoclinic lines. The remaining (linear)factor in(3.4.41) yields the third degenerate isocline which is either a hyperbola or a parabola consisting of two parallel real lines and in both second order critical points are tangent to the non-degenerate isoclines, which are transversal to the four isoclinic lines connecting the m_2^0 points.

The central conic of (3.4.39),(3.4.40) is given by

$$y(2c_{46}x + c_{56}y) = 0,$$

and is thus a degenerate hyperbola, one branch coinciding with the four isoclinic lines on y=0. The critical points in $x = \pm 1$ are saddle nodes if $b_{01} \pm (2a_{20} + b_{11}) \neq 0$ and cusps if $b_{01} \pm (2a_{20} + b_{11})$=0.

The situation with two complex second order critical points $c_2^0 c_2^0$ was already met in the proof of Theorem 3.5. It was concluded that all six isoclinic lines coincide as the real line connecting the c_2^0 critical points. They represent three coinciding degenerate parabolae. Since i_f=0 , A>0 so the central conic is of hyperbolic type. Using the expression (3.4.23) for the

Hessian determinant of the central conic as given by (3.4.22), it follows that $H=0$ if $c_{24} + \frac{1}{2}c_{25} = c_{26} + \frac{1}{2}c_{25}=0$, which are necessary conditions for system (3.4.3),(3.4.4) to have a c_2^0 point in (i,i) as well as $(-i,-i)$. The central conic is thus degenerate, consisting of two intersecting real lines, one of them coinciding with the three coinciding parabolae.

(ii)If the third order critical point is *semi-elementary* we may use (3.4.29) and (3.4.30) with $a_{20}=0$, yielding

$$\dot{x} = a_{11}xy + a_{02}y^2, \tag{3.4.42}$$
$$\dot{y} = y + b_{20}x^2 + b_{11}xy + b_{02}y^2, \tag{3.4.43}$$

and(0,0) is a saddle m_3^{-1} if $a_{11}b_{20} >0$ and a node m_3^1 if $a_{11}b_{20} <0$. The degenerate isoclines are determined by $\mu^3=0$ and consist of the three coinciding hyperbolae $y(a_{11}x + a_{02}y)=0$. The three coinciding branches on $y=0$ are in m_3 tangent to the center manifold(s) as are also the non-degenerate isoclines. The other three coinciding branches are transversal to $y=0$ and pass through the elementary critical point.

The central conic is given by

$$a_{11}y + 2c_{45}x^2 + 4c_{46}xy + 2c_{56}y^2 = 0, \tag{3.4.44}$$

non-degenerate since $H= \frac{1}{2}a_{11}^3 b_{20} \neq 0$.In m_3 it is tangent to the center manifold(s) with a curvature opposite to that of the non- degenerate isoclines as may be deduced from the approximation $y = 2b_{20}x^2+$h.o.t. $(b_{20} \neq 0)$. If $A>0$, so $i_f=0$, the central conic is a hyperbola and the elementary critical point is a saddle e^{-1} (antisaddle e^1) if m_3 is a node m_3^1 (saddle m_3^{-1}). If $A<0$ the central conic is of eliptic type and it is a real ellipse since for $c_{45} + c_{56} <0(>0),S_i <0(>0)$, (i=1,2) and $H=-\frac{1}{2}a_{11}^2 c_{45} >0(<0)$ so $AS_i/H >0(i=1,2)$ and (3.2.7) shows the result. For $c_{45}+c_{56} <0(>0)$, $i_f = -2(2)$ and the elementary critical point and third order critical point are both saddles (antisaddle and m_3^1 point).

If the third order point is *nilpotent*, we may use (2.3.5),(2.3.6) with $b_{20}=0$ yielding

$$\dot{x} = y + a_{20}x^2 + a_{11}xy + a_{02}y^2, \tag{3.4.45}$$
$$\dot{y} = b_{11}xy + b_{02}y^2, \tag{3.4.46}$$

and (0,0) is a saddle m_3^{-1} if $a_{20}b_{11} <0$ and a point with an elliptic and a hyperbolic sector m_3^1 if $a_{20}b_{11} >0$. The degenerate isoclines are determined by $\lambda^3=0$ and consist of the three coinciding hyperbolae $y(b_{11}x + b_{02}y)=0$. The three coinciding branches on $y=0$ are in m_3 tangent to the eigenvector of the linearized system in $(0,0)$ as are also the non-degenerate isoclines. The other three coinciding branches are transversal to $y=0$ and pass through the elementary critical point.

The central conic is given by

$$-b_{11}y + 2c_{45}x^2 + 4c_{46}xy + 2c_{56}y^2 = 0, \tag{3.4.47}$$

and is non-degenerate since $H=-\frac{1}{2}a_{20}b_{11}^3 \neq 0$. In m_3 it is tangent to the eigenvector of the linear part in $(3.4.45),(3.4.46)$ with a curvature opposite to the non- degenerate isoclines, as may be deduced from the approximation $y = 2a_{20}x^2+\text{h.o.t.}(a_{20} \neq 0)$. If A>0, then $i_f=0$, the central conic is a hyperbola and the elementary critical point is a saddle (antisaddle) if the third order critical point is of type $m_3^1(m_3^{-1})$. If $A <0$ the central conic is a real ellipse, under the same argument as given for the case with a semi- elementary critical point. Also follows the same conclusions with regard to the character of the elementary points: a saddle if $m_3 = m_3^{-1}$ and an antisaddle if $m_3 = m_3^1$. □

3.4.4   Systems with one finite critical point

Semi-elementary and nilpotent critical points

Theorem 3.8 *If a quadratic system of finite multiplicity 4 has a single critical point which is semi-elementary or nilpotent, then the point is a fourth order saddle node m_4^0. All three degenerate isoclines coincide and are degenerate parabolae consisting of two coinciding isoclinic lines. The six coinciding isoclinic lines are in m_4^0 tangent to all non-degenerate isoclines, the latter having a third order contact with each other. The non-degenerate isoclines are in m_4^0 also tangent to the center manifold(s) if m_4^0 is a semi-elementary critical point and the eigenvector of the linear part in m_4^0 if m_4^0 is a nilpotent saddle node. The central conic is a degenerate hyperbola, one branch of which coincides with the six isoclinic lines and the other branch is transversal to them.*

Proof. If the saddle node is *semi-elementary* we may write $(3.4.29),(3.4.30)$ as

$$\dot{x} = a_{02}y^2, \tag{3.4.48}$$
$$\dot{y} = y + b_{20}x^2 + b_{11}xy + b_{02}y^2, \tag{3.4.49}$$

with $a_{02}b_{20} \neq 0$. All non-degenerate isoclines are tangent to the center man-
ifold(s) in m_4^0 and have a third order contact there as follows from the approximation $y = -b_{20}x^2 + b_{20}b_{11}x^3 + h.o.t.$ To $\mu^3=0$ correspond the three
coinciding degenerate isoclines $y^2 \equiv 0$, thus forming six coinciding isoclinic
lines tangent to the non-degenerate isoclines in m_4^0 and also to the center
manifold(s) in m_4^0.

The central conic is given by

$$y(2b_{20}x + b_{11}y) = 0, \tag{3.4.50}$$

which is a degenerate hyperbola, one branch coinciding with the six isoclinic
lines and the other transversal to it since $b_{20} \neq 0$.

If the saddle node is *nilpotent* we may write (2.3.5),(2.3.6) as

$$\dot{x} = y + a_{20}x^2 + a_{11}xy + a_{02}y^2, \tag{3.4.51}$$
$$\dot{y} = b_{02}y^2, \tag{3.4.52}$$

with $a_{20}b_{02} \neq 0$. All non-degenerate isoclines are tangent in m_4^0 to the eigenvector of the linear part of (3.4.51) in m_4^0 and have a third order contact
there as follows from $y = -a_{20}x^2 + a_{20}a_{11}x^3+h.o.t.$ To $\lambda^3=0$ correspond
the three coinciding degenerate isoclines $y^2 \equiv 0$,which are three degenerate
parabolae together forming six coinciding isoclinic lines, also tangent to this
eigenvector.

The central conic is given by

$$y(2a_{20}x + a_{11}y) = 0, \tag{3.4.53}$$

which is a degenerate hyperbola, one branch coinciding with the six isoclinic
lines and the other transversal to it since $a_{20} \neq 0$.   □

Translated homogeneous systems

If a quadratic system with one finite critical point becomes a homogeneous
system when the critical point by a linear tansformation is transferred to the
origin, it will be called a *translated homogeneous system*. If the system is
non-degenerate the critical point is of fourth order as follows from Theorems
2.1 and 2.2 for a non-degenerate homogeneous system. So for $A \neq 0$, (and
$m_f=4$)it follows that $Ax^4 = Ay^4=0$, so (0,0) is an m_4^i point.

Theorem 3.9 *A non-degenerate translated homogeneous quadratic system belongs to the class $m_f = 4$ and all isoclines of such a system are degenerate. For $A < 0$ all isoclines are hyperbolic and the central conic is a degenerate ellipse, consisting of two conjugate complex lines intersecting in m_4. For $A > 0$ there exist hyperbolic, elliptic and parabolic isoclines. Of the latter there exist two different ones, each consisting of two coinciding lines and coinciding with the degenerate hyperbolic central conic.*

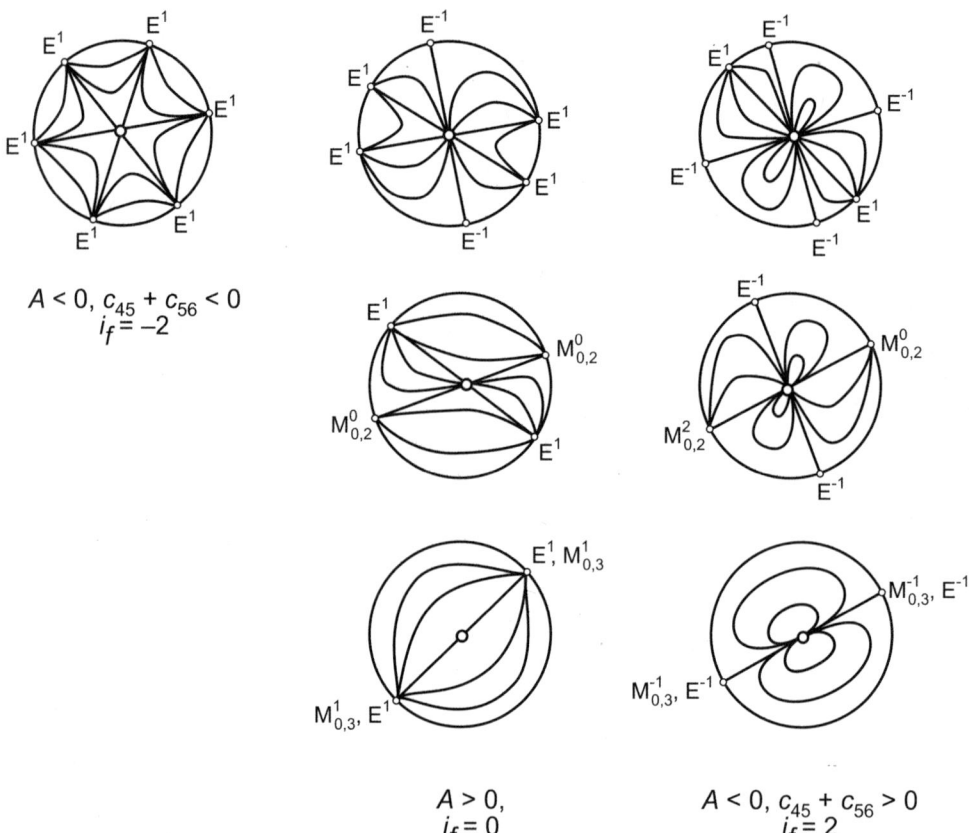

Figure 3.3 Topological characterization of the phase portraits of a non-degenerate translated homogeneous quadratic system

There exist seven topologically different phase portraits (see Figure 3.3). For $A<0$, $c_{45}+c_{56}<0$, the critical point is a point m_4^{-2} and has six hyperbolic sectors; at infinity there exist three elementary nodes E^1. For $A>0$ the critical point is an m_4^0 point; it has two, four or six sectors corresponding to one(three coinciding), two or three infinite critical points. In each case there exist two hyperbolic sectors; the remaining sectors are parabolic. At infinity the critical points are $E^{-1}E^1E^1$ and $C_{0,1}^0 C_{0,1}^0 E^1$ if there exist three infinite critical points, $E^1 M_{0,2}^0$ if there exist two such points and $M_{0,3}^1$ if only one. For $A<0$, $c_{45}+c_{56}>0$ the critical point is an m_4^2 point and also has two, four or six sectors corresponding to one (three), two or three infinite critical points. In each case there exist two elliptic sectors; the remaining sectors are parabolic. At infinity there exist the critical points $E^{-1}E^{-1}E^1$ and $C_{0,1}^0 C_{0,1}^0 E^{-1}$ if there exist three infinite critical points, $E^{-1} M_{0,2}^0$ if there exist two such points and $M_{0,3}^{-1}$ if only one.

Proof. The Lamé equation (3.2.1) is degenerate since $\alpha = \beta = \gamma = \delta = 0$ and all isoclines are degenerate. This follows from the fact that it is obvious for the homogeneous system and the invariance of this property under translation. From (3.1.2),(3.1.3) it follows that if $A<0$ all isoclines are hyperbolic, since if they would be all elliptic no real isoclines would exist as all isoclines would be complex. If $A>0$ there exist hyperbolic, elliptic as well as two different parabolic isoclines (see section 3.1 in the Introduction).Each parabolic isocline consists of two coinciding isoclinic lines and together they must form the degenerate central conic given by

$$c_{45}x^2 + 2c_{46}xy + c_{56}y^2 = 0, \qquad (3.4.54)$$

which thus is a degenerate hyperbola. If $A<0$ (3.4.54) is equally valid, the central conic is of elliptic type, consisting of two conjugate complex lines intersecting at the origin; this corresponds to the absence of parabolic isoclines for $A<0$.

The phase portraits of homogeneous quadratic systems were studied in several papers as a result of which it is known that there exist seven topologically different phase portraits; they are illustrated in Figure 3.3. We will discuss these in the present context.

Since $A \neq 0$, it follows from Theorem 2.5 that the critical points at infinity are transversally hyperbolic ($\lambda_z \neq 0$ in (2.1.10)). For an investigation of the possible infinite critical points we may rotate the axis such that there exists such a point at the ends of the x-axis. Thus $b_{20}=0$ in

$$\dot{x} = a_{20}x^2 + a_{11}xy + a_{02}y^2, \qquad (3.4.55)$$
$$\dot{y} = b_{20}x^2 + b_{11}xy + b_{02}y^2. \qquad (3.4.56)$$

If $a_{20}(a_{20} - b_{11}) > 0$ this point is an elementary node E^1 and if $a_{20}(a_{20} - b_{11}) < 0$ an elementary saddle E^{-1}. Since $\lambda_z \neq 0$ it follows that $a_{20} \neq 0$. If $a_{20} - b_{11}=0$, $a_{11} - b_{02} \neq 0$ the critical point is a second order saddle node M_{02}^0, and if $a_{20} - b_{11} = a_{11} - b_{02} = 0$ a third order node M_{03}^1 if $a_{20}a_{02} > 0$ and a third order saddle M_{03}^{-1} if $a_{20}a_{02} < 0$. (Note that $b_{20} = a_{20} - b_{11} = a_{11} - b_{02} = a_{02}=0$ cannot occur for $A \neq 0$). Equation (2.1.9) also allows two simple conjugate complex infinite critical points $C_{0,1}^0$.

For $A<0$, $c_{45} + c_{56} < 0$, we have $i_f = -2$ according to Theorem 3.4 so $i_i=3$ according to Theorem 2.8. This can only be achieved by three infinite nodes E^1, yielding six hyperbolic sectors in the finite critical point in accordance with Bendixsons formula $i=1+\frac{1}{2}(e - h)$ (section 2.2).

For $A>0$ we have $i_f=0$ so $i_i=1$. This admits the combinations $E^{-1}E^1E^1$ and $C_{0,1}^0 C_{0,1}^0 E^1$, $E^1 M_{0,2}^0$ and $M_{0,3}^1$ for three, two or one infinite critical points, respectively. Extending the flow near the Poincaré circle within a sector in a continuous way into the finite part of the plane yields two hyperbolic sectors in all four cases, as well as parabolic sectors in the remaining sectors. This is in accordance with Bendixson's formula.

For $A<0$, $c_{45} + c_{56} > 0$, we have $i_f=2$ so $i_i = -1$. This admits of the combinations $E^{-1}E^{-1}E^1$ and $C_{0,1}^0 C_{0,1}^0 E^{-1}$, $E^{-1}M_{0,2}^0$ and $M_{0,3}^{-1}$ for three, two or one infinite critical points, respectively. Extending in a continuous manner the flow near the Poincaré circle within a sector yields two elliptic sectors in all four cases , whereas the remaining sectors are parabolic. This is in accordance with Bendixson's formula.   □

3.4.5   The 65 classes of quadratic systems with finite multiplicity 4

Theorem 3.4 shows that within $m_f=4$ a further natural division in classes is obtained by the finite index i_f. Then there emerge the three classes:1 : $i_f = -2$, 2 : $i_f=0$, 3 : $i_f=2$. Within these classes a further division in classes is indicated by the various configurations of the degenerate isoclines and properties of the central conic, corresponding to the various combinations of finite and infinite critical points. This leads to five classes in class 1, 40 classes in class 2, 20 classes in class 3, yielding 65 classes in $m_f=4$. They are indicated in Table 1. Note that the possible combinations of infinite critical points in homogeneous quadratic systems are equally possible for the other systems in $m_f=4$. In Figure 3.4 the degenerate isoclines and the central conic are sketched for the various combinations of finite critical points.

$$e^{-1}e^{-1}e^{-1}e^1 \qquad e^{-1}e^{-1}c_1^0c_1^0 \qquad E^1E^1E^1$$
$$e^{-1}e^{-1}m_2^0$$
$$e^{-1}m_3^{-1}$$
$$m_4^{-2}$$

Class 1:$A < 0$, $c_{45} + c_{56} < 0$, $i_f = -2$, 5 classes.

$$e^{-1}e^{-1}e^1e^1 \qquad e^{-1}c_1^0c_1^0e^1 \qquad c_1^0c_1^0c_1^0c_1^0 \qquad\qquad E^{-1}E^1E^1 \qquad C_{0,1}^0C_{0,1}^0E^1$$
$$e^{-1}e^1m_2^0 \qquad c_1^0c_1^0m_2^0 \qquad\qquad\qquad E^1M_{0,2}^0$$
$$e^{-1}m_3^1 \qquad e^1m_3^{-1} \qquad m_2^0m_2^0 \qquad c_2^0c_2^0 \qquad\qquad M_{0,3}^1$$
$$m_4^0$$

Class 2:$A>0$, $i_f=0$, 40 classes.

$$e^{-1}e^1e^1e^1 \qquad c_1^0c_1^0e^1e^1 \qquad E^{-1}E^{-1}E^1 \qquad E^{-1}C_{0,1}^0C_{0,1}^0$$
$$e^1e^1m_2^0 \qquad\qquad\qquad E^{-1}M_{0,2}^0$$
$$e^1m_3^1 \qquad\qquad\qquad M_{0,3}^{-1}$$
$$m_4^2$$

Class 3:$A<0$, $c_{45} + c_{56} > 0$, $i_f=2$, 20 classes.

Table 1. Combinations of critical points in the 65 classes of class $m_f=4$.

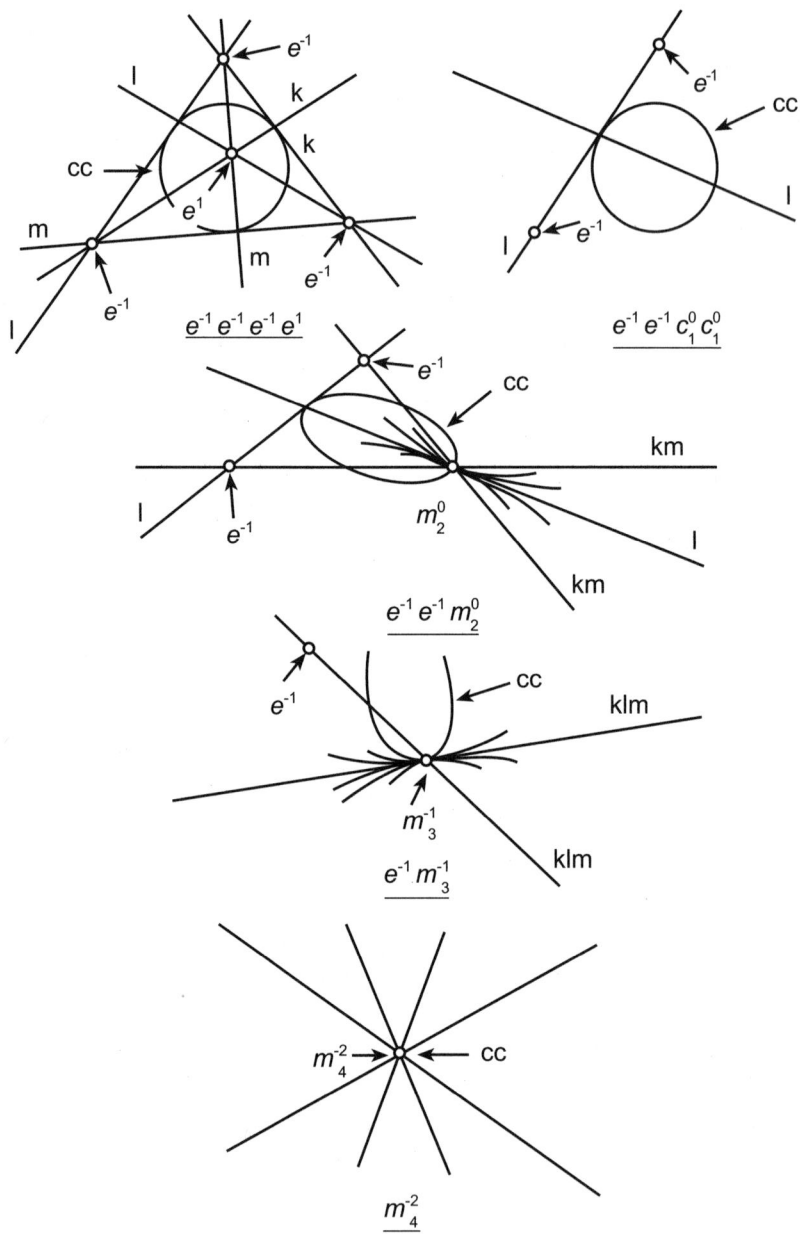

Figure 3.4a. Degenerate isoclines and central conic for class 1:m_f=4, $i_f = -2$

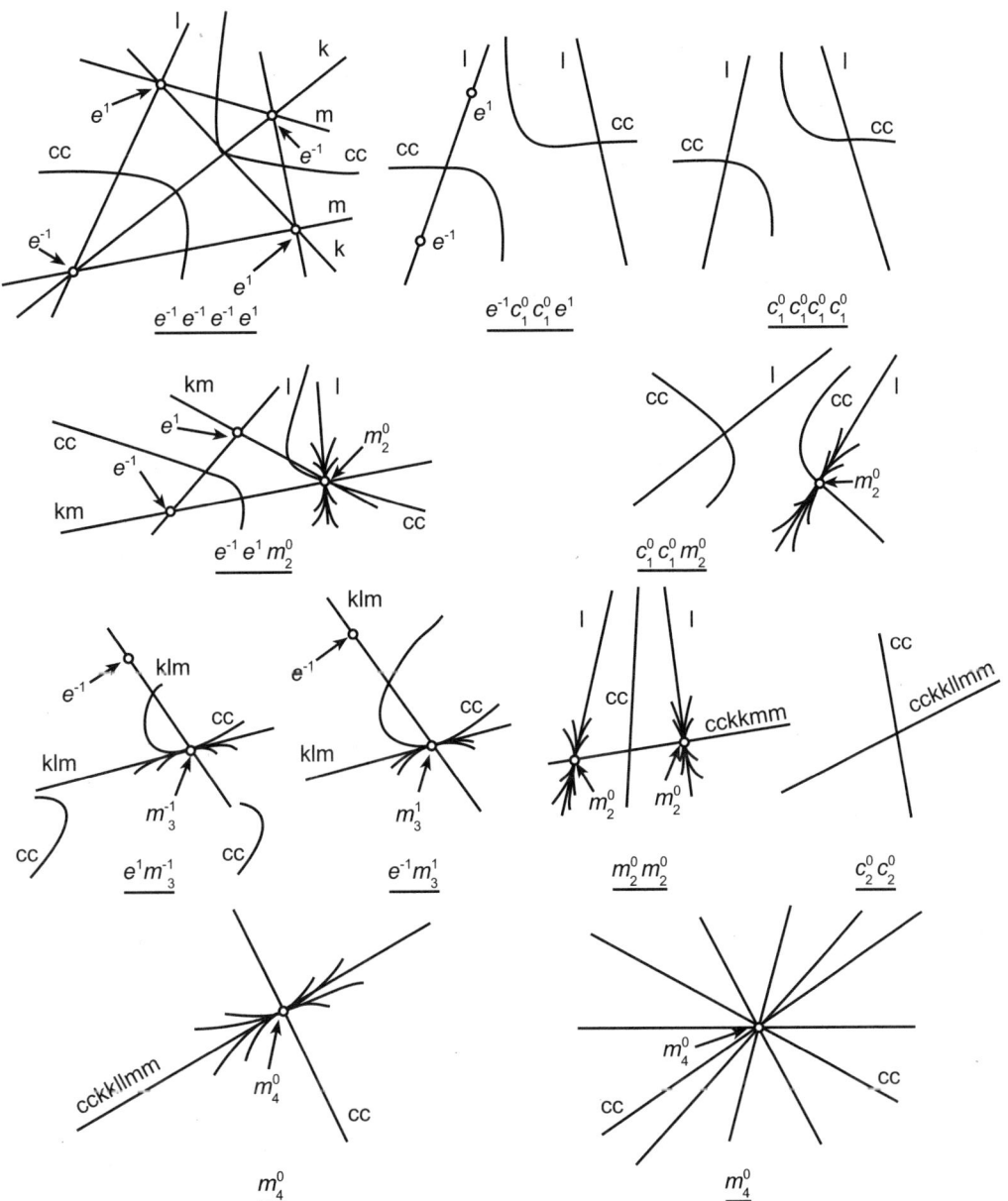

Figure 3.4b. Degenerate isoclines and central conic for class 2:$m_f=4$, $i_f=0$

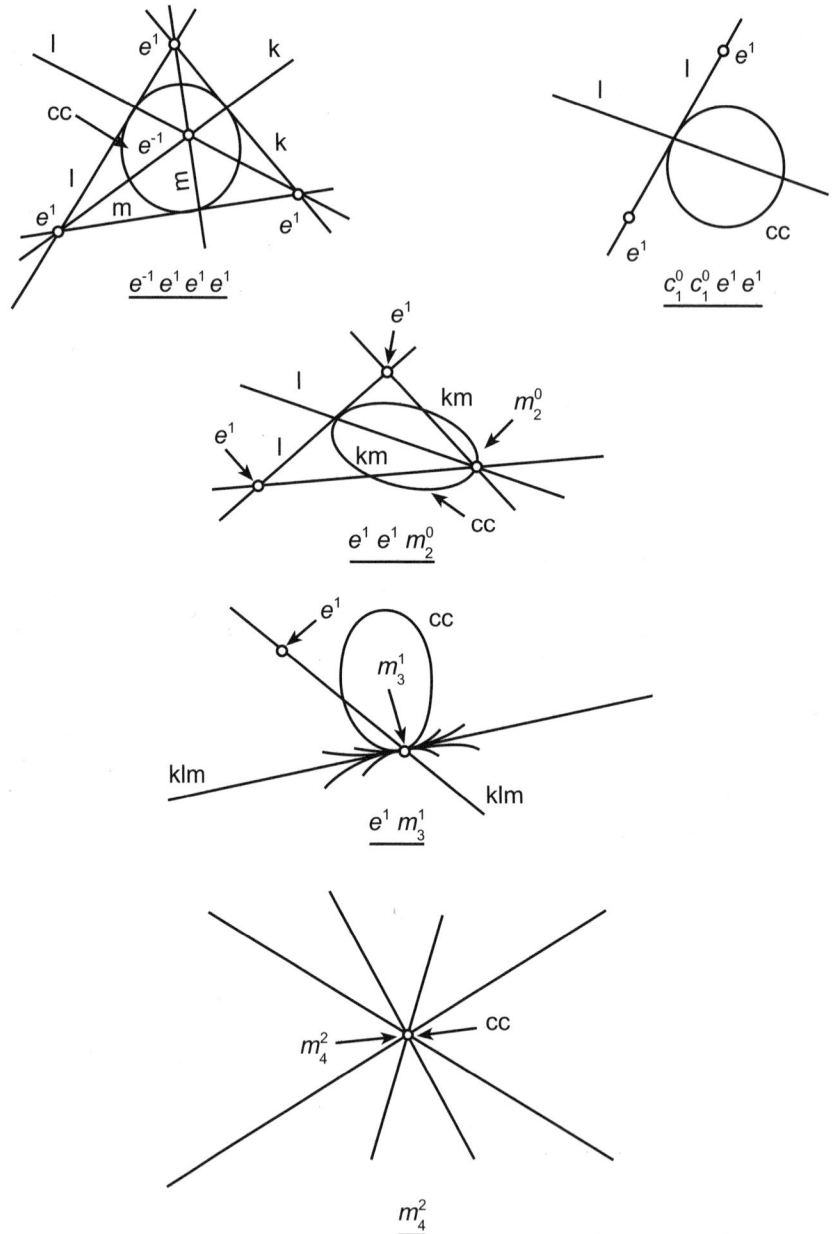

Figure 3.4c.Degenerate isoclines and central conic for class 3:m_f=4, i_f=2

3.5 Quadratic systems with $m_f{=}3$

3.5.1 Central conic

As Theorem 3.1 shows, a quadratic system in the class $m_f{=}3$ is characterized by the property that its central conic is a non-degenerate parabola. It may be seen as a limiting case of the class $m_f{=}4$. As was pointed out in section 3.4.1, the central conic in quadratic systems having four elementary critical points forming a non-degenerate quadrangle is a real ellipse, and the index of the critical point in the interior of the ellipse is equal to $1(-1)$if $i_f = -2(2)$. Using Theorem 3.4 this means that $(c_{45}+c_{56})\Lambda(x,y) < 0(> 0)$ in the interior (exterior) of the central conic. Sending one critical point to infinity seems to preserve this property, whereby the interior of the parabola is understood to be the region bounded by the parabola and containing its focus. In fact we have

Lemma 3.1 In the interior(exterior) of the central conic of a quadratic system in the class $m_f{=}3$ there is $(c_{45} + c_{56})\Lambda(x,y) < 0(> 0)$.

Proof First, let $c_{45} \neq 0$, then we may write

$$c_{45}\Lambda(x, y) = c_{23}c_{45} \tag{3.5.1}$$

$$-2c_{45}(c_{34} - \tfrac{1}{2}c_{25})x + 2c_{45}(c_{26} - \tfrac{1}{2}c_{35})y + 2c_{45}x^2 + 4c_{45}c_{46}xy + 2c_{45}c_{56}y^2,$$

or using $A = c_{46}^2 - c_{45}c_{56} = 0$,

$$c_{45}\Lambda(x, y) = c_{23}c_{45} - 2c_{45}(c_{34} - \tfrac{1}{2}c_{25})x + 2c_{45}(c_{26} - \tfrac{1}{2}c_{35})y + 2(c_{45}x + c_{46}y)^2. \tag{3.5.2}$$

Rotating the axes with respect to the coordinate system by putting $\bar{x} = c_{46}x - c_{45}y, \bar{y} = c_{45}x + c_{46}y$ yields

$$c_{45}\Lambda(\bar{x}, \bar{y}) = c_{23}c_{45} + \alpha\bar{x} + \beta\bar{y} + 2\bar{y}^2, \tag{3.5.3}$$

where

$$\alpha = \{-2c_{45}(c_{26} - \tfrac{1}{2}c_{35}) - 2c_{46}(c_{34} - \tfrac{1}{2}c_{25})\}/(c_{45} + c_{56}), \tag{3.5.4}$$

$$\beta = \{-2c_{45}(c_{34} - \tfrac{1}{2}c_{25}) + 2c_{46}(c_{26} - \tfrac{1}{2}c_{35})\}/(c_{45} + c_{56}), \tag{3.5.5}$$

and, where $c_{45} + c_{56} \neq 0$ as $c_{45} + c_{56}=0$ implies $c_{45}^2 + c_{46}^2 + c_{56}^2=0$. Let further (3.5.3) be written as

$$c_{45}\Lambda(\bar{x}, \bar{y}) = c_{23}c_{45} - \frac{1}{8}\beta^2 + \alpha\bar{x} + 2(\bar{y} + \frac{1}{4}\beta)^2. \qquad (3.5.6)$$

Then if $\alpha > 0(< 0)$, the interior of the central conic contains (part of) the negative (positive)$\bar{x}$ axis, whereas on the $\bar{x}$ axis $c_{45}\Lambda(\bar{x}, \bar{y})$ is an increasing (decreasing) function of $\bar{x}$, with $\Lambda(\bar{x}, \bar{y})=0$ on the central conic. So $c_{45}\Lambda(\bar{x}, \bar{y}) < 0(> 0)$ in the interior (exterior) of the central conic.

If $c_{56} \neq 0$, it may similarly be shown that $c_{56}\Lambda(\bar{x}, \bar{y}) < 0(> 0)$ in the interior(exterior) of the central conic. Adding these results leads to the statement in the lemma. □

Remark. According to (3.3.1) the condition $(c_{45} + c_{56})\Lambda(x, y) < 0(> 0)$ may be replaced by $H\Lambda(x, y) > 0(< 0)$.

3.5.2 Systems with three finite critical points

Systems with three real finite critical points

Theorem 3.10 *If a quadratic system in the class $m_f=3$ has three real finite critical points,then there exist six different real isoclinic lines formed by three degenerate hyperbolae. Three of these isoclinic lines form a triangle with the critical points as vertices. The other three isoclines are parallel lines through the critical points and directed towards the transversally non-hyperbolic infinite critical point. If $c_{45} + c_{56} <0$, there exist two saddles and one antisaddle and $i_f = -1$, whereas if $c_{45} + c_{56} >0$, there exist two antisaddles and one saddle and $i_f=1$. (See Figures 3.5,3.6)*

Proof. Starting with the system having four real elementary critical points (3.4.1),(3.4.2) and letting $\alpha, \beta \rightarrow \pm\infty$ such that $\gamma=\lim \alpha/\beta, \gamma \neq 0,1$, yields , with $D=a_1a_2 - b_1b_2 \neq 0$,

$$\dot{x} = a_1x(1 - x) + b_1y(1 - y) + (\gamma a_1 + \frac{1}{\gamma}b_1)xy, \qquad (3.5.7)$$

$$\dot{y} = a_2y(1 - y) + b_2x(1 - x) + (\frac{1}{\gamma}a_2 + \gamma b_2)xy, \qquad (3.5.8)$$

which has three finite critical points:A:(0,0),B(1,0) and C(0,1), and has a transversally non-hyperbolic critical point in the direction $x - \gamma y = 0$.

The central conic is given by

$$\gamma + (1 - 2\gamma)x + \gamma(-2 + \gamma)y - 2x^2 + 4\gamma xy - 2\gamma^2 y^2 = 0, \qquad (3.5.9)$$

which is a parabola with axis in the direction $x - \gamma y = 0$ and is non-degenerate since $H = \frac{1}{2}\gamma^2(1 + \gamma)^2 \neq 0$.

The degenerate isoclines are given by

$$(a_1\lambda + b_2\mu)(b_1\lambda + a_2\mu)[(b_1 + \gamma a_1)\lambda + (a_2 + \gamma b_2)\mu] = 0, \qquad (3.5.10)$$

which yields the three degenerate isoclines (hyperbolae)

$$k{:}\text{if } a_1\lambda + b_2\mu{=}0, \ y(-\gamma - x + \gamma y){=}0,$$
$$l{:}\text{if } b_1\lambda + a_2\mu{=}0, \ x(1 - x + \gamma y){=}0,$$
$$m{:}\text{if } (b_1 + \gamma a_1)\lambda + (a_2 + \gamma b_2)\mu{=}0, \ (1 - x - y)(-x + \gamma\, y){=}0.$$

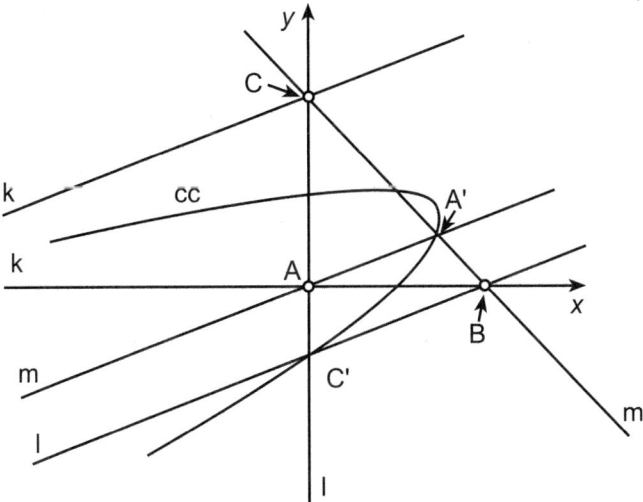

Figure 3.5 Degenerate isoclines and central conic in $m_f=3$ if there exist three elementary critical points

In figure 3.5 these degenerate isoclines and the central conic are sketched for $\gamma > 0$. The midpoints of the isoclines k,l,m are B', C' and A', respectively and given by $(-\gamma, 0)$, $(0, -\frac{1}{\gamma})$ and $(\frac{\gamma}{\gamma+1}, \frac{1}{\gamma+1})$,respectively.These points are on the central conic and form a triangle with A:(0,0) in its interior, the triangle in turn being situated in the interior of the central conic. Apart from A'

the point $(\frac{1}{2}, \frac{1}{2})$ is the other intersection point of BC with the central conic. The critical points B and C are therefore outside the central conic. Now $\Lambda(0,0) = D$, $\Lambda(1,0) = -\frac{\gamma+1}{\gamma}D$, $\Lambda(0,1) = -(\gamma+1)D$. Thus if $D > 0 (< 0)$ point A is an antisaddle (saddle) and points B and C are saddles (antisaddles) and $i_f = -1(1)$. According to Lemma 3.1 this corresponds to $c_{45} + c_{56} < 0 (> 0)$, as may also be concluded from the direct calculation of $c_{45} + c_{56}$ being equal to $-\frac{\gamma^2+1}{\gamma}D < 0 (> 0)$. If $\gamma < 0$ the conclusions may be reached similarly by noting that then point A is located outside of the central conic. Moreover also the relation $\Lambda(0,0)\Lambda(1,0)\Lambda(0,1) = -(c_{45} + c_{56})\frac{(\gamma+1)^2}{1+\gamma^2}D^2$ may be used. □

Systems having complex finite critical points

Theorem 3.11 *If a quadratic system in the class $m_f = 3$ has two complex and one (elementary) finite critical point, then there exist six different isoclinic lines, forming two complex degenerate isoclines each consisting of two non-conjugate complex lines and a (real) degenerate hyperbola. One of the branches of each degenerate isocline is in the direction of the transversally non-hyperbolic infinite critical point. If $c_{45} + c_{56} < 0 (> 0)$ the real critical point is a saddle (antisaddle) and $i_f = -1(1)$. (See Figure 3.6)*

Proof Rotate the axes such that the transversally non-hyperbolic infinite critical point is "at the ends of the x axis" and, by a shift, let $(0,0)$ be a real critical point.

This leads to the system

$$\dot{x} = a_{10}x + a_{01}y + a_{11}xy + a_{01}y^2, \qquad (3.5.11)$$
$$\dot{y} = b_{10}x + b_{01}y + b_{11}xy + b_{02}y^2, \qquad (3.5.12)$$

wherein $c_{23} \neq 0$ to make $(0,0)$ elementary and $c_{56} \neq 0$ because there exists precisely one direction wherein there exists a transversally non-hyperbolic infinite critical point.

Since $c_{56} \neq 0$, we may use (2.1.6) to determine the x coordinates of the critical points from

$$x(B_1x^2 + C_1x + D_1) = 0, \qquad (3.5.13)$$

with $B_1 = -c_{25}c_{56} \neq 0$, $C_1 = c_{26}^2 - c_{23}c_{56} - c_{25}c_{36}$ and $D_1 = -c_{23}c_{36}$.

The finite critical points outside of the origin are complex if

$$C_1^2 - 4B_1D_1 = (c_{26}^2 - c_{23}c_{56} - c_{25}c_{36})^2 - 4c_{23}c_{25}c_{36}c_{56}$$
$$= c_{26}^4 - 2c_{23}c_{26}^2c_{56} + c_{23}^2c_{56}^2 + c_{25}^2c_{36}^2 - 2c_{25}c_{26}^2c_{36} - 2c_{23}c_{25}c_{36}c_{56}$$

$$= c_{26}^2(c_{26}^2 - 2c_{23}c_{56} - 2c_{25}c_{36}) + (c_{23}c_{56} - c_{25}c_{36})^2$$
$$= c_{26}^2(c_{26}^2 - 2c_{23}c_{56} - 2c_{25}c_{36} + c_{35}^2) = c_{26}^2[(c_{35} - c_{26})^2 - 4c_{23}c_{56}] < 0, (3.5.14)$$

where use was made of the relations after Theorem 2.3.

For $a_{10}\lambda + b_{10}\mu = 0$ there exists the degenerate isocline l:

$$y(c_{23} + c_{25}x + c_{26}y) = 0, \tag{3.5.15}$$

where y ≡ 0 is the branch through (0,0) in the direction of the transversally non-hyperbolic infinite critical point and the other branch contains the complex critical points A and B. Isocline l is a hyperbola since $B_1 \neq 0$ implies $c_{25} \neq 0$ ($m_f=3$ implies $B_1 \neq 0$). The degenerate isocline k consists of the isoclinic line AO:$y_A x - x_A y = 0$ and B ∞ : y-y_B=0, which are non-conjugate complex lines, whereas the degenerate isocline m consists of the isoclinic lines BO:$y_B x - x_B y = 0$ and A ∞ : $y - y_A = 0$, which also are non-conjugate complex.

The central conic is given by

$$c_{23} + c_{25}x + 2(c_{26} - \frac{1}{2}c_{35})y + 2c_{56}y^2 = 0, \tag{3.5.16}$$

with the Hessian determinant H$= -\frac{1}{2}c_{25}^2 c_{56} \neq 0$.

Now if $c_{25}c_{56} > 0(< 0)$, the interior of the central conic contains (part of) the negative (positive)x axis. Moreover if $c_{23}c_{25} > 0(< 0)$ the midpoint of l,which is on the central conic, is on the negative (positive) x axis. As a result (0,0) is located outside (inside) the central conic if $c_{23}c_{56} > 0(< 0)$. In order to have complex critical points, however, $c_{23}c_{56} > 0$ as follows from (3.5.14). So (0,0) is located outside the central conic. From Lemma 3.1 it follows that for $c_{45} + c_{56} < 0(> 0), \Lambda(0,0) < 0(> 0)$ so (0,0) is a saddle (antisaddle) and $i_f = -1(1)$.   □

3.5.3   Systems with two finite critical points

Theorem 3.12 *If a quadratic system in the class $m_f=3$ has two finite critical points, then there exists one (real) elementary critical point and one second order critical point m_2^0, either a semi-elementary saddle node or a nilpotent cusp. There exist two coinciding degenerate isoclines, being hyperbolae of which one branch is in the direction of the transversally non-hyperbolic infinite critical point and the other branch contains the elementary critical point,* *the point m_2^0 being located at the intersection of the two branches which is on the central conic. The third degenerate isocline is also a degenerate hyperbola; one branch also contains the point m_2^0 and is tangent there to all non-degenerate isoclines. The other branch contains e and is in the direction of the transversally non-hyperbolic infinite critical point. If $c_{45}+c_{56} < 0(> 0)$, the elementary point is a saddle e^{-1} (antisaddle e^1) and $i_f = -1(1)$. (See Figure 3.6)*

Proof. The present case is a limiting case of the previous one, where there exist two complex and one real (elementary) critical point, and we may use (3.5.13) again with $c_{23}c_{56} \neq 0$. As before, there exist the degenerate isocline l as given by (3.5.15)

$$y(c_{23} + c_{25}x + c_{26}y) = 0, \tag{3.5.17}$$

where y $\equiv 0$ is the branch through (0,0) in the direction of the transversally non-hyperbolic infinite critical point and the other branch now contains the second order critical point m_2^0. Isocline l is a hyperbola since $B_1 \neq 0$, as $m_f = 3$,implies $c_{25} \neq 0$. For $\lambda = b_{11}, \mu = -a_{11}$ the non-degenerate isocline k given by

$$c_{25}x + c_{35}y - c_{56}y^2 = 0, \tag{3.5.18}$$

is obtained, what apart from (0,0) has the point m_2^0 in common with l, what implies that l and k are tangent there. In accordance with the properties of m_2^0 all non-degenerate isoclines are then tangent to k (and l) in m_2^0. From the tangency conditions it follows that $(c_{26} - c_{35})^2 - 4c_{23}c_{56}=0$, so $c_{23}c_{56} >0$ as $c_{23}c_{56} \neq 0$, in accordance with (3.5.14). Moreover, m_2^0 is given by

$$\frac{c_{26} - c_{35}}{2c_{56}}(\frac{c_{26} + c_{35}}{2c_{25}}, -1) \tag{3.5.19}$$

so that there exist the isocline(s)

$$(y + \frac{c_{26} - c_{35}}{2c_{56}})[2c_{25}x + (c_{26} + c_{35})y] = 0, \tag{3.5.20}$$

which not being tangent to k and l are the coinciding degenerate isoclines through m_2^0 of hyperbolic type.

The central conic is equally given by (3.5.16) and since $c_{23}c_{26} >0$ it may again be concluded that (0,0) is outside the central conic. From Lemma 3.1, the conclusion in the theorem may again be obtained.   □

3.5.4   Systems with one finite critical point

Theorem 3.13 *If a quadratic system in the class $m_f=3$ has one finite critical point, then there exists a third order critical point, which is a semi-linear or a nilpotent saddle m_3^{-1}, a semi-linear node or a nilpotent critical point with an elliptic and a hyperbolic sector m_3^1. There exist three coinciding degenerate isoclines; they are hyperbolae. One set of branches is tangent to the center manifold(s) or in the direction of the eigenvector of the linearized system in m_3 as well as to the non-degenerate isoclines, which osculate there. The other set of three coinciding branches is in the direction of the non-hyperbolic infinite critical point. The central conic is also tangent in m_3 to the non-degenerate isoclines. If $c_{45}+c_{56} < 0(> 0)$ the m_3 critical point is an $m_3^{-1}(m_3^1)$ point and $i_f = -1(1)$.(See Figure 3.6)*

   Proof. The statements with regard to the isoclines follow from the proof of Theorem 3.7 and the description of the properties of third order critical points in 2.3.1and 2.3.1. The last statement in the theorem follows applying Lemma 3.1.   □

3.5.5   The 36 classes of quadratic systems with finite multiplicity 3

For quadratic systems in the class $m_f=3$ the rôle of the invariants A,H and $c_{45} + c_{56}$ of the central conic can be formulated as follows.

   Theorem 3.14 *A quadratic system in the class $m_f=3$ is characterized by the condition $A=0$, $H \neq 0$. If $c_{45} + c_{56} < 0(>0)$, then $i_f = -1(1)$.*
   Proof. See Theorems 3.1,3.10-3.13.   □

   Remark. The formula given in Theorem 3.4 remains equally valid for $m_f=3$. Thus $i_f = [1 - sgn(c_{46}^2 - c_{45}c_{56})]sgn(c_{45} + c_{56})$.

   As for $m_f=4$, within $m_f=3$, a further natural division in classes is obtained by using the finite index i_f. Then, there emerge two classes:$1 : i_f = -1$ and $2 : i_f = 1$. At infinity there may exist the elementary critical points $E^{-1}, C_{0,1}^0$,and E^1,and the higher points $M_{1,1}^0, M_{0,2}^0, M_{1,2}^{-1}, M_{1,2}^1$ and $M_{1,3}^0$.

   Moreover all points at infinity are critical if $b_{20} = b_{11} - a_{20} = b_{02} - a_{11} = a_{02} = 0$,(see (2.3.11).

   The system (1.0.1),(1.0.2) then reads

$$\dot{x} = a_{00} + a_{10}x + a_{01}y + a_{20}x^2 + a_{11}xy, \qquad (3.5.21)$$
$$\dot{y} - b_{00} + b_{10}x + b_{01}y + a_{20}xy + a_{11}y^2, \qquad (3.5.22)$$

for which $c_{45} = a_{20}^2, c_{46} = a_{20}a_{11}, c_{56} = a_{11}^2$. So $A = a_{20}^2 a_{11}^2 - a_{20}^2 a_{11}^2 = 0$, and $H = -2[a_{20}(c_{26} - \frac{1}{2}c_{35}) + a_{11}(c_{34} - \frac{1}{2}c_{25})]^2 < 0$, so with a proper choice of coefficients we can make $H \neq 0$. Furthermore $c_{45} + c_{56} = a_{20}^2 + a_{11}^2 > 0$ and $i_f = 1$. Combining in each class the possible combinations of finite and infinite points yields eight classes in class 1 and 28 classes in class 2 yielding alltogether 36 classes in $m_f = 3$. They are indicated in Table 2. In Figure 3.6 the degenerate isoclines and the central conic are sketched for the various combinations of finite critical points in the classes 1 and 2.

$$e^{-1}e^{-1}e^1 \qquad e^{-1}c_1^0c_1^0 \qquad E^1 E^1 M_{1,1}^0$$
$$e^{-1}m_2^0 \qquad E^1 M_{1,2}^1$$
$$m_3^{-1}$$

Class 1: $c_{45} + c_{56} < 0, i_f = -1$, 8 classes

$$e^{-1}e^1e^1 \qquad c_1^0c_1^0e^1 \qquad E^{-1}E^1 M_{1,1}^0 \qquad C_{0,1}^0 C_{0,1}^0 M_{1,1}^0$$
$$e^1 m_2^0 \qquad E^{-1}M_{1,2}^1 \qquad E^1 M_{1,2}^{-1} \qquad M_{0,2}^0 M_{1,1}^0$$
$$m_3^1 \qquad \qquad M_{1,3}^0$$
$$\infty$$

Class 2: $c_{45} + c_{56} > 0, i_f = 1$, 28 classes

Table 2. Combinations of critical points in the 36 classes of class $m_f = 3$

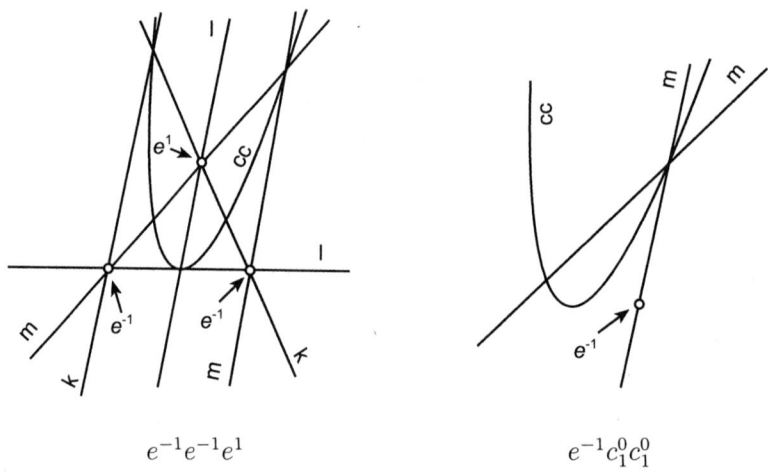

$$e^{-1}e^{-1}e^1 \qquad\qquad\qquad e^{-1}c_1^0c_1^0$$

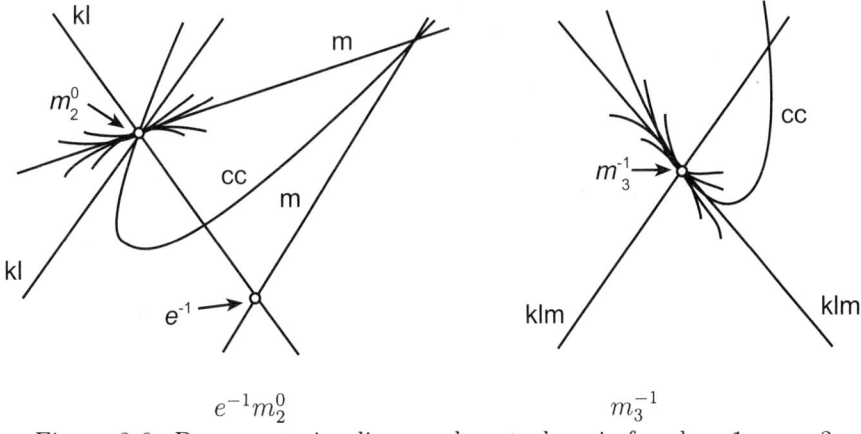

$$e^{-1}m_2^0 \qquad\qquad m_3^{-1}$$

Figure 3.6a Degenerate isoclines and central conic for class 1: $m_f=3$, $i_f = -1$

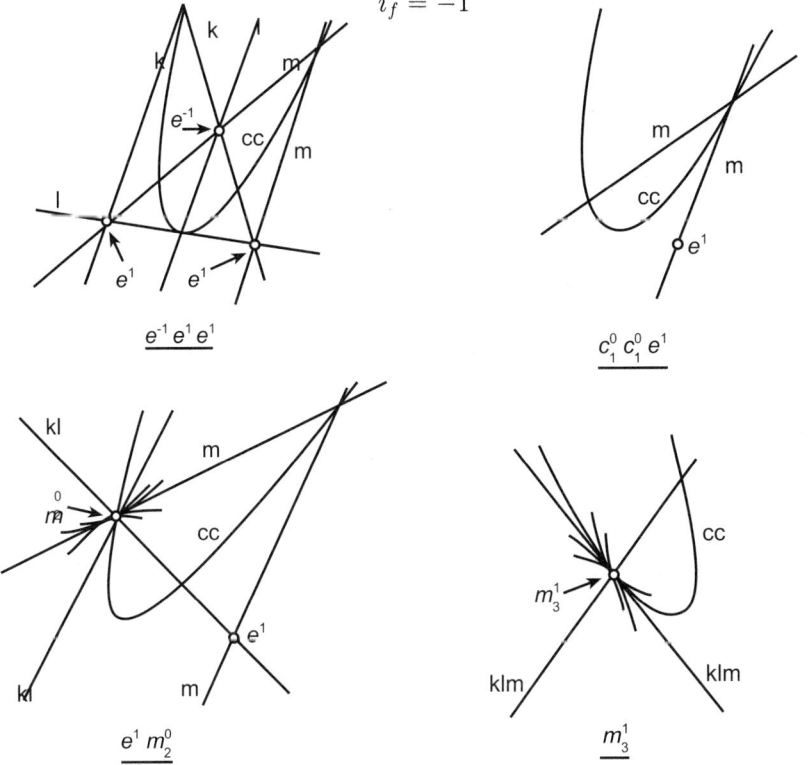

$$\underline{e^{-1}e^1e^1} \qquad\qquad \underline{c_1^0\,c_1^0\,e^1}$$

$$\underline{e^1\,m_2^0} \qquad\qquad \underline{m_3^1}$$

Figure 3.6b Degenerate isoclines and central conic for class 2: $m_f=3$, $i_f=1$

3.6   Quadratic systems with $m_f = 2$

For systems in the classes $m_f \leq 2$ we will make a distinction between one or two transversally non-hyperbolic infinite critical points. In the first case there exists for $m_f=2$ a critical point of type $M_{2,q}^i$, in the second either two points of type $M_{1,q}^i$ or $C_{1,1}^0$ or two coinciding directions of transversal non-hyperbolicity leading to one $M_{2,q}^i$ point.

3.6.1   Systems with one transversally non-hyperbolic infinite critical point

It follows from Theorem 3.2 that for $m_f=2$ the central conic is a degenerate parabola, consisting of two real parallel or coinciding lines. Let the region in between the parallel lines be indicated as the interior of the central conic and its complement as the exterior, then, since $m_f=2$ is a limiting case of $m_f=3$, Lemma 3.1 for $m_f=3$ may be thought to be equally valid for $m_f=2$. In fact we have

Lemma 3.2 In the interior(exterior) of the central conic of a quadratic system in class $m_f=2$ having one transversally non-hyperbolic infinite critical point there is $(c_{45} + c_{56})\Lambda(x, y) < 0 (> 0)$.

Proof. As in Lemma 3.1 let first $c_{45} \neq 0$ and calculate $c_{45}\Lambda(x, y)$; it then follows, with $\alpha = B_2/(c_{45} + c_{56})=0$, since $m_f=2$, that

$$c_{45}\Lambda(\bar{x}, \bar{y}) = c_{23}c_{45} + \beta\bar{y} + 2\bar{y}^2,$$

where β is given in (3.5.5). So $c_{45}\Lambda(\bar{x}, \bar{y}) < 0 (> 0)$ in the interior (exterior) of the central conic. Similarly it follows that $c_{56}\Lambda(\bar{x}, \bar{y}) < 0 (> 0)$ in the interior (exterior) of the central conic and the lemma follows from the invariance of sgnΛ under the used transformation.   □

Theorem 3.15 *If a quadratic system in the class $m_f=2$ has one transversally non-hyperbolic infinite critical point $M_{2,q}^i$, there exist six isoclinic lines of which two form a degenerate hyperbola and four form two coinciding degenerate parabolae consisting of two parallel lines either real distinct, coinciding or conjugate complex. One branch of the degenerate hyperbola and all branches of the parabolae point in the direction of $M_{2,q}^i$. There exist two elementary critical points, a second order point m_2^0 or two complex critical points, respectively. If there exist two real elementary critical points of the same type, they are located in the exterior of the central conic and are both saddles if*

$c_{45} + c_{56} < 0$, so if $i_f = -2$, and both antisaddles if $c_{45} + c_{56} > 0$, so if $i_f = 2$. In all other cases $i_f = 0$. This includes a saddle(antisaddle) in the interior and an antisaddle(saddle) in the exterior of the central conic if $c_{45} + c_{56} > 0 (< 0)$, a second order point m_2^0 on the central conic or two complex critical points. (See Figure 3.7)

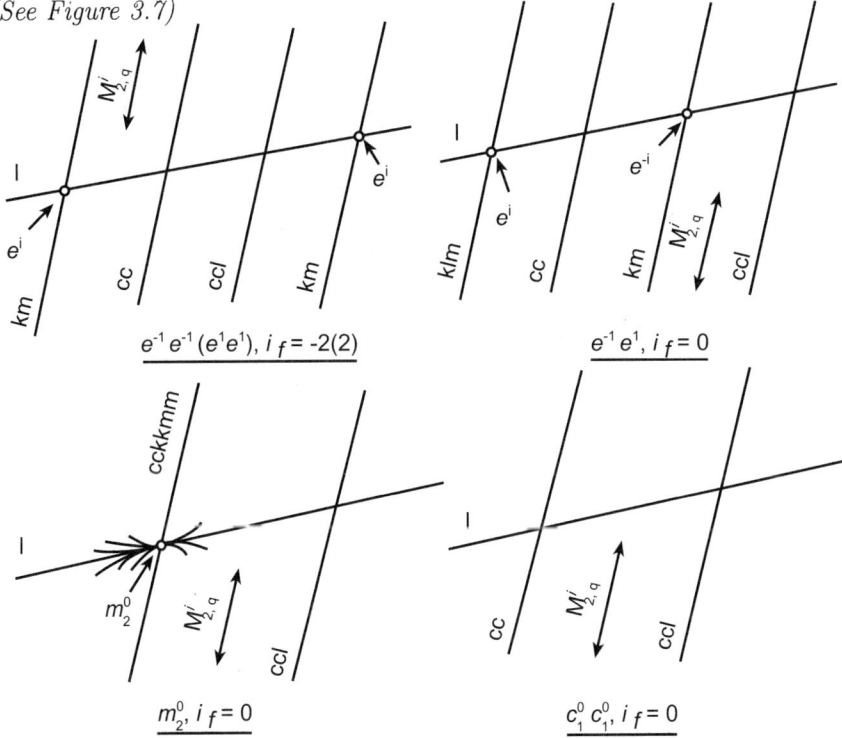

$e^{-1}e^{-1}(e^1e^1), i_f = -2(2)$ $e^{-1}e^1, i_f = 0$

$m_2^0, i_f = 0$ $c_1^0 c_1^0, i_f = 0$

Figure 3.7 Degenerate isoclines and central conic in $m_f = 2$. One transversally non-hyperbolic infinite critical point $M_{2,q}^i$

Proof. We may rotate the axes so that $M_{2,q}^i$ is at " the ends of the x axis" and y is a common factor of the quadratic terms in (1.0.1),(1.0.2). Then (1.0.1),(1.0.2) becomes

$$\dot{x} = a_{00} + a_{10}x + a_{01}y + a_{11}xy + a_{02}y^2, \tag{3.6.1}$$
$$\dot{y} = b_{00} + b_{10}x + b_{01}y + b_{11}xy + b_{02}y^2, \tag{3.6.2}$$

where $c_{56} \neq 0$ as $c_{45}^2 + c_{46}^2 + c_{56}^2 \neq 0$; from $B_1 = 0$ follows $c_{25} = 0$ and $C_1 = c_{26}^2 + c_{26}c_{35} - c_{15}c_{30} \neq 0$.

Since $c_{56} \neq 0$, it follows that $a_{11}^2 + b_{11}^2 \neq 0$, $a_{20}^2 + b_{02}^2 \neq 0$, so that there exist the isoclines

$$\bar{P}(x,y) \equiv c_{16} + c_{26}x + c_{36}y + c_{56}xy = 0, \qquad (3.6.3)$$
$$\bar{Q}(x,y) \equiv c_{15} + c_{35}y - c_{56}y^2 = 0, \qquad (3.6.4)$$

which we take as the basic conics of the pencil of isoclines of (3.6.1),(3.6.2): $\lambda\bar{P}(x,y) + \mu\bar{Q}(x,y)=0$. The degenerate isoclines of this pencil are given by

$$[(-c_{16}c_{56} + c_{26}c_{36})\lambda + C_1\mu]\lambda^2 = 0. \qquad (3.6.5)$$

To $\lambda^2=0$ correspond the two coinciding degenerate parabolae

$$k,m : c_{15} + c_{35}y - c_{56}y^2 = 0, \qquad (3.6.6)$$

which consist of two distinct lines if $c_{35}^2 + 4c_{15}c_{56} > 0$, two coinciding lines if $c_{35}^2 + 4c_{15}c_{56}=0$ and two complex lines if <0. The third degenerate isocline is given by

$$l : (c_{26} + c_{56}y)[c_{13}c_{56} - c_{16}c_{26} - C_1 x + (c_{16}c_{56} - c_{26}c_{36})y] = 0, \qquad (3.6.7)$$

which is a hyperbola since $C_1 \neq 0$.

The central conic may be written as

$$(c_{26} + c_{56}y)(c_{35} - 2c_{56}y) = 0, \qquad (3.6.8)$$

which is a degenerate parabola, consisting of two lines, distinct if $c_{26}+\frac{1}{2}c_{35} \neq 0$ and coinciding if $c_{26} + \frac{1}{2}c_{35}=0$.

For $D=c_{35}^2 + 4c_{15}c_{56} \neq 0$ there exist two elementary critical points (x_1, y_1) and (x_2, y_2) for which, with (3.6.3),(3.6.4) can be derived

$$\Lambda(x_i, y_i) = 2c_{15} + c_{23} + (2c_{26} + c_{35})y_i, i = 1, 2, \qquad (3.6.9)$$

and the relations

$$c_{56}^2\Lambda(x_1, y_1)\Lambda(x_2, y_2) = -DC_1, \qquad (3.6.10)$$
$$c_{56}[\Lambda(x_1, y_1) + \Lambda(x_2, y_2)] = D. \qquad (3.6.11)$$

Thus if $D>0$, there exist two real elementary critical points and if $C_1 <0$ they are of the same type. So for $c_{56} <0$ they are both saddles and for $c_{56} >0$ they are both antisaddles. From $\Lambda(x,y) = c_{23}+(2c_{26}-c_{35})y+2c_{56}y^2$ it follows further that for $c_{56} < 0(> 0)$, $\Lambda(x, y) > 0(< 0)$ in the interior of the central conic and $\Lambda(x, y) < 0(> 0)$ outside of it. So in both cases the critical points are outside of the central conic and Lemma 3.2 tells us that both are saddles if $c_{45} + c_{56} < 0$, so $i_f = -2$ and both are antisaddles if $c_{45} + c_{56} > 0$, so $i_f = 2$. If $C_1 > 0$ they are of different type; one saddle (antisaddle) in the interior and an antisaddle(saddle)in the exterior of the central conic if $c_{45} + c_{56} > 0(< 0)$ and $i_f = 0$.

If $D < 0$ there exist two complex critical points; $i_f = 0$.

If $D = 0$ there exists a second order point m_2^0, which is located on the central conic; $i_f = 0$.

Note that the branch of 1 pointing to $M_{2,q}^i$ coincides with one of the branches of the central conic. $\square$

3.6.2 Systems with two transversally non-hyperbolic infinite critical points

Two real distinct $M_{1,q}^i$ points

Theorem 3.16 *If a quadratic system in the class $m_f = 2$ has two transversally non-hyperbolic infinite critical points $M_{1,q}^i$ that are real and distinct, then there exist six isoclinic lines. One degenerate isocline is of parabolic type and consists of one finite line and one line at infinity. The other two degenerate isoclines are degenerate hyperbolae in the case of real critical points; distinct if there exist two finite critical points and coinciding if there exists a single m_2^0 point. The two branches of the hyperbola point in the directions of the $M_{1,q}^i$ points. In the case of complex critical points the degenerate isocline consists of two non-conjugate complex lines. If there exist two real critical points, there exist a saddle and a antisaddle. For all cases $i_f = 0$. The central conic is of parabolic type and consists of one finite line and one line at infinity. See Figure 3.8.*

Proof. Let the $M_{1,q}^i$ points be "at the ends of the x and y axes" then we may study the system

$$\dot{x} = a_{00} + a_{10}x + a_{01}y + a_{11}xy, \qquad (3.6.12)$$
$$\dot{y} = b_{00} + b_{10}x + b_{01}y + b_{11}xy, \qquad (3.6.13)$$

with $a_{11}^2 + b_{11}^2 \neq 0$, and which, according to Theorem 2.3, is of class $m_f = 2$ if $c_{25}c_{35} \neq 0$.

The Lamé equation reads

$$(a_{11}\lambda + b_{11}\mu)[(-a_{00}a_{11} + a_{10}a_{01})\lambda^2$$
$$+(-a_{00}b_{11} - a_{11}b_{00} + a_{10}b_{01} + a_{01}b_{10})\lambda\mu + (-b_{00}b_{11} + b_{10}b_{01})\mu^2] = 0. \quad (3.6.14)$$

To $a_{11}\lambda + b_{11}\mu = 0$ corresponds the degenerate isocline

$$k : c_{15} + c_{25}x + c_{35}y = 0, \quad (3.6.15)$$

which is of parabolic type and consists of a line at infinity and a finite line connecting the critical points, which are real distinct, coinciding or complex.

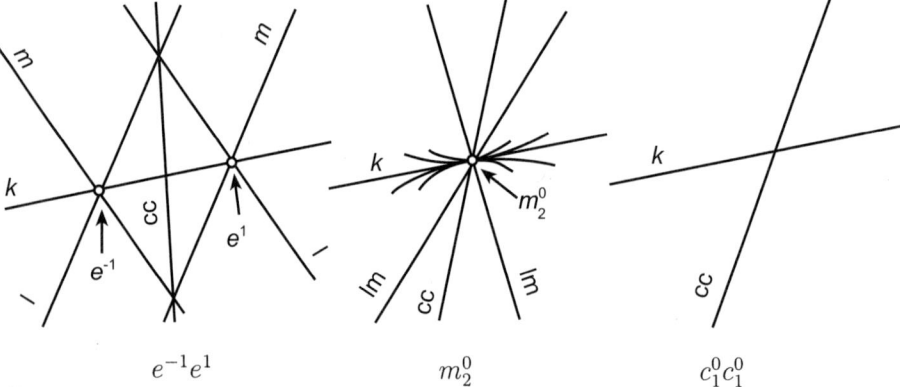

$$e^{-1}e^1 \qquad\qquad m_2^0 \qquad\qquad c_1^0 c_1^0$$

Figure 3.8 Degenerate isoclines and central conic in m_f=2. Two real distinct $M_{1,q}^i$ critical points: i_f=0

Since $c_{35} \neq 0$ there exists the isocline

$$c_{13} + c_{23}x - c_{35}xy = 0. \quad (3.6.16)$$

From (3.6.15),(3.6.16) then follows for the x coordinates of the critical points

$$c_{13} + (c_{15} + c_{23})x + c_{25}x^2 = 0, \quad (3.6.17)$$

and, with D=$(c_{15} + c_{23})^2 - 4c_{13}c_{25}$, if D>0 there exist two elementary critical points , if D=0 a second order point m_2^0. If D<0 there are two complex critical points which then are given by

$$(x_\pm, y_\pm) = [\frac{-c_{15} - c_{23} \pm \sqrt{D}}{2c_{25}}, \frac{-c_{15} + c_{23} \mp \sqrt{D}}{2c_{35}}]. \quad (3.6.18)$$

Since the discriminant of the quadratic factor in the Lamé equation also equals D, there exist two real, distinct solutions (λ, μ) if D>0, two coinciding if D=0 and two complex solutions if D<0. Accordingly, there exist the degenerate isoclines

$$l : [c_{25}x + \frac{1}{2}(c_{15} + c_{23} + \sqrt{D})][c_{35}y + \frac{1}{2}(c_{15} - c_{23} + \sqrt{D})] = 0, \quad (3.6.19)$$

$$m : [c_{25}x + \frac{1}{2}(c_{15} + c_{23} - \sqrt{D})][c_{35}y + \frac{1}{2}(c_{15} - c_{23} - \sqrt{D})] = 0, \quad (3.6.20)$$

which are two distinct degenerate hyperbolae, with pairwise parallel branches in the directions of the $M^i_{1,q}$ points, if D>0, two coinciding degenerate hyperbolae if D=0, and two pairs of non-conjugate complex lines if D<0.

The central conic is given by

$$c_{23} + c_{25}x - c_{35}y = 0, \quad (3.6.21)$$

being of parabolic type and consisting of one line at infinity and one finite line. If D>0 this line separates the two elementary critical points, which then must be a saddle and an antisaddle. This may also be concluded from the relation $\Lambda(x_-, y_-)\Lambda(x_+, y_+) = -D < 0$. In all cases there is $i_f = 0$.   □

Two coinciding transversally non-hyperbolic infinite critical points: an $M^i_{2,q}$ point

Theorem 3.17 *If a quadratic system in the class $m_f = 2$ has two coinciding transversally non-hyperbolic infinite critical points, then there exist six isoclinic lines. One degenerate isocline is of parabolic type and consists of one finite line and one line at infinity. The other two degenerate isoclines coincide and are also of parabolic type ; they consist of two parallel lines , real and distinct if there exist two real critical points, coinciding if there exists a second order point m^0_2 and complex conjugate if there exist two complex critical points. They point in the direction of $M^i_{2,q}$. The real critical points are a saddle and a antisaddle. In all cases $i_f = 0$. The central conic is of parabolic type and consists of a finite line pointing in the direction of $M^i_{2,q}$ and a line at infinity. See Figure 3.9.*

    Proof. The critical point $M^i_{2,q}$ may be thought to be located "at the ends of the x axis"; then (1.0.1),(1.0.2) may be written as

$$\dot{x} = a_{00} + a_{10}x + a_{01}y + a_{02}y^2, \quad (3.6.22)$$

$$\dot{y} = b_{00} + b_{10}x + b_{01}y + b_{02}y^2, \quad (3.6.23)$$

where $a_{02}^2 + b_{02}^2 \neq 0$ and if $c_{26} \neq 0$ is in the class $m_f{=}2$ according to Theorem 2.1.

The Lamé equation reads

$$(\lambda a_{10} + \mu b_{10})^2(\lambda a_{02} + \mu b_{02}) = 0. \tag{3.6.24}$$

To $a_{02}\lambda + b_{02}\mu{=}0$ corresponds the degenerate isocline

$$k : c_{16} + c_{26}x + c_{36}y = 0, \tag{3.6.25}$$

which is of parabolic type and consists of a line at infinity and a finite line connecting the critical points, which may be real , coinciding or complex.

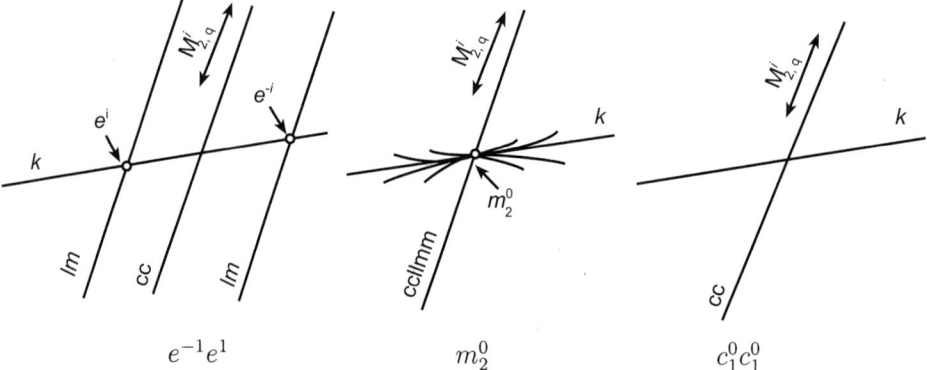

$$e^{-1}e^1 \qquad\qquad m_2^0 \qquad\qquad c_1^0c_1^0$$

Figure 3.9 Degenerate isoclines and central conic in $m_f{=}2$.Two coinciding transversally non-hyperbolic infinite critical points

To $(a_{10}\lambda + b_{10}\mu)^2{=}0$ corresponds the coinciding degenerate isoclines

$$l, m : c_{12} - c_{23}y - c_{26}y^2 = 0, \tag{3.6.26}$$

which are of parabolic type and consists of two parallel lines, real distinct if $D{>}0$, coinciding if $D{=}0$ and conjugate complex if $D{<}0$; here $D{=}c_{23}^2 + 4c_{12}c_{26}$. Correspondingly , there exist two elementary critical points , a second order point m_2^0 or two complex critical points c_1^0.

The central conic is given by

$$c_{23} + 2c_{26}y = 0, \tag{3.6.27}$$

being of parabolic type and consisting of one line at infinity and one finite line. If $D{>}0$ this line separates the two elementary critical points,which then must be a saddle and an antisaddle. This may also be concluded from the relation $\Lambda(x_-,y_-)\Lambda(x_+,y_+) = -D < 0$. In all cases there is $i_f{=}0$.   □

Two complex transversally non-hyperbolic infinite critical points $C_{1,1}^0$

Theorem 3.18 *If a quadratic system in the class $m_f = 2$ has two complex transversally non-hyperbolic infinite critical points $C_{1,1}^0$, then there exist six isoclinic lines. One degenerate isocline is of parabolic type and consists of one finite line and one line at infinity. The other two degenerate isoclines are both complex. If there exist two complex critical points, then both of these isoclines are represented by a conic with real coefficients and consist of two conjugate complex lines. If there exists a second order point m_2^0 these isoclines coincide. If there exist two critical points, then both of these isoclines are represented by a conic with complex coefficients and consist of two non-conjugate complex lines. Then there exist a saddle and an antisaddle. In all cases $i_f = 0$. The isoclinic lines are in the direction of the points $C_{1,1}^0$. The central conic is of parabolic type and consists of one finite line and one line at infinity. See Figure 3.10*

Proof. If the complex infinite critical points $C_{1,1}^0$ are located in the direction $y = (\alpha \pm i\beta)x$, the quadratic terms in (1.0.1),(1.0.2) have the common factor $(\alpha^2 + \beta^2)x^2 - 2\alpha xy + y^2, \alpha, \beta \in \mathbb{R}, \beta \neq 0$. Then $P_2(x,y) = a_{02}[(\alpha^2 + \beta^2)x^2 - 2\alpha xy + y^2]$, $Q_2(x,y) = b_{02}[(\alpha^2 + \beta^2)x^2 - 2\alpha xy + y^2]$. Locating the third (real) infinite critical point at the ends of the x axis yields $b_{02}(\alpha^2 + \beta^2) = 0$, thus $b_{02} = 0$ and $Q_2(x,y) = 0$. Then replace x by $a_{02}(x + \alpha y)$, y by $a_{02}(\alpha^2 + \beta^2)y$ and we obtain the system

$$D\dot{x} \equiv x' = \alpha_0 + \alpha_1 x + \alpha_2 y + \alpha_3 x^2 + \alpha_4 y^2, \tag{3.6.28}$$
$$D\dot{y} \equiv y' = \beta_0 + \beta_1 x + \beta_2 y, \tag{3.6.29}$$

where D$=a_{02}^2(\alpha^2 + \beta^2) \neq 0$, as $a_{02} \neq 0$ to avoid (1.0.1),(1.0.2) to be linear and $\alpha_3 \alpha_4 > 0$. Stretching x and y and returning to the original notation yields the system

$$\dot{x} = a_{00} + a_{10}x + a_{01}y + x^2 + y^2, \tag{3.6.30}$$
$$\dot{y} = b_{00} + b_{10}x + b_{01}y, \tag{3.6.31}$$

which, using $\bar{x} = x + \frac{1}{2}a_{10}, \bar{y} = y + \frac{1}{2}a_{01}$ and returning to the old notation, yields

$$\dot{x} = a_{00} + x^2 + y^2, \tag{3.6.32}$$
$$\dot{y} = b_{00} + b_{10}x + b_{01}y. \tag{3.6.33}$$

According to theorem 2.1, this system is in the class $m_f=2$ if $C_1 \equiv c_{26}^2 + c_{34}c_{36} = b_{10}^2 + b_{01}^2 \neq 0$.

From (2.1.6),(2.1.7) follows, for the coordinates of the critical points,

$$(b_{10}^2 + b_{01}^2)x^2 + 2b_{00}b_{10}x + a_{00}b_{01}^2 + b_{00}^2 = 0, \qquad (3.6.34)$$
$$(b_{10}^2 + b_{01}^2)y^2 + 2b_{00}b_{01}y + a_{00}b_{10}^2 + b_{00}^2 = 0. \qquad (3.6.35)$$

So, with $D = -(a_{00}b_{10}^2 + a_{00}b_{01}^2 + b_{00}^2)$ there exist two real elementary points if $D>0$, a second order point m_2^0 if $D=0$ and two complex critical points if $D<0$.

Obviously, to $\lambda=0$ corresponds the degenerate isocline

$$k : b_{00} + b_{10}x + b_{01}y = 0, \qquad (3.6.36)$$

which is of parabolic type and consists of a finite line and a line at infinity.

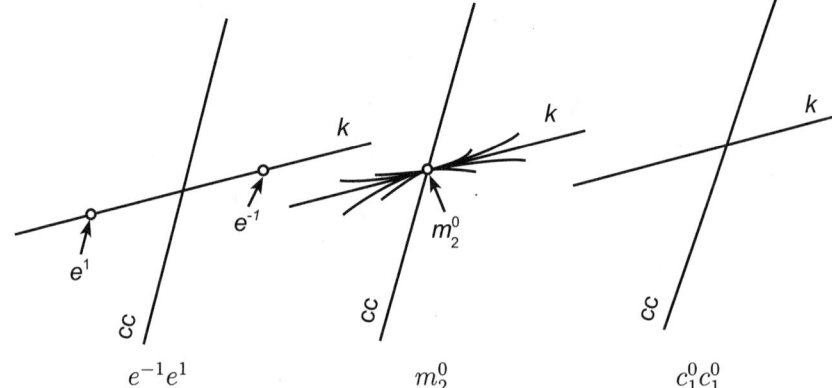

Figure 3.10 Degenerate isoclines and central conic in $m_f=2$. Two complex transversally non-hyperbolic infinite critical points:$C_{1,1}^0$

If $\lambda \neq 0$, let $\lambda=1$ and the isoclines of (3.6.32),(3.6.33) may be written as

$$a_{00} + \mu b_{00} + \lambda b_{10}x + \mu b_{01}y + x^2 + y^2 = 0, \qquad (3.6.37)$$

or

$$(x + \frac{1}{2}\mu b_{10})^2 + (y + \frac{1}{2}\mu b_{01})^2 = R, -\infty < \mu < \infty, \qquad (3.6.38)$$

where

$$R^2 = -a_{00} - b_{00}\mu + \frac{1}{4}(b_{10}^2 + b_{01}^2)\mu^2. \qquad (3.6.39)$$

The circle is degenerate if it is a point circle, so if R=0. As a result $R^2=0$ yields two real solutions μ_1 and μ_2 if D<0, so if there exist two complex critical points. There exist then two distinct degenerate isoclines l and m, consisting of two complex conjugate lines

$$l, m : [(x + \frac{1}{2}\mu_i b_{10}) + i(y + \frac{1}{2}\mu_i b_{01})][(x + \frac{1}{2}\mu_i b_{10}) - i(y + \frac{1}{2}\mu_i b_{01})] = 0, (i = 1, 2),$$
(3.6.40)

forming together a conic with real coefficients.

If D=0 these two solutions coincide as do the two isoclines l and m;there exists an m_2^0 point.

If D<0, $R^2=0$ yields two complex solutions $\mu = \alpha \pm i\beta$, and (3.6.32),(3.6.33) has two real critical points . Equation (3.6.38) then represents a point circle with a complex midpoint and consisting of two non-conjugate complex lines

$$l : [\frac{1}{2}\alpha b_{10} - \frac{1}{2}\beta b_{01} + x + i(\frac{1}{2}\alpha b_{01} + \frac{1}{2}\beta b_{10} + y)]$$
$$[\frac{1}{2}\alpha b_{10} + \frac{1}{2}\beta b_{01} + x + i(-\frac{1}{2}\alpha b_{01} + \frac{1}{2}\beta b_{10} - y)] = 0,$$

$$m : \frac{1}{2}\alpha b_{10} + \frac{1}{2}\beta b_{01} + x + i(\frac{1}{2}\alpha b_{01} - \frac{1}{2}\beta b_{10} + y)]$$
$$[\frac{1}{2}\alpha b_{10} - \frac{1}{2}\beta b_{01} + x + i(-\frac{1}{2}\alpha b_{01} - \frac{1}{2}\beta b_{10} - y)] = 0.$$

Each of the branches of l and m point in the direction of a $C_{1,1}^0$ point $(x \pm iy=0)$.

The central conic of (3.6.32),(3.6.33)is given by

$$b_{01}x - b_{10}y = 0, \qquad (3.6.41)$$

it is of parabolic type and consists of a finite line and a line at infinity.

Moreover $\Lambda(x_-, y_-)\Lambda(x_+, y_+) = -4D$ <0 if D>0 and there exist two critical points (x_-, y_-) and (x_+, y_+) of which one is a saddle and the other an antisaddle.

So $i_f=0$ in all cases. $\square$

Conclusion

From Theorems 3.16-3.18 it follows that if a quadratic system in the class $m_f=2$ has two (distinct, coinciding or complex)transversally non-hyperbolic infinite critical points, then $i_f=0$ and the central conic consists of a finite line and a line at infinity.

3.6.3   The 48 classes in quadratic systems with finite multiplicity 2

Within systems with one transversally non-hyperbolic infinite critical point we distinguish the classes $1 : i_f = -2, 2 : i_f = 0$ and $3 : i_f=2$. The systems with two transversally non-hyperbolic infinite critical points form class 4 : $i_f=0$. At infinity there can occur the elementary points $E^{-1}, C^0_{0,1}$ and E^1 and the higher order points $C^0_{1,1}, M^0_{0,2}, M^0_{1,1}; M^1_{1,2}, M^{-1}_{2,1}, M^1_{2,1}; M^{-2}_{2,2}, M^0_{2,2}, M^2_{2,2}$ and $M^{-1}_{2,3}, M^1_{2,3}$. Moreover, all points at infinity are critical if $b_{20} = b_{11} - a_{20} = b_{02} - a_{11} = a_{02} = 0$ which leads to the system (3.5.21),(3.5.22)

$$\dot{x} = a_{00} + a_{10}x + a_{01}y + a_{20}x^2 + a_{11}xy, \qquad (3.6.42)$$
$$\dot{y} = b_{00} + b_{10}x + b_{01}y + b_{20}x^2 + b_{11}xy, \qquad (3.6.43)$$

where $a^2_{20} + a^2_{11} \neq 0$ as otherwise the system would be linear. The quadratic terms in (3.6.42),(3.3.43) have the common linear factor $a_{20}x + a_{11}y$ which makes the analysis in section 3.6.1 applicable. In fact, let $a_{11} \neq 0$,otherwise interchange x and y , and put $\bar{x} = x, \bar{y} = a_{20}x + a_{11}y$, then (3.6.42),(3.6.43) becomes, after returning to the original notation , the system

$$\dot{x} = a_{00} + a_{10}x + a_{01}y + xy, \qquad (3.6.44)$$
$$\dot{y} = b_{00} + b_{10}x + b_{01}y + y^2, \qquad (3.6.45)$$

which is a special case of (3.6.42),(3.6.43). Now $c_{45} = c_{46} = 0, c_{56} = 1$ so A=0, and $B_1=0$ if $c_{25}=0$ or $b_{10}=0$. In order for system (3.6.44),(3.6.45) to be in the class $m_f=2$ it is moreover required that $C_1 \equiv c^2_{26} + c_{26}c_{35} - c_{15}c_{56} = a^2_{10} - a_{10}b_{01} + b_{00} \neq 0$. Let $D \equiv b^2_{01} - 4b_{00}$, then for D>0 there are two real distinct elementary critical points , for D=0 a second order point and for D<0 two complex critical points. If $D> 0, C_1 <0$ it follows from the analysis in section 3.6.1 that the two critical points are both antisaddles so $i_f=2$ and the system belongs to class 3. Similarly , if $D> 0, C_1 >0$, there exist a saddle and an antisaddle,so $i_f=0$ and the system belongs to class 2.If $D \leq 0, i_f=0$ and the system also belongs to class 2.

Combining in each class the possible combinations of finite and infinite critical points yields two classes in class 1, 24 classes in class 2, seven classes in class 3 and 15 classes in class 4, yielding 48 classes in $m_f=2$. They are given in Table 3.

$$e^{-1}e^{-1} \qquad E^1 E^1 M_{2,1}^1$$
$$E^1 M_{2,2}^2$$

Class 1:$c_{45} + c_{56} < 0, i_f = -2$, 2 classes.

$$e^{-1}e^1 \qquad c_1^0 c_1^0 \qquad E^1 E^1 M_{2,1}^{-1} \qquad E^{-1}E^1 M_{2,1}^1 \qquad C_{0,1}^0 C_{0,1}^0 M_{2,1}^1$$
$$m_2^0 \qquad E^1 M_{2,2}^0 \qquad E^{-1} M_{2,2}^2 \qquad M_{0,2}^0 M_{2,1}^1$$
$$M_{2,3}^1$$
$$\infty$$

Class 2:$c_{45} + c_{56} < 0(> 0) : e^1(e^{-1})$ in c.c.,$e^{-1}(e^1)$ out c.c.,i_f=0, 24 classes.

$$e^1 e^1 \qquad E^{-1}E^1 M_{2,1}^{-1} \qquad C_{0,1}^0 C_{0,1}^0 M_{2,1}^{-1}$$
$$E^1 M_{2,2}^{-2} \qquad E^{-1} M_{2,2}^0 \qquad M_{0,2}^0 M_{2,1}^{-1}$$
$$M_{2,3}^{-1}$$
$$\infty$$

Class 3:$c_{45} + c_{56} > 0, i_f$=2, 33 classes.

a. Systems with one transversally non-hyperbolic infinite critical point.

$$e^{-1}e^1 \qquad c_1^0 c_1^0 \qquad E^1 M_{1,1}^0 M_{1,1}^0 \qquad E^1 C_{1,1}^0 C_{1,1}^0$$
$$m_2^0 \qquad M_{1,1}^0 M_{1,2}^1 \qquad E^1 M_{2,2}^0$$
$$M_{2,3}^1$$

Class 4:i_f=0, 15 classes.

b. Systems with two transversally non-hyperbolic infinite critical points.

Table 3. Combinations of critical points in the 48 classes of class m_f=2.

3.7 Quadratic systems with $m_f = 1$

3.7.1 Systems with one transversally non-hyperbolic infinite critical point

According to Theorem 3.2 the central conic of a quadratic system with $m_f \leq 2$ is a degenerate parabola, consisting of two real parallel or coinciding lines. However for m_f=1 the latter cannot occur.

Theorem 3.19 *The central conic of a quadratic system in the class m_f=1 and having one transversally non-hyperbolic infinite critical point is of parabolic type and consists of two real distinct parallel lines.*

Proof. Rotating the coordinate axes, such that the $M_{3,q}^i$ point is "at the ends of the x axis" and shifting the origin to the elementary critical point yields the system

$$\dot{x} = a_{10}x + a_{01}y + a_{11}xy + a_{02}y^2, \qquad (3.7.1)$$
$$\dot{y} = b_{10}x + b_{01}y + b_{11}xy + b_{02}y^2, \qquad (3.7.2)$$

where $c_{23} \neq 0, c_{56} \neq 0$. Obviously A=0, since $c_{45} = c_{46} = 0, B_1=0$ yields $c_{25} = 0, C_1=0$ yields $c_{26}(c_{26} + c_{35})=0$ and $D_1 \neq 0$ gives $c_{23}c_{36} \neq 0$. From $c_{23}c_{56} + c_{26}c_{35} = c_{25}c_{36} = 0$ it follows further that $c_{26}c_{35} \neq 0$ so $C_1=0$ yields $c_{26} + c_{35}=0$.

The central conic of (3.7.1) may be written

$$c_{23} + 2(c_{26} - \frac{1}{2}c_{35})y + 2c_{56}y^2 = 0, \qquad (3.7.3)$$

which is of parabolic type, consisting of two real distinct parallel lines, since the discriminant D may be written as $D=4[(c_{26} - \frac{1}{2}c_{35})^2 - 2c_{23}c_{56}] = 4(c_{26} + \frac{1}{2}c_{35})^2 = c_{26}^2 > 0$. □

Lemma 3.3 In the interior (exterior) of the central conic of a quadratic system in the class $m_f=1$ and one transversally non-hyperbolic infinite critical point there is $(c_{45} + c_{56})\Lambda(x, y) < 0(> 0)$.

Proof. See proof of Lemma 3.2. □

Lemma 3.4 The critical point of a quadratic system in the class $m_f=1$ and one transversally non-hyperbolic infinite critical point is located in the exterior of the central conic. It is a saddle point if $c_{45} + c_{56} < 0$, then $i_f = -1$ and an antisaddle if $c_{45} + c_{56} > 0$, then $i_f=1$

Proof. From $c_{23}c_{56} = -c_{26}c_{35} = c_{26}^2 > 0$ it follows that the roots of (3.7.3) are of the same sign, and the critical point in (0,0) is located in the exterior of the central conic. Then use Lemma 3.3. □

Theorem 3.20 *If a quadratic system in the class $m_f=1$ has one transversally non-hyperbolic infinite critical point $M_{3,q}^i$, then there exist six real isoclinic lines all pointing in the direction of $M_{3,q}^i$. They represent three coinciding isoclines, of parabolic type, and each consisting of two real parallel lines. One of the three coinciding isoclinic lines contains the finite critical point, whereas the other three coincide with one branch of the central conic. The other branch of the central conic is halfway between the coinciding isoclinic lines. See Figure 3.11*

Proof. Since $c_{56} \neq 0$, as there exist only one transversally non-hyperbolic infinite critical point, it follows that $a_{11}^2 + b_{11}^2 \neq 0$, $a_{02}^2 + b_{02}^2 \neq 0$, so there exist the isoclines of (3.7.1),(3.7.2)

$$\bar{P}(x,y) \equiv c_{26}x + c_{36}y + c_{56}xy = 0, \qquad (3.7.4)$$
$$\bar{Q}(x,y) \equiv c_{35}y - c_{56}y^2, \qquad (3.7.5)$$

which we take as the basic conics of the pencil of isoclines of (3.7.1),(3.7.2): $\lambda\bar{P}(x,y) + \mu\bar{Q}(x,y)=0$. The Lamé equation for this pencil is given by $\lambda^3=0$, yielding $\bar{Q}(x,y)=0$, so that there exist three isoclinic lines through the finite critical point in (0,0) whereas the other three coincide with $c_{35} - c_{56}y=0$, which is one of the branches of the central conic given by $(c_{35} - c_{56}y)(c_{35} - 2c_{56}y)=0$. $\qquad\square$

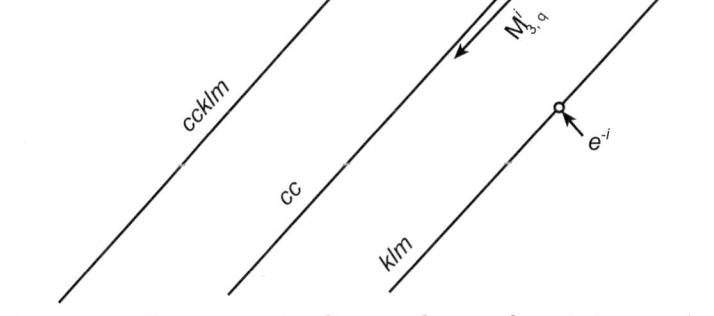

Figure 3.11 Degenerate isoclines and central conic in $m_f=1$. One transversally non-hyperbolic infinite critical point $M_{3,q}^i$

3.7.2 Systems with two transversally non-hyperbolic infinite critical points

Two distinct points : an $M_{1,q}^i$ point and an $M_{2,q}^i$ point

Theorem 3.21 *The central conic of a quadratic system with $m_f=1$ and two distinct real transversally non-hyperbolic infinite critical points $M_{1,q}^i$ and $M_{2,q}^i$ is of parabolic type and consists of one finite line in the direction of $M_{2,q}^i$ and one line at infinity. There exist six isoclinic lines. One isocline is a degenerate hyperbola ; one branch coincides with the finite branch of the central conic and the other contains the finite critical point and points in* the direction of $M_{1,q}^i$. The other two degenerate isoclines coincide and are of parabolic type; they consist of one finite line in the direction of $M_{2,q}^i$ and contain the critical point, whereas the other line is located at infinity. See Figure 3.12

Proof. As in the sections 3.6 and 2.1 we may use (3.6.11) with $a_{00} = b_{00}=0$;

$$\dot{x} = a_{10}x + a_{01}y + a_{11}xy, \qquad (3.7.6)$$
$$\dot{y} = b_{10}x + b_{01}y + b_{11}xy, \qquad (3.7.7)$$

with $a_{11}^2 + b_{11}^2 \neq 0$, which in accordance with Theorem 2.3 is of class $m_f=1$ if either $c_{25} = 0, c_{35} \neq 0, c_{15} + c_{23} \neq 0$ or $c_{25} \neq 0, c_{35} = 0, c_{15} - c_{23} \neq 0$.

The latter condition may be obtained from the first by interchanging x and y, a_{10} and b_{01}, a_{01} and b_{10}, a_{11} and b_{11}; so we will work with the first conditions; then $M_{2,q}^i$ is "at the ends of the x axis".

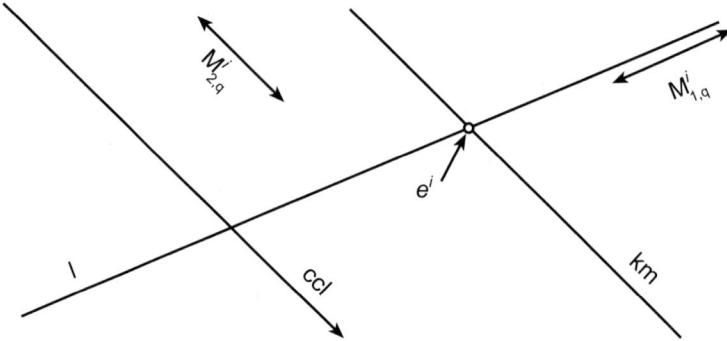

Figure 3.12 Degenerate isoclines and central conic in $m_f=1$. Two real distinct infinite critical points $M_{1,q}^i$ and $M_{2,q}^i$.

The Lamé equation reads

$$(a_{10}\lambda + b_{10}\mu)(a_{01}\lambda + b_{01}\mu)(a_{11}\lambda + b_{11}\mu) = 0, \qquad (3.7.8)$$

yielding the degenerate isoclines

$\quad\quad$ · $\quad$ k:$y \equiv 0$,for $a_{11}\lambda + b_{11}\mu=0$,
$\quad\quad\quad$ l:$x(c_{23} - c_{35}y)=0$,for $a_{01}\lambda + b_{01}\mu =0$,
$\quad\quad\quad$ m:$y \equiv 0$,for $a_{10}\lambda + b_{10}\mu=0$.
For the central conic is obtained

$$c_{23} - c_{35}y = 0. \qquad (3.7.9)$$

The critical point is at the intersection of l and k,m. $\qquad\qquad$ □

Two coinciding transversally non-hyperbolic infinite critical points : a point $M_{3,q}^i$

Theorem 3.22 *If a quadratic system in class $m_f=1$ has two coinciding transversally non- hyperbolic infinite critical points, then the central conic is a degenerate parabola consisting of two coinciding lines at infinity. The three degenerate isoclines coincide and are also of parabolic type; they consist of a finite line and a line at infinity. See Figure 3.13*

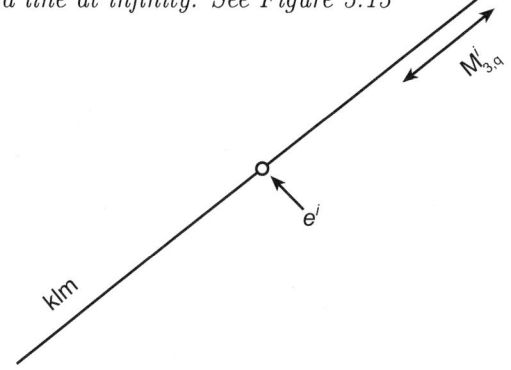

Figure 3.13 Degenerate isoclines and central conic in $m_f=1$. Two coinciding infinite critical points:an $M_{3,q}^i$ point

Proof. As in section 3.6.2 we may work with (3.6.20), $a_{00} = b_{00}=0$;then

$$\dot{x} = a_{10}x + a_{01}y + a_{02}y^2, \qquad (3.7.10)$$
$$\dot{y} = b_{10}x + b_{01}y + b_{02}y^2, \qquad (3.7.11)$$

with $a_{02}^2+b_{02}^2 \neq 0$, so we can use Theorem 2.1. It follows that $A=B_1=0$,whereas $C_1=0$ yields $c_{26}=0$ and $D_1 \neq 0$ yields $c_{23}c_{36} \neq 0$.

For the central conic follows the equation

$$c_{23} = 0. \qquad (3.7.12)$$

The Lamé equation reads

$$(\lambda a_{10} + \mu b_{10})^2(\lambda a_{02} + \mu b_{02}) = 0, \qquad (3.7.13)$$

yielding

$$y^3 \equiv 0. \qquad (3.7.14)$$

$\square$

3.7.3 Concluding remark

If a quadratic system in class $m_f=1$ has one transversally non-hyperbolic infinite critical point, then, as for $m_f=3$ and $m_f=4$, there is $i_f=[1-sgn(c_{46}^2-c_{45}c_{56})]sgn(c_{45}+c_{56})$.

3.7.4 The 19 classes in quadratic systems with finite multiplicity 1

We distinguish the classes $1 : i_f = -1$, and $2 : i_f = 1$; each class being again divided in a) with one, b) with two distinct, and c) with two coinciding transversally non-hyperbolic infinite critical points. At infinity there can occur the elementary critical points $E^{-1}, C_{0,1}^0$ and E^1 and the higher order points $M_{0,2}^0, M_{1,1}^0; M_{1,2}^1, M_{2,1}^{-1}, M_{2,2}^0, M_{2,2}^2; M_{3,2}^{-1}, M_{3,2}^1$ and $M_{3,3}^0, M_{3,3}^2$. Moreover all infinite points are critical if $b_{20} = b_{11} - a_{20} = b_{02} - a_{11} = a_{02} = 0$, which leads to (3.6.44),(3.6.45), with $a_{00} = b_{00} = 0$. So

$$\dot{x} = a_{10}x + a_{01}y + xy, \qquad (3.7.15)$$
$$\dot{y} = b_{10}x + b_{01}y + y^2, \qquad (3.7.16)$$

with $c_{23} \neq 0$. Now $c_{45} = c_{46} = 0, c_{56} = 1$ so A=0 and $B_1=0$ if $c_{25} =0$ or $b_{10}=0$. In order for system (3.7.15),(3.7.16) to be in class $m_f=1$, it is moreover required that $C_1 = a_{10}(a_{10} - b_{01}) = 0$ so $a_{10} = b_{01}$ as $a_{10}=0$ means $c_{23} = a_{10}b_{01} - a_{01}b_{10} =0$. So $c_{23} = a_{10}^2 >0$ and $i_f=1$; the critical point is an antisaddle . Furthermore $D_1 \neq 0$ yields $D_1 = -a_{10}^2a_{01} \neq 0$ so $a_{01} \neq 0$. From $c_{23} \neq 0$ follows also $b_{01} \neq 0$. System (3.7.15),(3.7.16) will be considered to be of class 2a.

Combining in each class the possible combinations of finite and infinite critical points yields two classes in class 1a ; three classes in class 1b ; two classes in class 1c; seven classes in class 2a, three classes in class 2b and two classes in class 2c yielding all together 19 classes in $m_f=1$.They are given in Table 4 below.

e^{-1}	$E^1E^1M_{3,1}^0$	e^{-1}	$E^1M_{1,1}^0M_{2,1}^1$	e^{-1}	$E^1M_{3,2}^1$
	$E^1M_{3,2}^1$		$M_{2,1}^1M_{1,2}^1$ $M_{1,1}^0M_{2,2}^2$		$M_{3,3}^2$
	1a:2 classes		1b:3 classes		1c:2classes

Class 1:$c_{45} + c_{56} <0$, $i_f = -1$.

$$e^1 \quad \begin{matrix} E^{-1}E^1M^0_{3,1} \\ E^1M^{-1}_{3,2} \quad E^{-1}M^1_{3,2} \\ M^0_{3,3} \\ \infty \end{matrix} \qquad \begin{matrix} C^0_{0,1}C^0_{0,1}M^0_{3,1} \\ M^0_{0,2}M^0_{3,1} \end{matrix} \qquad e^1 \quad \begin{matrix} E^1M^0_{1,1}M^{-1}_{2,1} \\ M^0_{1,1}M^0_{2,2} \quad M^{-1}_{2,1}M^1_{1,2} \end{matrix} \qquad e^1 \quad \begin{matrix} E^1M^{-1}_{3,2} \\ M^0_{3,3} \end{matrix}$$

$$\text{2a:7 classes} \qquad\qquad\qquad\qquad \text{2b:3 classes} \qquad\qquad \text{2c:2 classes}$$

Class 2:$c_{45} + c_{56} > 0$, $i_f = 1$.

Table 4. Combinations of critical points in the 19 classes of class $m_f = 1$.

3.8 Quadratic systems with $m_f = 0$

3.8.1 Systems with one transversally non-hyperbolic infinite critical point

According to Theorem 3.2 the central conic of a quadratic system with $m_f \leq 2$ is a degenerate parabola,consisting of two real parallel or coinciding lines. However, for $m_f = 0$ the first possibilty cannot occur.

Theorem 3.23 *The central conic of a quadratic system in the class $m_f = 0$ having one transversally non-hyperbolic infinite critical point is of parabolic type and consists of two (real finite) coinciding lines.*

Proof. Rotating the coordinate axes such that the $M^i_{4,q}$ point is at the ends of the x axis leads to the system

$$\dot{x} = a_{00} + a_{10}x + a_{01}y + a_{11}xy + a_{02}y^2, \qquad (3.8.1)$$

$$\dot{y} = b_{00} + b_{10}x + b_{01}y + b_{11}xy + b_{02}y^2, \qquad (3.8.2)$$

where $c_{56} \neq 0$. Obviously A=0 since $c_{45} = c_{46}=0, B_1=0$ yields $c_{25}=0$. If $a_{11} \neq 0$ let $\bar{y} = y + a_{10}/a_{11}$, if $a_{11}=0$, then $b_{11} \neq 0, a_{10}=0$ and let then $\bar{y} = y + b_{10}/b_{11}$. In both cases $a_{10} = b_{10} = 0$ in the new coordinates . From $C_1=0$ then follows $c_{15}=0$. Moreover $D_1 \equiv -c_{23}c_{36} - 2c_{13}c_{56} + 2c_{16}(c_{26} - \frac{1}{2}c_{35}) = -2c_{13}c_{56} - c_{16}c_{35} = -c_{13}c_{56}$ by using $c_{15}=0$ and $c_{13}c_{56} + c_{16}c_{35} = c_{15}c_{36}$. Thus $D_1=0$ yields $c_{13}=0$. As $E_1 \equiv c_{16}^2 - c_{13}c_{36} = c_{16}^2 \neq 0$ there follows further from $c_{13}c_{56} + c_{16}c_{35} = c_{15}c_{36}$ that $c_{35}=0$. If $a_{11} \neq 0$ let $\bar{x} = x + a_{01}/a_{11}$, if $a_{11}=0$, then $b_{11} \neq 0, a_{01}=0$ and let $\bar{x} = x + b_{01}/b_{11}$. In both cases we may take $a_{01} = b_{01}=0$. So (3.8.1),(3.8.2) can now be written as

$$\dot{x} = a_{00} + a_{11}xy + a_{02}y^2, \qquad (3.8.3)$$

$$\dot{y} = b_{00} + b_{11}xy + b_{02}y^2, \tag{3.8.4}$$

with $c_{15} = 0, c_{16} \neq 0, c_{56} \neq 0$.

The central conic of (3.8.1),(3.8.2), is given by $y^2 \equiv 0$, which is a degenerate parabola consisting of two coinciding lines.   □

Theorem 3.24 *If a quadratic system in the class $m_f = 0$ has one transversally non-hyperbolic infinite critical point $M_{4,q}^i$, then there exist six coinciding lines, all pointing in the direction of $M_{4,q}^i$ and coinciding with the central conic. They represent three coinciding isoclines, being of parabolic type and consisting of two coinciding lines. See Figure 3.14*

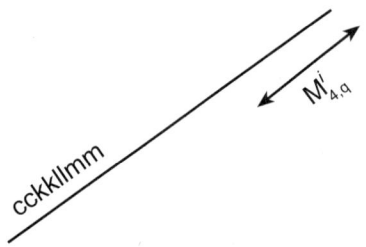

Figure 3.14 Degenerate isoclines and central conic in $m_f = 0$. a)One transversally non-hyperbolic infinite critical point $M_{4,q}^i$.b)An infinite number of infinite critical points

Proof. The Lamé equation of (3.8.3),(3.8.4) reads

$$(a_{00}\lambda + b_{00}\mu)(a_{11}\lambda + b_{11}\mu)^2 = 0, \tag{3.8.5}$$

which corresponds to the six coinciding isoclinic lines given by $y^6 \equiv 0$.   □

3.8.2 Systems with two transversally non-hyperbolic infinite critical points

Two distinct transversally non-hyperbolic infinite critical points

Let these points be located at the ends of the x axis and y axis, then (1.0.1),(1.0.2) may be written as

$$\dot{x} = a_{00} + a_{10}x + a_{01}y + a_{11}xy, \tag{3.8.6}$$
$$\dot{y} = b_{00} + b_{10}x + b_{01}y + b_{11}xy, \tag{3.8.7}$$

where $a_{11}^2 + b_{11}^2 \neq 0$. According to Theorem 2.3 there exist three cases: a)$c_{25} = 0, c_{35} \neq 0, c_{15} + c_{23} = 0, c_{13} \neq 0$; then, there exists an $M_{3,q}^i$ point at the ends of the x axis and an $M_{1,q}^i$ point at the ends of the y axis; b) $c_{25} = c_{35} = 0, c_{15} \neq 0$; then there exist an $M_{2,q}$ point at both ends of the x axis and the y axis; c) $c_{25} \neq 0, c_{35} = 0, c_{15} - c_{23} = 0, c_{12} \neq 0$; this case is related to case a) through the transformation $x \leftrightarrow y, a_{00} \leftrightarrow b_{00}, a_{10} \leftrightarrow b_{01}, a_{01} \leftrightarrow b_{10}, a_{11} \leftrightarrow b_{11}$, so that we can restrict to the cases a and b.

Lemma 3.5 If there exists an $M_{3,q}^i$ point at the ends of the x axis and an $M_{1,q}^i$ point at the ends of the y axis, (3.8.6),(3.8.7) can be written as

$$\dot{x} = a_{00} + a_{01}y + a_{11}xy, \qquad (3.8.8)$$
$$\dot{y} = b_{00} + b_{01}y + b_{11}xy, \qquad (3.8.9)$$

with $c_{15} = 0, c_{13} \neq 0, c_{35} \neq 0$.

If there exist an $M_{2,q}^i$ point at both ends of the x axis and the y axis, (3.8.8),(3.8.9) can be written as

$$\dot{x} = a_{00} + a_{11}xy, \qquad (3.8.10)$$
$$\dot{y} = b_{00} + b_{11}xy, \qquad (3.8.11)$$

with $c_{15} \neq 0$.

Proof. In the first case let $\bar{y} = y + a_{10}/a_{11}$ if $a_{11} \neq 0$ or $\bar{y} = y + b_{10}/b_{11}$ if $a_{11} = 0$ (then $b_{11} \neq 0, a_{10} = 0$). Then (3.8.8),(3.8.9) follows. In the second case , shifting along both the x axis and the y axis and using $c_{25} = c_{35} = 0$ leads to (3.8.10),(3.8.11).   □

Theorem 3.25 *If a quadratic system in the class $m_f = 0$ has two distinct transversally non-hyperbolic infinite critical points, then there are two cases:*

a)There exists an $M_{3,q}^i$ point and an $M_{1,q}^i$ point. Then the three degenerate isoclines coincide. A degenerate isocline is of the parabolic type and consists of a line at infinity and a line pointing in the direction of $M_{3,q}^i$. The central conic is also of parabolic type and consists of a line at infinity and a line coinciding with the three coinciding isoclinic lines. See Figure 3.15a

b)There exist two $M_{2,q}^i$ points. Then two degenerate isoclines coincide, each consisting of two coinciding lines at infinity. The third isocline is a degenerate hyperbola, the branches of which point in the direction of the $M_{2,q}^i$ points. The central conic degenerates into the midpoint of the degenerate hyperbolic isocline, which is also the midpoint of all other isoclines in the finite plane. See Figure 3.15b

Proof.a)The Lamé equation of (3.8.8),(3.8.9) reads

$$(a_{00}\lambda + b_{00}\mu)(a_{11}\lambda + b_{11}\mu)^2 = 0, \qquad (3.8.12)$$

which corresponds to the three coinciding isoclines each given by y≡0. The central conic of (3.8.8),(3.8.9) is also given by y ≡0.

b) The Lamé equation of (3.8.10),(3.8.11) also reads

$$(a_{00}\lambda + b_{00}\mu)(a_{11}\lambda + b_{00}\mu)^2 = 0. \qquad (3.8.13)$$

To $a_{00}\lambda + b_{00}\mu$=0 corresponds the degenerate hyperbola l: xy=0 with center in x=y=0. To $a_{11}\lambda + b_{11}\mu$=0 corresponds c_{15}=0, representing two lines at infinity. The equation for the central conic is degenerate. However, all isoclines of (3.8.10),(3.8.11) have their center in (0,0).   □

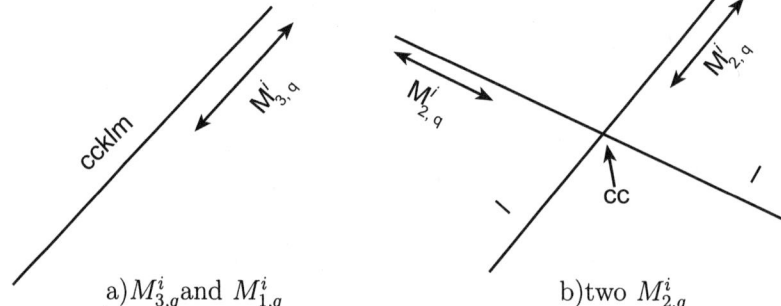

a)$M_{3,q}^i$and $M_{1,q}^i$ b)two $M_{2,q}^i$

Figure 3.15 Degenerate isoclines and central conic in m_f=0. Two real distinct transversally non-hyperbolic critical points

Two coinciding transversally non-hyperbolic infinite critical points: a point $M_{4,q}^i$

Theorem 3.26 *If a quadratic system within class m_f=0 has two coinciding transversally non-hyperbolic infinite critical points in the point $M_{4,q}^i$, then there exist two cases: a) All six isoclinic lines are located at infinity and the central conic is degenerated into the point $M_{4,q}^i$, b) All lines pointing in the direction of $M_{4,q}^i$ are isoclinic lines and all points of the real plane belong to the central conic.*

Proof. As in section 6.2.2 we can work with

$$\dot{x} = a_{00} + a_{10}x + a_{01}y + a_{02}y^2, \qquad (3.8.14)$$

$$\dot{y} = b_{00} + b_{10}x + b_{01}y + b_{02}y^2, \qquad (3.8.15)$$

with $a_{20}^2 + b_{02}^2 \neq 0$. Obviously, A=0 since $c_{45} = c_{46} = c_{56} = 0$ and also $B_1 = 0$. Furthermore $C_1 = 0$ yields $c_{26} = 0$ and $D_1 = 0$ yields $c_{23}c_{36} = 0$, whereas $E_1 \equiv c_{16}^2 - c_{13}c_{36} \neq 0$ in order to let (3.8.14),(3.8.15) belong to $m_f = 0$. From this it follows that $a_{02}u^3 - b_{02}u^2 = 0$, so $M_{4,q}^i$ is $M_{4,2}^i$ if $b_{02} \neq 0$ and $M_{4,3}^i$ if $b_{02} = 0$, $a_{02} \neq 0$

(i)If there exists a point $M_{4,2}^i$, so $b_{02} \neq 0$, then $a_{10} = a_{02}b_{10}/b_{02}$ as $c_{26} = 0$, whereas $c_{23} = a_{10}b_{01} - a_{01}b_{10} = a_{02}b_{10}b_{01}/b_{02} - a_{01}b_{10} = -b_{10}c_{36}/b_{02}$ so $c_{23}c_{36}$ becomes $-b_{10}c_{36}^2/b_{02} = 0$. We distinguish between $b_{10} \neq 0$ and $b_{10} = 0$. This corresponds to the cases a) and b), respectively.

a)$b_{10} \neq 0$. Then $c_{36} = 0$ and also $c_{23} = 0$. Furthermore $c_{12} = a_{00}b_{10} - a_{10}b_{00} = a_{00}b_{10} - a_{02}b_{00}b_{10}/b_{02} = b_{10}c_{16}/b_{02} \neq 0$ since $E_1 = c_{16}^2 \neq 0$.

For the Lamé equation of (3.8.14),(3.8.15) then follows

$$(a_{10}\lambda + b_{10}\mu)^2(a_{02}\lambda + b_{02}\mu) = 0. \qquad (3.8.16)$$

To $(a_{10}\lambda + b_{10}\mu)^2 = 0$ corresponds $c_{12} = 0$, yielding four coinciding isoclinic lines at infinity. To $a_{20}\lambda + b_{02}\mu = 0$ corresponds $c_{16} = 0$, representing two isoclinic lines at infinity; thus yielding six isoclinic lines at infinity alltogether.

The equation for the central conic is degenerate. However all isoclines of (3.8.14),(3.8.15) in the finite part of the plane are non-degenerate parabolae, having their centers in $M_{4,2}^i$, which can therefore be considered as the central conic (of the finite isoclines).

b)$b_{10} = 0$. Then also $a_{10} = 0$ since $c_{26} = 0$ and (3.8.14),(3.8.15) reads

$$\dot{x} = a_{00} + a_{01}y + a_{02}y^2, \qquad (3.8.17)$$
$$\dot{y} = b_{00} + b_{01}y + b_{02}y^2, \qquad (3.8.18)$$

and all isoclines consist of lines y≡constant and the Lamé equation is degenerate. The equation for the central conic is also degenerate. For each pair (λ, μ) the isocline is of parabolic type and consists of two real ,coinciding or complex lines. The midpoints of each isocline are on a line halfway. As a result all points of the real plane are on the central conic.

ii)If there exists a point $M_{4,3}^i$ so $b_{20} = 0, a_{02} \neq 0, c_{26} = 0$ then $b_{10} = 0$, whereas $c_{23}c_{36} = 0$ yields $a_{10}b_{01} = 0$. We distinguish between $b_{01} = 0$ and $a_{10} = 0$. This corresponds to case a and case b, respectively.

a) $b_{01} = 0$. Since $E_1 \equiv c_{16}^2 - c_{13}c_{36} = a_{02}^2b_{00}^2 \neq 0$ it follows that $b_{00} \neq 0$. System (3.8.14),(3.8.15) then reads

$$\dot{x} = a_{00} + a_{10}x + a_{01}y + a_{02}y^2, \qquad (3.8.19)$$
$$\dot{y} = b_{00}, \qquad (3.8.20)$$

with $a_{02}b_{00} \neq 0$.

If $a_{10} \neq 0$, then the Lamé equation reads $\lambda^3=0$ yielding the six coinciding isoclinic lines at infinity given by $b_{00}=0$. All isoclines corresponding to $\lambda \neq 0$ are non-degenerate parabolae having their centers in $M^1_{4,3}$.

b) $a_{10}=0$. Now eqs.(3.8.14),(3.8.15) read

$$
\begin{aligned}
\dot{x} &= a_{00} + a_{01}y + a_{02}y^2, & (3.8.21)\\
\dot{y} &= b_{00} + b_{01}y, & (3.8.22)
\end{aligned}
$$

with $a_{02}b_{01} \neq 0$. All lines $y\equiv$ constant are isoclinic lines and the Lamé equation is degenerate. The equation for the central conic is also degenerate. For each pair (λ, μ) the isocline is of parabolic type and consists of two real, distinct or two complex lines. The midpoints of each isocline are on a line halfway. As a result all points in the real plane are on the central conic. $\square$

Two complex transversally non-hyperbolic infinite critical points

Theorem 3.27 *If a quadratic system in the class $m_f=0$ has two complex transversally non-hyperbolic infinite critical points : $C^0_{2,1}$, the degenerate isoclines consist of two degenerate parabolae and one degenerate ellipse. The degenerate parabolae yield four isoclinic lines at infinity. The degenerate ellipse consists of two conjugate complex isoclinic lines intersecting in one real point, which coincides with the central conic.*

Proof. We may again work with (3.6.32),(3.6.33) in section 3.6.2

$$
\begin{aligned}
\dot{x} &= a_{00} + x^2 + y^2, & (3.8.23)\\
\dot{y} &= b_{00} + b_{01}x + b_{01}y, & (3.8.24)
\end{aligned}
$$

which has $C^0_{p,q}$ points in $u=\pm i$ and a real infinite critical point in $u=0$. Since $c_{45} = c_{46} = c_{56}=0$ it follows that $A=B_1=0$. Furthermore $C_1 = b^2_{10} + b^2_{01}=0$ yields $b_{10} = b_{01} =0$, whereas $D_1=0$ is then also satisfied. Finally $E_1 \equiv c^2_{16} - c_{13}c_{36} = b^2_{00} \neq 0$ and (3.8.23),(3.8.24) may be written as

$$
\begin{aligned}
\dot{x} &= a_{00} + x^2 + y^2, & (3.8.25)\\
\dot{y} &= b_{00}, & (3.8.26)
\end{aligned}
$$

with $b_{00} \neq 0$.

The Lamé equation of (3.8.25),(3.8.26),reads

$$
\lambda^2(a_{00}\lambda + b_{00}\mu) = 0. \qquad (3.8.27)
$$

To $\lambda^2 = 0$ correspond the four isoclinic lines at infinity $b_{00} = 0$. To $a_{00}\lambda + b_{00}\mu = 0$ corresponds the point circle $x^2 + y^2 = 0$, or the conjugate complex lines $y = \pm ix$. All isoclines $\lambda a_{00} + \mu b_{00} + \lambda(x^2 + y^2) = 0, \lambda \neq 0$ are circles with center in $(0,0)$, which can therefore be considered as the central conic degenerated into one point. The equation for the central conic is degenerate. $\square$

3.8.3   The five classes in quadratic systems with finite multiplicity 0

If there exists a finite number of transversally non-hyperbolic infinite critical points, there exists the classes: 1: with one, 2: with two distinct, 3: with two coinciding and 4: with two complex such points. Moreover there exists class 5 with infinitely many infinite critical points. There appear [95,RK] to be the following combinations of critical points in the various classes:

in 1: $E^{-1}E^1 M_{4,1}^1, C_{0,1}^0 C_{0,1}^0 M_{4,1}^1, E^1 E^1 M_{4,1}^{-1}, E^1 M_{4,2}^0, M_{0,2}^0 M_{4,1}^1, E^{-1}M_{4,2}^2$ and $M_{4,3}^1$;

in 2: $E^1 M_{1,1}^0 M_{3,1}^0, M_{1,1}^0 M_{3,2}^1, M_{1,2}^1 M_{3,1}^0, E^1 M_{2,1}^{-1} M_{2,1}^1, M_{2,1}^{-1} M_{2,2}^2$ and $M_{2,1}^1 M_{2,2}^1$;

in 3: $E^1 M_{4,2}^0, M_{4,3}^1$; and in 4: $E^1 C_{2,1}^0 C_{2,1}^0$.

For class 5 we work again with system (3.6.44),(3.6.45)

$$\dot{x} = a_{00} + a_{10}x + a_{01}y + xy, \qquad (3.8.28)$$
$$\dot{y} = b_{00} + b_{01}x + b_{01}y + y^2. \qquad (3.8.29)$$

Now $c_{45} = c_{46} = 0$ so A=0, whereas $B_1 = 0$ yields $c_{25} = 0$ so $b_{10} = 0$. Also $C_1 = a_{10}^2 - a_{10}b_{01} + b_{00} = 0$ so $b_{00} = a_{10}b_{01} - a_{10}^2$ which may be used in $D_1 = -a_{10}a_{01}b_{01} - a_{00}b_{01} + 2a_{00}a_{10} + 2a_{01}b_{00} = 0$ to yield $(2a_{10} - b_{01})(a_{00} - a_{10}a_{01}) = 0$. Since $E_1 \equiv c_{16}^2 - c_{13}c_{36} = (a_{00} + a_{10}a_{01} - a_{01}b_{01})(a_{00} - a_{10}a_{01}) \neq 0, D_1 = 0$ yields $b_{01} = 2a_{10}$ and $E_1 = (a_{00} - a_{10}a_{01})^2 \neq 0$. As a result (3.8.28),(3.8.29) becomes

$$\dot{x} = a_{00} - a_{10}a_{01} + (a_{01} + x)(a_{10} + y), \qquad (3.8.30)$$
$$\dot{y} = (a_{10} + y)^2, \qquad (3.8.31)$$

or with $\bar{x} = a_{01} + x, \bar{y} = a_{10} + y$ and returning to the previous notation results in the system

$$\dot{x} = \alpha + xy, \qquad (3.8.32)$$
$$\dot{y} = y^2, \qquad (3.8.33)$$

with $\alpha = a_{00} - a_{10}a_{01} \neq 0$, which after scaling may be taken as $\alpha = 1$.

Theorem 3.28 *If a quadratic system in the class $m_f=0$ has an infinite number of infinite critical points, then the three degenerate isoclines are formed by six coinciding finite lines. The central conic consists of two coinciding lines also coinciding with these lines, and is also an orbit of the system. See Figure 3.14*

Proof. The Lamé equation of (3.8.32),(3.8.33) reads $\lambda^3=0$ giving six lines coinciding with the x axis, which is also an orbit. The central conic is given by $y^2=0$. □

Remark. System (3.8.32),(3.8.33) may be seen as a special case of (3.8.28)and(3.8.29).

Chapter 4

Analyzing phase portraits of quadratic systems

4.1 Introduction

As was mentioned in Chapter 1, this book aims at giving a survey of what
has become known so far about the phase portraits of quadratic systems,
thereby indicating what still has to be done to complete the picture on the
subject. In order to approach this goal systematically, in the previous chapter
non-degenerate quadratic systems were subdivided into 173 classes. For each
class we now investigate what is known about its phase portraits and how
to proceed to give a complete classification. In order to get an insight into
the arguments and methods useful to obtain the topological character of a
phase portrait, it is of interest to take into account general properties of
quadratic systems and results obtained from studying classes of quadratic
systems, characterized by a specific property.

General properties can be obtained, e.g. by investigating the various
possible types of flow over arbitrary straight lines in quadratic systems. From
them follows, for instance, that in a quadratic system there exist at most two
finite critical points on a straight line; that the total number of critical points
which are either a focus or a center is at most two and if there exist two, the
flow near them is clockwise around one point and anticlockwise around the
other point. Moreover, it follows that a closed orbit is a convex curve; it has
either a center or a focus in its interior and no other critical points, and the
flow in the interior of the closed orbit is oriented in the same sense as the

flow on the closed orbit. Another result is that there exist at most two nests of closed orbits in a quadratic system.

Classes of quadratic systems, characterized by a specific property, studied in the literature are, for instance, systems containing center points or systems obtained from them by slight perturbations of the coefficients of these systems; systems with a starlike node, a cusp, a third or fourth order finite critical point or a weak critical point (wherein $P_x + Q_y=0$); systems having algebraic orbits, including those with a straight line orbit; Volterra−Lotka , Hamiltonian, gradient or homogeneous systems; structurally stable quadratic systems without limit cycles and systems containing separatrix cycles. Particular attention has been given in the literature to the limit cycle problem in quadratic systems. This involves the question whether, for a given quadratic system, there exist limit cycles and if so how many and how distributed: in one or two nests. Finally it should be remarked that for some values of the finite multiplicity m_f there are already classifications that can be found in the literature.

A further point of concern is the partition of the space of coefficients a_{ij}, b_{ij} (i,j=0,1,2), corresponding to the topologically different phase portraits, resulting from the various analyses, and using various forms for differential equations. Attention should be given to the possibility to determine this partition in terms of algebraic invariants.

Although examples of the analysis of phase portraits of quadratic systems may be found already in the writings of Poincaré [P] on the curves determined by differential equations, a more extensive, but still tentative investigation seems to be given for the first time in 1904 by Büchel [04,B]. Thereafter, interest in quadratic systems arose in relation to the center problem, asking for conditions on the coefficients such that a system contains a center point. The papers that appeared first on this problem , starting a flow of papers continuing throughout the twentieth century , were of Dulac in 1908 [08,D], and of Kapteyn in 1911 and 1912 [11,K],[12,K]. Using quadratic systems as models in various fields of applications also started around that time. A quadratic system occurring frequently in applications is the Volterra-Lotka equation (1.0.3),(1.0.4). An early paper using this equation appeared in 1920 by A.J.Lotka [20,L] and considers chemical reactions of an oscillatory character. Various other fields of application of this equation were exploited in the thirties and forties of the twentieth century and in these years the literature on quadratic systems is almost completely devoted to applications. Although in many cases limit cycles were encountered in these applications a general result was first given by Bautin [39,B], who showed that from a focus or a center point, up to three limit cycles may bifurcate through perturbations that leave the system quadratic. A further result was given in 1955 by Petrovskii and Landis [55,PL], stating that three was the maximimum number of limit cycles a quadratic system could contain. The limit cycle problem has since that time been given ample attention in the theory of quadratic systems. In particular this problem was taken up in China where it was given much attention over many decades thereafter. The first papers appeared from Qin Yuanxun and Ye Yanqian, both in 1957 and from Dong Jinzhu in 1958. As a result general properties of quadratic systems were discovered by considering flow over staight lines in quadratic systems. The next section presents a discussion on this issue based on the papers by Ye Yanqian [58,Y1], Dong Jinzhu [59,D] and subsequent discussions by Coppel [66,C] and Chicone and Tian Jinghuang [82,CT].

4.2   Flow over straight lines

To characterize the flow over a straight line L in a quadratic system we use the notation $L_{q,r}^p$, where p is the number of critical points, q the number of zero cross flow points in the finite part of the plane and r the number of critical points at infinity on L. There exist the following possibilities.

Lemma 4.1 (i) If L is not in the direction of an infinite critical point, then $r=0$ and there exist the possibilities: $L_{0,0}^0$, $L_{1,0}^0$, $L_{1,0}^1$, $L_{2,0}^0$, $L_{2,0}^1$ and $L_{2,0}^2$, (ii) If L is in the direction of a transversally hyperbolic infinite critical point, then $r=1$ and there exist the possibilities $L_{0,1}^0$, $L_{1,1}^0$, $L_{1,1}^1$, L_{∞}^0, $L_{\infty,1}^1$, and $L_{\infty,1}^2$, (iii)If L is in the direction of a transversally non-hyperbolic infinite critical point, then $r=1$ and there exist the possibilities $L_{0,1}^0$, $L_{1,1}^0$, $L_{1,1}^1$, $L_{\infty,1}^0$, $L_{\infty,1}^0$ and $L_{\infty,1}^\infty$. The various possibilities are sketched in Figure 4.1. Flow reversal , mirroring the flow around L and changing the angle of L with respect to a fixed direction does not lead to flow types that will be considered to be different in the present context.

Proof. We may use a shift and/or a rotation so that L coincides with the x axis. Then on $L : \dot{y} = b_{00} + b_{10}x + b_{01}x^2$.(i)If $b_{20} \neq 0$,there exist no infinite critical point at the ends of L and depending on the sign of $D_1 \equiv b_{10}^2 - 4b_{00}b_{20}$ there exist three possibilities.If $D_1 < 0$, there exist no zero cross flow points on L and $L = L_{0,0}^0$ is a line without contact. If $D_1 = 0$ there exist(s) one (two coinciding) zero cross flow point(s) on L, a contact point, then $L \equiv L_{1,0}^0$ or a critical point then $L \equiv L^1_{1,0}$. If $D_1 > 0$, then there exist two distinct zero cross flow points; if there are two contact points then $L \equiv L^0_{2,0}$, one contact point

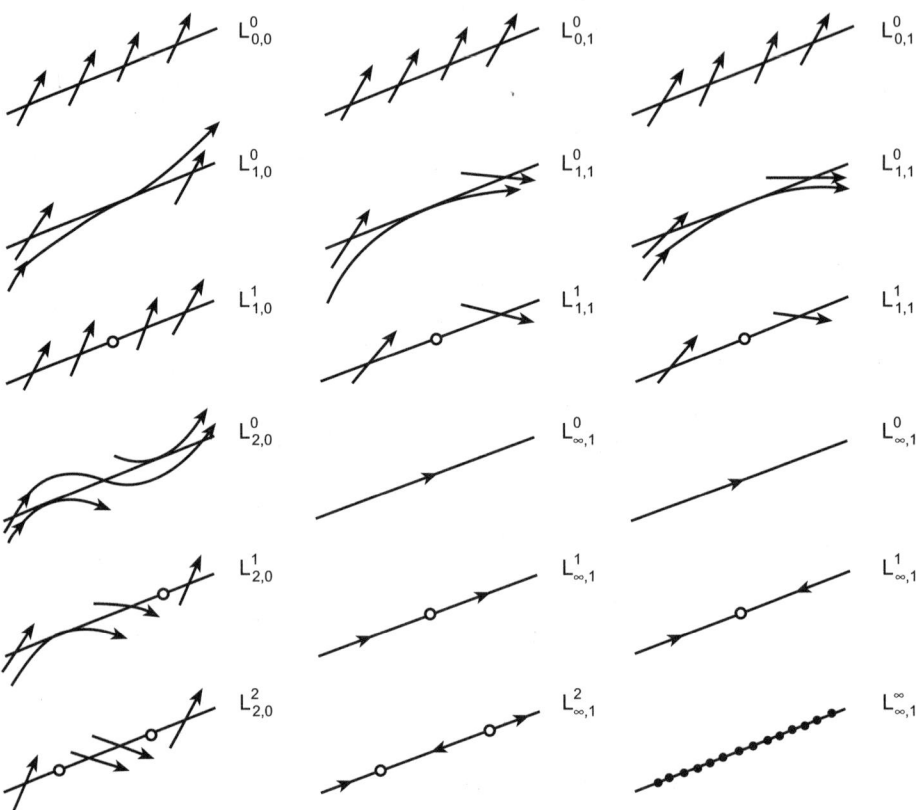

Figure 4.1 Possible flow types over straight lines in quadratic systems

and one critical point then $L \equiv L^1_{2,0}$ and if there are two critical points then $L \equiv L^2_{2,0}$. (ii) If $b_{20} = 0, a_{20} \neq 0$ there exists a transversally hyperbolic infinite critical point on L which is also a zero cross flow point, so $r = 1$. There exist three possibilities. If $b_{10} \neq 0$ then there exists one finite zero cross flow point, if a contact point then $L \equiv L^0_{1,1}$, if a critical point then $L \equiv L^1_{1,1}$. If $b_{10} = 0, b_{00} \neq 0$ then there exist no finite zero cross flow points on L and $L \equiv L_{0,1}^0$ is a line without contact. If $b_{20} = b_{10} = b_{00} = 0$ then L consists of orbit(s). Now on L there is $\dot{x} = a_{00} + a_{10}x + a_{20}x^2, a_{20} \neq 0$ and the number of critical points on L depends on $D_2 = a_{10}^2 - 4a_{00}a_{20}$. If $D_2 < 0$ there exist no critical points on L and $L \equiv L_{\infty,1}^1$, if $D_2 = 0$ there exist two coinciding critical points and $L \equiv L_{\infty,1}^1$ and if $D_2 > 0$ there exist two distinct critical points and $L \equiv L_{\infty,1}^2$. (iii) If $b_{20} = a_{20} = 0$ there exists a transversally non-hyperbolic infinite critical point at the ends of L, which is also a zero cross flow point, so $r=1$. If $b_{10}^2 + b_{00}^2 \neq 0$ we again have the flow types $L_{0,1}^0, L_{1,1}^0$ and $L_{1,1}^1$. If $b_{20} = b_{10} = b_{00} = 0$ then L consists of orbit(s). Now on L there is $\dot{x} = a_{00} + a_{10}x$ and there exists one critical point if $a_{10} \neq 0$. If $a_{10} = 0, a_{00} \neq 0$ there exist no critical points on L. If $a_{20} = a_{10} = a_{00} = 0$, L consists of critical points and the system is degenerate.   □

From Lemma 4.1 there follow two statements known as lemmas of Tung Chinchu [59,D].

Lemma 1(Tung). The sum of the number of contacts with an orbit and the number of critical points on an arbitrary straight line not being an orbit is at most equal to 2.

Lemma 2(Tung). The line through two critical points P_1 and P_2 will be either composed of orbits or of three line segments without contact ∞P_1, $P_1 P_2$ and $P_2 \infty$; the flow over $P_1 P_2$ being in opposite sense to the flow over ∞P_1 and $P_2 \infty$.

It was remarked by Coppel [66,C] that in a quadratic system three critical points can never be collinear . Although this statement sharpens Lemma 1 of Tung Chinchu, Lemma 4.1 allows to state:

Theorem 4.1 *There exist at most two critical points on an arbitrary straight line L that is not directed towards a transversally non-hyperbolic infinite critical point. If L is in the direction of a transversally non-hyperbolic infinite critical point, there exists at most one critical point on L if the system is non-degenerate or all points on L are degenerate and the system is degenerate.*

It could be a useful procedure to scan a particular quadratic system or class of systems with respect to the flow over the straight lines in the plane. Consider the straight line through the point $(x,y)=(a,b)$, $a,b \in \mathbb{R}$ and rewrite (1.0.1),(1.0.2) using the transformation $x=a+r\cos\theta, y=b+r\sin\theta, 0 \leq \theta < 2\pi$ and we obtain

$$\dot{r} - \bar{A}_1(\theta) + r\bar{B}_1(\theta) + r^2 C_1(\theta), \qquad (4.2.1)$$

$$r\dot{\theta} = \bar{A}_2(\theta) + r\bar{B}_2(\theta) + r^2 C_2(\theta), \tag{4.2.2}$$

where

$$\bar{A}_1(\theta) = \bar{a}_{00}\cos\theta + \bar{b}_{00}\sin\theta,$$
$$\bar{B}_1(\theta) = \bar{a}_{10}\cos^2\theta + (\bar{a}_{01} + \bar{b}_{10})\sin\theta\cos\theta + \bar{b}_{01}\sin^2\theta,$$
$$C_1(\theta) = a_{20}\cos^3\theta + (a_{11} + b_{20})\sin\theta\cos^2\theta + (a_{02} + b_{11})\sin^2\theta\cos\theta + b_{02}\sin^3\theta,$$
$$\bar{A}_2(\theta) = \bar{b}_{00}\cos\theta - \bar{a}_{00}\sin\theta,$$
$$\bar{B}_2(\theta) = \bar{b}_{10}\cos^2\theta + (\bar{b}_{01} - \bar{a}_{10})\sin\theta\cos\theta - \bar{a}_{01}\sin^2\theta,$$
$$C_2(\theta) = b_{20}\cos^3\theta + (b_{11} - a_{20})\sin\theta\cos^2\theta + (b_{02} - a_{11})\sin^2\theta\cos\theta - a_{02}\sin^3\theta,$$

and

$$\bar{a}_{00} = a_{00} + aa_{10} + ba_{01} + a^2 a_{20} + aba_{11} + b^2 a_{02},$$
$$\bar{a}_{10} = a_{10} + 2aa_{20} + ba_{11},$$
$$\bar{a}_{01} = a_{01} + aa_{11} + 2ba_{02},$$
$$\bar{b}_{00} = b_{00} + ab_{10} + bb_{01} + a^2 b_{20} + abb_{11} + b^2 b_{02},$$
$$\bar{b}_{10} = b_{10} + 2ab_{20} + bb_{11},$$
$$\bar{b}_{01} = b_{01} + ab_{11} + 2bb_{02}.$$

Obviously, zero cross flow points on a line L having a direction given by $\theta = \theta_0$ are obtained from the condition $\dot{\theta} = 0$ or $\bar{A}_2(\theta_0) + r\bar{B}_2(\theta_0) + r^2 C_2(\theta_0) = 0$ and there exists such a point at infinity on L if $C_2(\theta_0) = 0$, thus if there exists an infinite critical point for $\theta = \theta_0$. This is also a necessary condition for L to consists of orbit(s). Moreover, the infinite critical point is transversally non-hyperbolic if also $C_1(\theta_0)=0$. This may be seen as follows. With $C_2(\theta_0)=0$ there follows $C_1(\theta_0) = \cos\theta_0(a_{20} + a_{11}\tan\theta_0 + a_{02}\tan^2\theta_0)$, which for $\theta_0 \neq \frac{\pi}{2}, \frac{3\pi}{2}$, using $u = \tan\theta_0$ yields with (2.3.13) $C_1(\theta_0) = -\lambda_z \cos\theta_0$, and $\lambda_z=0$ if $C_1(\theta_0)=0$. For $\theta_0 = \frac{\pi}{2}, \frac{3\pi}{2}, C_2(\theta_0)=0$ implies $a_{20}=0$ and $C_1(\theta_0)=0$ implies $b_{20}=0$, so x is a common factor in the quadratic terms.

Lemma 4.2 If a line L through a finite critical point is aligned along a (real) eigenvector of the locally linearized system and not in the direction of an infinite critical point, then $L \equiv L_{1,0}^1$. If L is also pointing in the direction of an infinite critical point, L consists of orbits. If the infinite critical point is transversally hyperbolic there may exist another finite critical point on L, if transversally non-hyperbolic,there exists no other finite critical point on L if the system is non-degenerate.

Proof. Let the system have real eigenvalues, then we may use the normal form

$$\dot{x} = \lambda_1 x + \alpha y + a_{20}x^2 + a_{11}xy + a_{02}y^2, \tag{4.2.3}$$
$$\dot{y} = \lambda_2 y + b_{20}x^2 + b_{11}xy + b_{02}y^2, \tag{4.2.4}$$

and the various cases may be considered.

(i)$\lambda_1 \neq \lambda_2, \alpha=0$. The origin is a saddle point or a node with two directions wherein the orbits approach. Eigenvectors are $(1,0)$ and $(0,1)$. Along $(1,0)$ take the x axis as the straight line L. If $b_{20} \neq 0$ there is no infinite critical point at the ends of L and $L \equiv L_{1,0}^1$. If $b_{20}=0$ there exists such a point and L consists of orbits. If $a_{20} \neq 0$, the infinite critical point is transversally hyperbolic and there exists another finite critical point in $x = -\frac{\lambda_1}{a_{20}} \neq 0$. If $a_{20}=0$ there exists no other finite critical point apart from $(0,0)$ on L. For the eigenvector $(0,1)$ the arguments run the same.

(ii)$\lambda_1 = \lambda_2 = \lambda \neq 0, \alpha = 0$. The origin is a starlike node and orbits approach in all directions; eigenvectors are given by $(a,b), a^2 + b^2=1$. Use (4.2.1),(4.2.2), then $\bar{A}_2(\theta) = \bar{B}_2(\theta) \equiv 0$ and $\dot{\theta} = C_2(\theta)r$. Take L through $(0,0)$ along (a,b), then since $C_2(\theta) = -C_2(\theta + \pi)$ there follows $L \equiv L_{1,0}^1$ if $C_2(\theta) \neq 0$. If $C_2(\theta)=0$, L consists of orbits. Then $\bar{A}_1(\theta)=0, \bar{B}_1(\theta) = \lambda \neq 0$ and $\dot{r} = \lambda r + r^2 C_1(\theta)$. If $C_1(\theta) \neq 0$ then the infinite critical point is transversally hyperbolic and there exists another finite critical point in $r = -\frac{\lambda}{C_1(\theta)}$, so $L \equiv L_{\infty,1}^2$. If $C_1(\theta) =0$, the infinite critical point on L is transversally non-hyperbolic and there exist no other critical point on L if the system is non-degenerate, so $L \equiv L_{\infty,1}^1$.

(iii) $\lambda_1 = \lambda_2 = \lambda, \alpha = 1$. The origin is a node, in one direction approached by orbits; eigenvector given by $(1,0)$. The analysis for $\lambda_1 \neq \lambda_2, \alpha=0$ applies.

(iv) $\lambda_1 = \alpha = 0, \lambda_2 = 1$. The origin is a semi-elementary critical point; a second or fourth order saddle node or a third order saddle or node. The eigenvector is given by $(0,1)$ so take L to coincide with the y axis. If $a_{02} \neq 0$ then $L \equiv L_{1,0}^1$. If $a_{02}=0$ then there exists an infinite critical point at the ends of L. If $b_{02} \neq 0$ this point is transversally hyperbolic and there exists another finite critical point in $y = -\frac{1}{b_{02}}$, so $L \equiv L_{\infty,1}^2$. If $b_{20}=0$ this point is transversally non-hyperbolic and there exists no other critical point on L if the system is non-degenerate, so $L \equiv L_{\infty,1}^1$.

(v) $\lambda_1 = \lambda_2 = 0, \alpha = 1$. The origin is a nilpotent critical point; a second order cusp, a third order saddle or point with an elliptic and a hyperbolic sector, or a fourth order saddle node. The eigenvector is given by $(1,0)$. If $b_{20} \neq 0$, the origin is a cusp and $L \equiv L_{1,0}^1$; L is not pointing towards an infinite critical point. If $b_{20}=0$, L is pointing into the direction of an infinite critical point and L coincides with the orbits on $y=0$. Since the system is non-degenerate $a_{20} \neq 0$, the infinite critical point is hyperbolic transversally to the Poincaré circle and there exist no other finite critical points on L, so

$L \equiv L_{\infty,1}^1$. The critical point is of third or fourth order.

(vi) $\lambda_1 = \lambda_2 = \alpha = 0$. The system is homogeneous and all vectors $(a,b), a^2+b^2=1$ are eigenvectors. In (4.2.1),(4.2.2) then $\bar{A}_2(\theta) = \bar{B}_2(\theta) \equiv 0$ and $\dot{\theta} = C_2(\theta)r$. Take L through (0,0) along (a,b), then since $C_2(\theta) = -C_2(\theta+\pi)$ there follows that $L \equiv L_{1,0}^1$ if $C_2(\theta) \neq 0$. If $C_2(\theta) = 0$, L consists of orbits. Then $\bar{A}_1(\theta) = \bar{B}_1(\theta) \equiv 0$, whereas $C_1(\theta) \neq 0$ since infinite critical points are transversally hyperbolic in homogeneous quadratic systems. So $\dot{r} = r^2 C_1(\theta)$ and (0,0) is the only critical point.

4.3 General properties of quadratic systems.

On the basis of the various types of flow over straight lines in quadratic systems we may derive some general properties of quadratic systems.

4.3.1 Critical points of center or focus type.

A center point is a critical point having a neighborhood such that through any point in that neighborhood there exists a closed orbit encircling the critical point. The direction of increasing t is the same for all orbits in that neighborhood. As a result, the periodic motion in a certain region around the center point is clockwise or anticlockwise. Similarly, the motion around a focus is also clockwise or anticlockwise, whereas the orbits are now spiraling towards or away from the critical point upon increasing t.

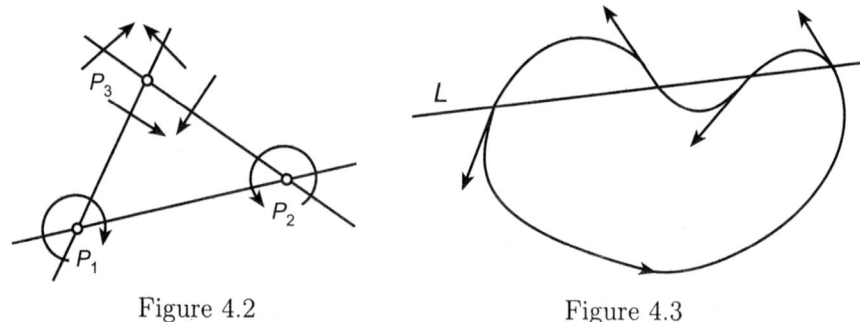

Figure 4.2   Figure 4.3

Theorem 4.2 *If a quadratic system contains two critical points, which are either a focus or a center point, one critical point is oriented clockwise and the other anticlockwise.*

Proof. The theorem is an immediate consequence of Lemma 2 of Tung Chinchu. □

An immediate consequence of Theorem 4.2 is

Theorem 4.3 *The total number of critical points that are either a focus or a center point is at most equal to 2.*

Proof. Consider three critical points of focus or center type P_1, P_2 and P_3 as in Figure 4.2. Assume that P_1 is clockwise oriented, so P_2 is anticlockwise oriented, according to Theorem 4.2. From Theorem 4.1 it follows that $P_1 P_3$ and $P_2 P_3$ are $L_{2,0}^2$ lines, and P_3 should be both clockwise and anticlockwise oriented, which is a contradiction. □

4.3.2　Closed orbits

Theorem 4.4 *A closed orbit in a quadratic system is convex.*

Proof. Suppose, as sketched in Figure 4.3 that there exists a closed orbit which is not, convex. Then there exists a line L intersecting this orbit in at least four points. Consider the normal component to L of the vector field $(P(x,y), Q(x,y))$. Since this vector field is continuous there exist on L at least three zeros of this component, so three points which are either contact points or critical points. This contradicts Lemma 1 of Tung Chinchu. □

Theorem 4.5 *In the interior of a closed orbit of a quadratic system there exists precisely one critical point and this point is either a center point or a focus.*

Proof. Since a closed orbit is convex it has unique points with minimum abscissa P_1 and maximum abscissa P_2 ; they are located on the ∞ isocline $P(x,y)=0$. See Figure 4.4. Also there exist unique points with minimum ordinate Q_1 and maximum ordinate Q_2 ; they are located on the 0 isocline $Q(x,y)=0$. Any line $L : x = x_0$, where $x_{min} < x_0 < x_{max}$, intersects the closed orbit in two points, having opposite cross flow over L. Since $(P(x,y), Q(x,y))$ is continuous there exists at least one point on L in the interior of the closed orbit with zero cross flow , thus being a point on the ∞ isocline $P(x,y)=0$. So there exists a continuous curve connecting P_1 and P_2, which is part of the ∞ isocline. Similarly , there exists a continuous curve connecting Q_1 and Q_2, which is part of the o isocline. As a result there exists at least one critical point in the interior of the closed orbit.

Now suppose there exists more than one critical point in the interior of the closed orbit and consider the line through two critical points A and B. According to Lemma 2 of Tung Chinchu, the flow over ∞A and B∞ are then in the same sense in contradiction to what follows for the flow on the closed orbit. There exists thus precisely one critical point in the interior of the closed orbit.

According to Lemma 4.2, if a critical point has real (possibly zero) eigenvalues, thus real eigenvector(s), then there exists either an $L_{1,0}^1$ line, leading to contradictory conclusions with regard to the flow direction on the closed orbit or a line $L_{\infty,1}^p$ consisting of orbits intersecting with the closed orbit, which contradicts the uniqueness theorem. This leaves the focus or the center point as possible character for the critical point in the interior of the closed orbit. $\qquad\qquad\qquad\qquad\qquad\qquad\qquad\qquad\qquad\qquad\qquad\square$

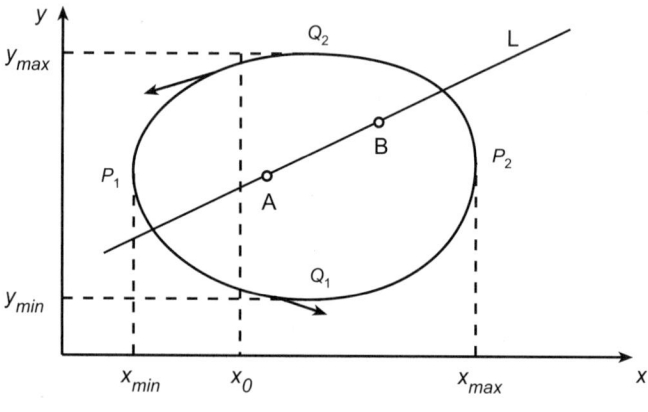

Figure 4.4

Remark. Since the closed orbits enclose a convex region it is simple to deduce that the change of the angle that $(P(x,y), Q(x,y))$ makes with a fixed direction when the closed orbit is traversed in an anti-clockwise direction equals 2π. As a result the Poincaré index of a closed orbit equals 1, a result known for a wider class of systems. Letting a curve that initially coincides with the closed orbit contract onto the critical point shows that the Poincaré index of the critical point in the interior of the closed orbit also equals 1, in accordance with it being a focus or a center point.

Theorem 4.6 *The motion along a closed orbit and along the orbits in the interior of the closed orbit of a quadratic system is either both in clockwise or both in anticlockwise direction.*

Proof. Consider, as in Figure 4.5, the closed orbit γ and a line L through the critical point P in the interior of γ. The line L intersects γ in two points S_1 and S_2 having opposite cross flow with respect to L, since γ is a convex curve whereon the flow has a unique direction, and let this be anticlockwise. Near the critical point the flow is then also anticlockwise as otherwise both on S_1P and S_2P there would be at least one point of contact on L, which together with the critical point P would contradict Lemma 1 of Tung Chinchu. Now let in some point Q on S_1P the orbit through Q cross L in the clockwise direction. As the orbit through Q cannot approach P in a fixed direction and cannot remain for all t on one side of L in the interior of γ, it must cross L either over S_1P or PS_2. In both cases there would exist at least two points of contact on S_1S_2 which plus the critical point P, again would be in contradiction with Lemma 1 of Tung Chinchu. □

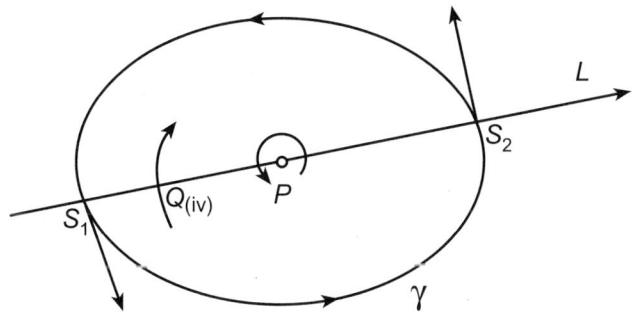

Figure 4.5

Remark. Note that in the interior of a closed orbit other closed orbits may occur, either that the whole interior is filled with orbits around a center point or that there exists a nest of limit cycles encircling a common critical point; the motion on all closed orbits being oriented in the same sense in one nest.

Theorem 4.7 *Two closed orbits of a quadratic system are oppositely oriented if external to each other and similarly oriented if one is inside the other.*

Proof. The last statement follows directly from the previous theorem and the remark thereafter.

Consider now two closed orbits external to each other as in Figure 4.6. According to Lemma 2 of Tung Chinchu, the direction of the flow on the segment ∞P_1 of the line L through the two critical points inside the closed

orbits γ_1 and γ_2, and that are on the segment $P_2\infty$, are in the same sense. This determines the opposite orientation of γ_1 and γ_2.   □
    From Theorem 4.7 follows directly

Theorem 4.8 *In a quadratic system there exist at most two nests of closed orbits.*

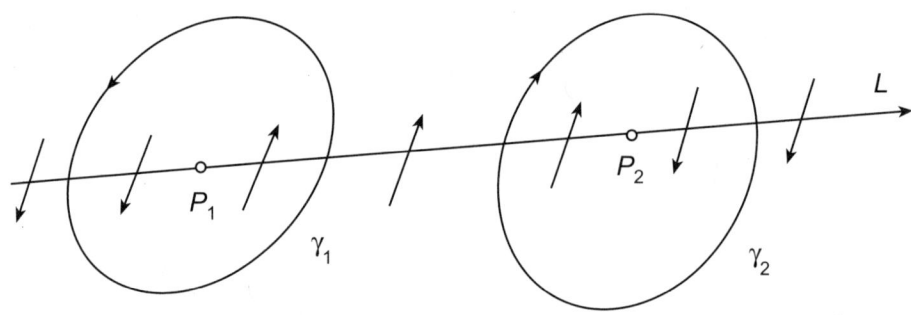

Figure 4.6

Chapter 5

Phase portraits of quadratic systems in the class $m_f = 0$

5.1 Introduction

A system having no real finite critical points cannot have closed orbits. As a result, orbits in these systems must connect points on the Poincaré circle,which is why they are sometimes indicated in the literature as chordal systems.

Attention to chordal systems occurred rather late in the development of the qualitative theory of quadratic systems, probably because no closed orbits are present and no applications are known for these systems.There appeared a paper in 1986 by Gasull, Shen Liren and Llibre [86,GSL] and in 1987 by Camacho and Palmeira [87,CP], followed by two additional papers of these authors, commenting on the previous ones [88,CP],[88,GL]. In [86,GSL] it was first shown that by a linear transformation and a scaling of the variable t, if necessary,any quadratic system could be brought into one of ten simpler forms. Although this reduces the number of coefficients remaining as parameters in the systems to be investigated, the relation between phase portraits and corresponding conditions on the coefficients of (1 0 1),(1 0 2) in $\mathbb{R}^{12}$ is actually not given in [86,GSL]. In [87,CP] the phase portraits are studied starting from the possible combinations of infinite critical points; examples are given but also the relation between the resulting phase portraits and the coefficients of (1.0.1),(1.0.2) is not given.

Quadratic systems in the class $m_f=0$ are a subclass of chordal quadratic systems, since in $m_f=0$ also complex finite critical points are not allowed to occur. These systems were classified by Reyn [96,R] following the approach of Camacho and Palmeira [87,CP] by ordering first the possible combinations of infinite critical points, yet giving also the corresponding conditions on the coefficients in (1.0.1),(1,.0.2). Apart from considering phase portraits as determined by their topological structure, characterized by the homeomorphic equivalence class with given separatrix structure, as is done in [86,GSL],[87,CP], in [96,R] attention is also given to the bifurcation properties of the infinite critical points; they are indicated by their notation. The affine invariant partition of the coefficient space $\mathbb{R}^{12}$ of (1.0.1),(1.0.2) corresponding to the phase portraits given in [96,R] were determined in [95,VV].

In the present account we will follow the approach in [96,R], using the equations developed in section 3.8 of Chapter 3.

5.2 One transversally non-hyperbolic critical point

For these systems we use (3.8.3),(3.8.4)

$$\dot{x} = a_{00} + a_{11}xy + a_{02}y^2, \tag{5.2.1}$$
$$\dot{y} = b_{00} + b_{11}xy + b_{02}y^2, \tag{5.2.2}$$

with $c_{15} = 0, c_{16} \neq 0, c_{56} \neq 0$, and

$$z' = -z(a_{11}u + a_{00}z^2 + a_{02}u^2) \equiv P(z,u), \tag{5.2.3}$$
$$u' = b_{11}u + b_{00}z^2 + (-a_{11} + b_{02})u^2 - a_{00}z^2u - a_{02}u^3, \tag{5.2.4}$$

so that the infinite critical points are located in $u = \bar{u}$ where $\bar{u}$ is a solution of $f(u) = b_{11}u + (-a_{11} + b_{02})u^2 - a_{02}u^3 = 0$. There exist the following cases:

(i) $b_{11} \neq 0$. If $b_{11} \neq 0$, q=1 so $M_{4,q}^i = M_{4,1}^i$ and the critical point at the ends of the x axis is a semi-elementary critical point with (the) center manifold(s) transversal to the Poincaré circle. It follows that $b_{00} \neq 0$, since $b_{00}=0$ leads with $a_{00}=0$ to $c_{16}=0$ and with $a_{00} \neq 0$ to $b_{11}=0$ as $c_{15}=0$, both being contradictory to the data. The center manifold(s) may be approximated by $u = \varphi(z) = -\frac{b_{00}}{b_{11}}z^2 - \frac{b_{00}^2 b_{02}}{b_{11}^3}z^4$+h.o.t.which results in a flow over them given approximately by $z' = \psi(z) = P(z, \varphi(z)) = \frac{b_{00}^2 c_{56}}{b_{11}^3}z^5$ +h.o.t. According to

[ALGM 1], theorem 65 (p.340) or [95,RK], then if $b_{11}c_{56} > 0$, $M_{4,1}^i$ is a node $M_{4,1}^1$ and if $b_{11}c_{56} < 0$, a saddle $M_{4,1}^{-1}$

Apart from $M_{4,1}^i$ there exist two other infinite critical points, located in the non-zero roots of $f(u)=0$. Let D$\equiv (a_{11} - b_{02})^2 + 4a_{02}b_{11} = (a_{11} + b_{02})^2 - 4c_{56}$, then for D>0 these points are both elementary nodes E^1 if $M_{4,1}^i$ is a saddle $M_{4,1}^{-1}$, an elementary saddle E^{-1} and a node E^1 if $M_{4,1}^i$ is a node $M_{4,1}^1$; for D=0 these points coincide into a second order tangentially non-hyperbolic saddle node $M_{0,2}^0$, whereas for D<0 they are complex critical points $C_{0,1}^0$. This leads to the phase portraits $E^1 E^1 M_{4,1}^{-1}, E^{-1}E^1 M_{4,1}^1, C_{0,1}^0 C_{0,1}^0 M_{4,1}^1, M_{0,2}^0 M_{4,1}^1$ given in Figure 5.1 (1,2,3 and 4, respectively). They may be obtained by starting in the infinite critical points and proceeding continuously into the finite part of the plane. It may be seen that the phase portraits in Figure 5.1 (1) and 5.1(2) may be considered to belong to the same homeomorphic equivalent class; they are different, however, in their bifurcational properties.

(ii) $b_{11} = 0, a_{11} - b_{02} \neq 0$. Then $c_{56} = a_{11}b_{02} \neq 0$, so $c_{15} = -b_{00}a_{11}=0$ yields $b_{00}=0$, whereas $c_{16} = a_{00}b_{02} \neq 0$ and (5.2.3),(5.2.4) may be written as

$$z' = -z(a_{11}u + a_{00}z^2 + a_{02}u^2), \tag{5.2.5}$$
$$u' = -(a_{11} - b_{02})u^2 - a_{00}z^2u - a_{02}u^3. \tag{5.2.6}$$

So $M_{4,2}^i$ is a critical point at the ends of the x axis of type (iii)c in section 2.3.2 of Chapter 2, being a critical point in a cubic system that cannot occur in quadratic systems and wherein the quadratic terms have a common factor. An analysis of these points can be made using polar coordinates $z = r\cos\theta, u = r\sin\theta, 0 \leq \theta < 2\pi$, use of which in (5.2.5),(5.2.6) gives the system

$$\dot{r} = -[(a_{11} - b_{02}) + b_{02}\cos^2\theta]r\sin\theta - [a_{00}\cos^2\theta + a_{02}\sin^2\theta]r^2 \tag{5.2.7}$$
$$\dot{\theta} = b_{02}\sin^2\theta\cos\theta. \tag{5.2.8}$$

An analysis in the (r, θ) plane of the critical points at $(0,0), (0, \frac{\pi}{2}), (0, \pi), (0, \frac{3\pi}{2})$ and $(0, 2\pi)$, which are elementary or of the type occurring in homogeneous quadratic systems, yields the topological structure of $M_{4,2}^i$. This was done in [95,RK]. Using these results, in particular those given in figure 8 of [95,RK] it follows that $M_{4,2}^i$ is a point $M_{4,2}^2$ having two elliptic and one parabolic sector if $b_{02}^2 < a_{11}b_{02}$, a point $M_{4,2}^0$(b)with two parabolic and two hyperbolic sectors if $-b_{02}^2 \leq a_{11}b_{02} < b_{02}^2$, and a point $M_{4,2}^0$ (a) with four parabolic sectors and two hyperbolic sectors if $a_{11}b_{02} < -b_{02}^2$. Apart from $M_{4,2}^2$ there exists a saddle

E^{-1}; the phase portrait $E^{-1}M_{4,2}^2$ is given in Figure 5.1(7).Apart from $M_{4,2}^0$ there exists a node E^1; the phase portrait for $E^1M_{4,2}^0$(b) is given in Figure 5.1(5) and for $E^1M_{4,2}^0$(a) in Figure 5,1 (6),

(iii) $b_{11} = a_{11} - b_{02} = 0, a_{02} \neq 0$. Then $c_{56} = a_{11}^2 \neq 0$, so $c_{15} = -b_{00}a_{11}=0$ yields $b_{00}=0$, whereas $c_{16} = a_{00}a_{11} \neq 0$ and (5.3.3),(5.3.4) may be written as

$$z' = -z(a_{11}u + a_{00}z^2 + a_{02}u^2), \tag{5.2.9}$$
$$u' = -a_{00}z^2u - a_{02}u^3. \tag{5.2.10}$$

So $M_{4,3}^i$ is a critical point at the ends of the x axis of type (iii)c in section 2.3.2 of Chapter 2. An analysis near (0,0), using polar coordinates z=rcos θ, u = $r\sin\theta, 0 \leq \theta < 2\pi$ was given in [95,RK]. In particular, using figure 12 in [95,RK] yields that $M_{4,3}^i$ is a point $M_{4,3}^1$(a) of nodal type if $a_{00}a_{02} > 0$ and a point $M_{4,3}^1$(b) with two elliptic and two hyperbolic sectors if $a_{00}a_{02} < 0$. The phase portraits for $M_{4,3}^1$(a) and $M_{4,3}^1$(b) are given in Figures 5.1 (8) and 5.1(9), respectively.

5.3 Two transversally non-hyperbolic critical points

5.3.1 Two real distinct transversally non-hyperbolic infinite critical points

Systems with an $M_{3,q}$point and an $M_{1,q}$point

For these systems we use (3.8.8),(3.8.9)

$$\dot{x} = a_{00} + a_{01}y + a_{11}xy, \tag{5.3.1}$$
$$\dot{y} = b_{00} + b_{01}y + b_{11}xy, \tag{5.3.2}$$

with $c_{15} = 0, c_{13} \neq 0, c_{35} \neq 0$ and

$$z' = -z(a_{11}u + a_{00}z^2 + a_{01}zu) \equiv P(z,u), \tag{5.3.3}$$
$$u' = b_{11}u + b_{00}z^2 + b_{01}zu - a_{11}u^2 - a_{01}zu^2 - a_{00}z^2u \equiv Q(z,u) \tag{5.3.4}$$

so that the infinite critical points are located in u=$\bar{u}$, where $\bar{u}$ is the solution of f(u)=$b_{11}u - a_{11}u^2$=0.

(i) $b_{11} \neq 0$. Then the critical point at the ends of the x axis is a semi-elementary critical point $M_{p,1}^i$ with center manifold(s) transversal to the Poincaré circle. It follows that $b_{00} \neq 0$, since $b_{00}=0$ leads with $c_{15} = a_{00}b_{11} - a_{11}b_{00}=0$ to $a_{00}=0$ and $c_{13} = a_{00}b_{01} - a_{01}b_{00}=0$ in contradiction with $c_{13} \neq 0$. The center manifold(s) may be approximated by $u = \varphi(z) = -b_{00}z^2 + \frac{b_{00}b_{01}}{b_{11}}z^3$ +h.o.t.,which results in a flow over them given approximately by $\dot{z} = \psi(z) = P(z, \varphi(z)) = \frac{b_{00}c_{35}}{b_{11}^3}z^4$+h.o.t. . According to [ALGM1], theorem 65 (p.340) or [95,RK], [96,R] then $M_{p,1}^i$ is a fourth order saddle node $M_{3,1}^0$ at the ends of the x axis.

The critical point at the ends of the y axis can be investigated by interchanging x and y in (5.3.1),(5.3.2) so that we obtain

$$\dot{x} = b_{00} + b_{01}x + b_{11}xy, \qquad (5.3.5)$$
$$\dot{y} = a_{00} + a_{01} + a_{01}x + a_{11}xy, \qquad (5.3.6)$$

with $c_{15} = 0, c_{13} \neq 0, c_{35} \neq 0$ and in (5.3.3),(5.3.4), so

$$z' = -z(b_{01}z + b_{11}u + b_{00}z^2), \qquad (5.3.7)$$
$$u' = a_{01}z + a_{11}u + a_{00}z^2 - b_{01}zu - b_{11}u^2 - b_{00}z^2u, \qquad (5.3.8)$$

and the infinite critical points are located in $u=\bar{u}$,where $\bar{u}$ is a solution of $a_{11}u - b_{11}u^2=0$.

If $a_{11} \neq 0$, it can be shown in a similar way as shown above that the critical point at the ends of the (original) y axis is a second order semi-elementary saddle node $M_{1,1}^0$.The third infinite critical point is an elementary node E^1 in $u = \frac{a_{11}}{b_{11}}$ (or $\frac{a_{11}}{b_{11}}$ for (5.3.4)). This leads to the phase portraits in the class $E^1M_{1,1}^0M_{3,1}^0$. For $a_{11}b_{11} >0$, the phase portrait is indicated by $E^1M_{1,1}^0M_{3,1}^0$(a) and given in Figure 5.1(10), for $a_{11}b_{11} <0$ by $E^1M_{1,1}^0M_{3,1}^0$(b), given in Figure 5.1(11).

If $a_{11}=0$, the critical point at the ends of the (original) y axis is a nilpotent third order critical point $M_{1,2}^1$ with an elliptic and a hyperbolic sector, as may be deduced from Figure 14 in [95,RK]. The phase portrait for $M_{1,2}^1M_{3,1}^0$ is given in Figure 5.1(15).

(ii)$b_{11}=0$ Then $a_{00}a_{11}b_{01} \neq 0, b_{00}=0$, and from Figure 11 in [95,RK] we obtain that the critical point at the ends of the x axis is a point $M_{3,2}^1$ having an elliptic, a hyperbolic and two parabolic sectors. At the ends of the y axis there is a semi-elementary saddle node $M_{1,1}^0$ with center manifold(s) transversal to the Poincaré circle. The phase portrait for $M_{1,1}^0M_{3,2}^1$ is given in Figure 5.1(16).

Systems with two $M_{2,q}^i$ points

For these systems we use (3.8.10),(3.8.11)

$$\dot{x} = a_{00} + a_{11}xy, \tag{5.3.9}$$
$$\dot{y} = b_{00} + b_{11}xy, \tag{5.3.10}$$

with $c_{15} \neq 0$, and

$$z' = -z(a_{11}u + a_{00}z^2), \tag{5.3.11}$$
$$u' = b_{11}u + b_{00}z^2 - a_{11}u^2 - a_{00}z^2u, \tag{5.3.12}$$

so that the infinite critical points are located in $u = \bar{u}$, where $\bar{u}$ is a solution of $f(u) = b_{11}u - a_{11}u^2 = 0$.

(i) $a_{11}b_{11} \neq 0$. At both ends of the x axis as well as of the y axis there exists a semi- elementary infinite critical point with center manifold(s) transversal to the Poincaré circle an $M_{p,1}^i$. From figure 2 in [95,RK] there follows that at the ends of the x axis there exists an $M_{2,1}^1$ nodal point for $c_{15} < 0$ and an $M_{2,1}^{-1}$ saddle if $c_{15} > 0$; at the ends of the y axis the reverse occurs. The third infinite critical point is an elementary critical point E^1 at $u = \frac{b_{11}}{a_{11}}$. The phase portraits for $E^1 M_{2,1}^{-1} M_{2,1}^1$ can now easily be constructed. Further analysis shows that there exist three possible phase portraits according as $a_{00}b_{00} <, =$ or > 0; they are shown in Figures 5.1(12), 5.1(13) and 5.1(14), respectively.

(ii) $a_{11}b_{11} = 0$, $(a_{11}^2 + b_{11}^2 \neq 0)$. Then, the nodal point E^1 coincides with either the third order node $M_{2,1}^1$ or the third order saddle $M_{2,1}^{-1}$, giving rise to a point $M_{2,2}^2$ or $M_{2,2}^0$, respectively. Let $a_{11} \neq 0, b_{11} = 0$, then $b_{00} \neq 0$, since $c_{15} = -a_{11}b_{00} \neq 0$ and the critical point at the ends of the x axis is of type (iii)b in section 2.3.2 of Chapter 2, being a critical point of a homogeneous quadratic system, wherein the quadratic terms have no common factor. Using Figure 11 in [95,RK] it follows that, if $a_{11}b_{00} < 0$, this point is a point $M_{2,2}^0$ with two hyperbolic sectors and, if $a_{11}b_{00} > 0$, it is a point $M_{2,2}^2$ with two elliptic sectors. For $a_{11}b_{00} < 0$ at the ends of the y axis there exists a third order node $M_{2,1}^1$ yielding the class $M_{2,1}^1 M_{2,2}^0$ of which the phase portrait is given in Figure 5.1(19). If $a_{11}b_{00} > 0$ there exists a third order saddle $M_{2,1}^{-1}$ at the ends of the y axis yielding the class $M_{2,1}^{-1} M_{2,2}^2$. For $a_{00} \neq 0$ the phase portrait is given in Figure 5.1(17) and the class indicated by $M_{2,1}^{-1} M_{2,2}^2$ (a). For $a_{00} = 0$ in Figure 5.1(18) and the class by $M_{2,1}^{-1} M_{2,2}^2$(b).

5.3.2 Two coinciding transversally non-hyperbolic infinite critical points

For these systems we use (3.8.14),(3.8.15)

$$\dot{x} = a_{00} + a_{10}x + a_{01}y + a_{02}y^2, \qquad (5.3.13)$$
$$\dot{y} = b_{00} + b_{10}x + b_{01}y + b_{02}y^2, \qquad (5.3.14)$$

with $a_{02}^2 + b_{02}^2 \neq 0$, $c_{26} = c_{23}c_{36} = 0$, $c_{16}^2 - c_{13}c_{36} \neq 0$ and

$$z' = -z(a_{10}z + a_{00}z^2 + a_{01}zu + a_{02}u^2), \qquad (5.3.15)$$
$$u' = b_{10}z + b_{00}z^2 + (-a_{10} + b_{01})zu + b_{02}u^2$$
$$-a_{01}zu^2 - a_{00}z^2u - a_{02}u^3, \qquad (5.3.16)$$

so that the infinite critical points are located in $u = \bar{u}$, where $\bar{u}$ is a solution of $-b_{02}u^2 + a_{02}u^3 = 0$.

Systems with an $M_{4,2}^i$ point

Then $b_{02} \neq 0$ and it follows from $c_{26} = c_{23}c_{36} = 0$ that $b_{10}c_{36}^2 = 0$.

(i) $b_{10} \neq 0$. Then $c_{23} = c_{26} = c_{36} = 0$ and $c_{12}c_{16} \neq 0$. The critical point at the ends of the x axis is a nilpotent critical point. From Figure 5 in [95,RK] it is seen that $M_{4,2}^i$ is a six order nilpotent saddle node $M_{4,2}^0$. There exists one phase portrait containing such a point; it also contains an elementary node E^1 and is given as class $E^1 M_{4,2}^0(\text{a})$ in Figure 5.1(21).

(ii)$b_{10} = 0$. Then also $a_{10} = 0$ since $c_{26} = 0$, and the critical point at the ends of the x axis may be determined using Figure 7-10 in [95,RK]. Then, it follows from Figure 7 that if $\Delta \equiv b_{01}^2 - 4b_{00}b_{02} < 0$ that the critical point is a point $M_{4,2}^0$ with two hyperbolic sectors; it occurs in class $E^1 M_{4,2}^0(\text{b})$ given in Figure 5.1(22). If $\Delta = 0$ it follows from Figure 8 that the point is a point $M_{4,2}^0$ with two hyperbolic and two parabolic sectors. It should be noted that $\Delta = 0$ implies that $y = -\frac{b_{01}}{2b_{02}}$ solves $b_{00} + b_{01}y + b_{02}y^2 = 0$ and should not solve $a_{00} + a_{01}y + a_{02}y^2 = 0$ since $c_{16}^2 - c_{13}c_{36} \neq 0$. This yields $-4b_{02}c_{16} + 2b_{01}c_{36} \neq 0$. Furthermore $b_{02}(a_{11} + b_{02}) > 0$. The point occurs in class $E^1 M_{4,2}^0(\text{c})$ given in Figure 5.1(23). If $\Delta > 0$, Figure 10 in [95,RK] shows there exist two cases:(i)if $E_1 \equiv c_{16}^2 - c_{13}c_{36} > 0$, then there exists a point $M_{4,2}^0$ with two hyperbolic and two parabolic sectors; it occurs in the class $E^1 M_{4,2}^0(\text{d})$ given in Figure 5.1(24),(ii) if $E_1 < 0$ the point $M_{4,2}^0$ has three hyperbolic, two parabolic and one elliptic sector and occurs in the class $E^1 M_{4,2}^0(\text{e})$, given in Figure 5.1(25).

Systems with an $M_{4,3}^1$ point

Then $b_{20}=0$, so $a_{02} \neq 0$, whereas $c_{26}=0$ yields $b_{10}=0$ and $c_{23}c_{36}=0$ yields $a_{10}b_{01}=0$.

(i)$b_{01}=0$. Then (5.3.13),(5.3.14) become

$$\dot{x} = a_{00} + a_{10}x + a_{01}y + a_{02}y^2, \qquad (5.3.17)$$
$$\dot{y} = b_{00}, \qquad (5.3.18)$$

with $a_{02}b_{00} \neq 0$. Figure 15 in [95,RK] shows that there exist two cases; (i) if $a_{10} \neq 0$, $M_{4,3}^1$ has an elliptic, a hyperbolic and a parabolic sector and occurs in the class $M_{4,3}^1$(c) given in Figure 5.1(28);(ii) if $a_{10} = 0$, $M_{4,3}^1$ has one parabolic sector and is a seventh order node; it occurs in class $M_{4,3}^1$(d) and is given in Figure 5.1(29).

(ii)$a_{10}=0$. Then (5.3.13),(5.3.14) become

$$\dot{x} = a_{00} + a_{01}y + a_{02}y^2, \qquad (5.3.19)$$
$$\dot{y} = b_{00} + b_{01}y, \qquad (5.3.20)$$

with $a_{02}b_{01} \neq 0$. Figure 15 in [95,RK] shows that there exist two cases; (i) if $E_1 \equiv c_{16}^2 - c_{13}c_{36} < 0$, $M_{4,3}^1$ has two hyperbolic and two elliptic sectors and occurs in class $M_{4,3}^1$(a), given in Figure 5.1(26); (ii) if $E_1 > 0$, $M_{4,3}^1$ is a seventh order node and occurs in class $M_{4,3}^1$(b), given in Figure 5.1(27).

5.3.3 Two complex transversally non-hyperbolic infinite critical points

It may easily be seen that the third infinite critical point is an elementary node E^1. The phase portrait of class $E^1C_{2,1}^0C_{2,1}^0$ is given in Figure 5.1(20).

5.4 Systems with infinitely many infinite critical points

In this case we can work with (3.8.32),(3.8.33)

$$\dot{x} = 1 + xy, \qquad (5.4.1)$$
$$\dot{y} = y^2, \qquad (5.4.2)$$

and

$$z' = -z(u + z^2), \qquad (5.4.3)$$
$$u' = -z^2 u, \qquad (5.4.4)$$

or, after changing the parameter, with

$$\dot{z} = u + z^2, \qquad (5.4.5)$$
$$\dot{u} = zu, \qquad (5.4.6)$$

which in (0,0) has a nilpotent critical point with an elliptic and a hyperbolic sector. The phase portrait in the x, y plane is given in Figure 5.1(30). It confirms that given in the paper of Gasull and Prohens on quadratic and cubic systems with a degenerate infinity [96,GP1].

5.5 Conclusion

Theorem 5.1 There exist 30 possible phase portraits in the class of quadratic systems with finite multiplicity $m_f=0$. They are given in Figure 5.1.

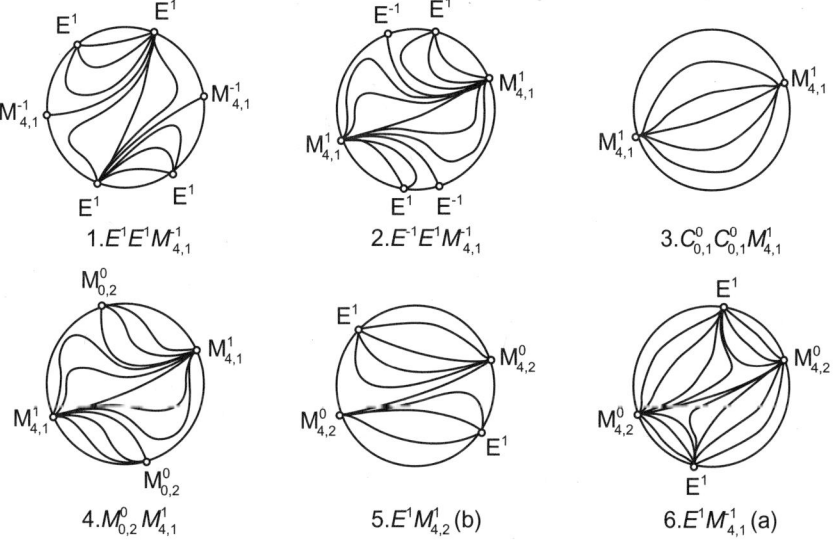

Figure 5.1 (i) Phase portraits of the systems in m_f-0

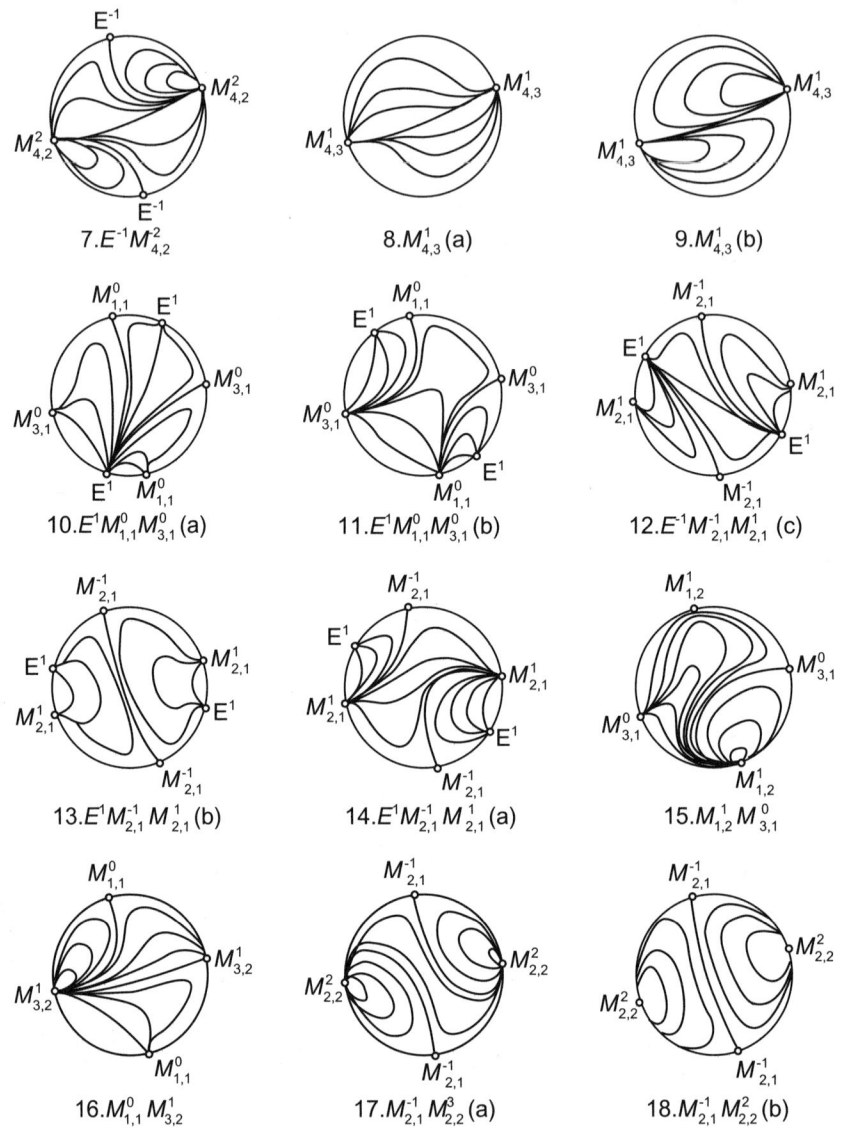

Figure 5.1 (ii)Phase portraits of the systems in $m_f = 0$

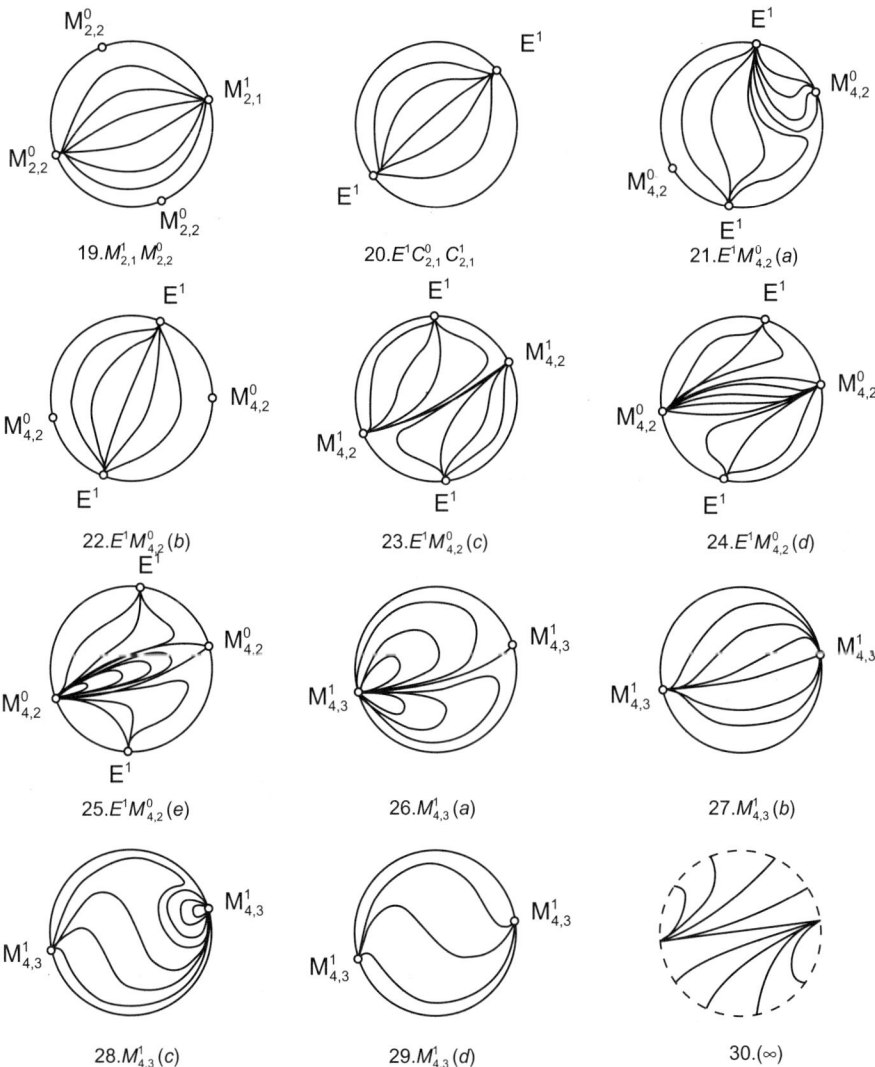

Figure 5.1 (iii)Phase portraits of the systems in $m_f=0$

Chapter 6

Quadratic systems with center points

6.1 Integrability and center conditions

When proceeding to analyze the phase portraits of quadratic systems for $m_f > 0$, it should be realized that the presence of critical points in systems with $m_f > 0$ induces the possible occurrence of closed orbits. Closed orbits represent periodic solutions and they may be either imbedded in an annular region without other closed orbits (limit cycles) or be members of a family of closed orbits surrounding a center point and filling out a region containing precisely one center point in its interior.

In this chapter we consider quadratic systems with (a) center point(s). Examples of such systems were already given in 1904 by Büchel [04,B]; some of them having two center points, the maximum number that can occur in a quadratic system.

The study of quadratic systems with center points illuminates the problem of finding an integrating factor for such a system. As can be concluded from the existence and uniqueness statements for the initial problem, mentioned in Chapter 1, quadratic systems belong to the wide class of systems $P(x,y)\frac{dy}{dx} - Q(x,y) = 0$ for which there exists a (first) integral $F(x,y)=C$, where C is constant on an orbit. So $F(x,y) = \frac{\partial F}{\partial x}\dot{x} + \frac{\partial F}{\partial y}\dot{y} = \frac{\partial F}{\partial x}P(x,y) + \frac{\partial F}{\partial y}Q(x,y) = 0$ or $\frac{\partial F}{\partial x} = Q(x,y)I(x,y)$, $\frac{\partial F}{\partial y} = -P(x,y)I(x,y)$ for some function $I(x,y)$ satisfying the condition $\frac{\partial^2 F}{\partial x \partial y} - \frac{\partial^2 F}{\partial y \partial x} = \frac{\partial P(x,y)I(x,y)}{\partial x} + \frac{\partial Q(x,y)I(x,y)}{\partial y} = 0$; $I(x,y)$ is called an integrating factor. The problem that, usually, an integrating factor, if it at all can be expressed by well-known functions, does not have a simple form, is one of the reasons for the development of the qualitative theory of differential equations, which aims at finding properties of the solutions of these equations, without actually solving them. Since multiplying the vector field $P(x, y), Q(x, y)$ by $I(x, y)$ amounts to making the divergence of the vector field equal to zero, $I(x, y)$ must have a singular behavior in critical points for which the sum of the eigenvalues is unequal to zero, such as, for instance, for nodes or non-weak foci. The same applies for limit cycles, as such orbits can not occur in a divergence free vector field. Since, moreover, in general, the shape of a limit cycle cannot be represented by a simple expression, the integrating factor, being singular there, can hardly be expected to be expressible in a simple form as well.

However, integrating factors and first integrals for quadratic systems with center points are of much simpler form and easier to find. Obviously, at a center point (x_0, y_0) of a quadratic system there is $\mathrm{div}(P(x_0, y_0), Q(x_0, y_0))=0$. If we assume the contrary, then div $(P(x, y), Q(x, y)) \neq 0$ in a neighborhood of (x_0, y_0) and Green's formula for a closed orbit C enclosing a region S_C within this neighborhood yields the contradiction

$$0 = \oint P(x, y), Q(x, y) . \underline{n} ds = \int \int \mathrm{div}(P(x, y), Q(x, y)) dS \neq 0.$$

Here $\underline{n}$ is the outward pointing unit vector normal to C. Inspecting all critical points (x_0, y_0) possible in a quadratic system, having the property that div $(P(x_0, y_0), Q(x_0, y_0))=0$ shows that a center point should also be a center point in the linear approximation. Making a linear transformation and a scaling of t, if necessary, then shows that, if (1.0.1),(1.0.2) has a center point in the origin, it can be written as

$$
\begin{aligned}
\dot{x} &= -y + a_{20}x^2 + a_{11}xy + a_{02}y^2 \equiv -y + P_2(x, y), & (6.1.1) \\
\dot{y} &= x + b_{20}x^2 + b_{11}xy + b_{02}y^2 \equiv x + Q_2(x, y). & (6.1.2)
\end{aligned}
$$

Determination of an integrating factor and first integral for this system in case it has a center point at (0,0) may subsequently be obtained by using a result of Lyapunov stating that (6.1.1),(6.1.2) belongs to the class for which the necessary and sufficient conditions to have this property is satisfied if $F(x, y)$ is an analytic function in a neighborhood of (0,0). This means that for $F(x, y)$ may be written the power series

$$F(x, y) = \frac{1}{2}x^2 + \frac{1}{2}y^2 + F_3(x, y) + F_4(x, y) + \cdots + F_k(x, y) + \cdots, \quad (6.1.3)$$

where $F_k(x, y)$ is a homogeneous polynominal of degree k, and which converges in a neighborhood of the origin. Substituting this expression in

$$(-y + P_2(x, y))\frac{\partial F}{\partial x} + (x + Q_2(x, y))\frac{\partial F}{\partial y} = 0 \qquad (6.1.4)$$

yields the infinite set of equations

$$x\frac{\partial F_3}{\partial y} - y\frac{\partial F_3}{\partial x} = -xP_2(x, y) - yQ_2(x, y), \qquad (6.1.5)$$

$$x\frac{\partial F_4}{\partial y} - y\frac{\partial F_4}{\partial x} = -P_2(x, y)\frac{\partial F_3}{\partial x} - Q_2(x, y)\frac{\partial F_3}{\partial y}, \qquad (6.1.6)$$

$$\dotfill \qquad (6.1.7)$$

$$x\frac{\partial F_k}{\partial y} - y\frac{\partial F_k}{\partial x} = -P_2(x, y)\frac{\partial F_{k-1}}{\partial x} - Q_2(x, y)\frac{\partial F_{k-1}}{\partial y}, \qquad (6.1.8)$$

$$\dotfill \qquad (6.1.9)$$

with k $\to \infty$.

Solving these equations is possible only if certain conditions on a_{ij}, b_{ij} are satisfied; these conditions already appear when constructing $F_k(x, y)$ for low values of $k(k \le 6)$. Once these conditions are obtained and substituted in (6.1.1),(6.1.2) systems are yielded that can simply be integrated, so that for higher values of k the function $F_k(x, y)$ need not be determined and $F(x, y)$ can be expressed in elementary functions, such as polynomials and exponentials. So, all quadratic systems in the normal form (6.1.1),(6.1.2) having a center point in the origin as well as their solution can be determined.The remaining systems in (6.1.1),(6.1.2)are said to have a weak focus at the origin.

The center point problem for quadratic systems was first discussed in 1908, by Dulac [08,Du].Dulac allowed the variables x and y to take complex values, using a wider definition for a center point, and did not concern himself with the consequences of his results for real quadratic systems. The first study to obtain the center conditions and integrals by working directly in the real plane was made in 1911 by Kapteyn [11,K]. In order to simplify the calculations he took $b_{20} + b_{02}=0$, obtainable, if needed, by a rotation of the coordinate system. Moreover, one of the four alternative conditions to have a center point in the origin of (6.1.1),(6.1.2) appears to be too restrictive, as a result of which not all quadratic systems with a center are covered in [11,K]. This was noticed by Bautin in 1939, who gave the correct center conditions for (6.1.1),(6.1.2) in [39,B1], using also the condition $b_{20} + b_{02}=0$. In fact he considered the system

$$\dot{x} = \lambda_1 x - y - \lambda_3 x^2 + (2\lambda_2 + \lambda_5)xy + \lambda_6 y^2, \qquad (6.1.10)$$
$$\dot{y} = x + \lambda_1 y + \lambda_2 x^2 + (2\lambda_3 + \lambda_4)xy - \lambda_2 y^2, \qquad (6.1.11)$$

and found that the origin is a center point if and only if $\lambda_1=0$ and at least one of the following conditions is satisfied:(i)$\lambda_2 = \lambda_5 = 0$, (ii)$\lambda_3 - \lambda_6 = 0$, (iii)$\lambda_4 = \lambda_5=0$, or $\lambda_5 = 5\lambda_3 + \lambda_4 - 5\lambda_6 = \lambda_2^2 - \lambda_3\lambda_6 + 2\lambda_6^2=0$. As a result, the phase portraits for quadratic systems with a center point can be found by analyzing the correct equations.

6.2 Phase portraits for quadratic systems with center points

6.2.1 The symmetric case

We rewrite (6.1.10),(6.1.11) as

$$\dot{x} = -y + ax^2 + by^2, \qquad (6.2.1)$$
$$\dot{y} = x + cxy, \qquad (6.2.2)$$

which may be analyzed on the unit sphere $a^2 + b^2 + c^2=1$, since $\bar{x} = \lambda x, \bar{y} = \lambda y, \lambda = (a^2 + b^2 + c^2)^{-\frac{1}{2}}$ leads to (6.2.1),(6.2.2) with $\bar{a} = \lambda a, \bar{b} = \lambda b, \bar{c} = \lambda c$ and $\bar{a}^2 + \bar{b}^2 + \bar{c}^2=1$. The bifurcation diagram will be drawn on the lower half of this unit sphere and projected onto the unit disk $a^2 + b^2 \leq 1$ in $c\equiv 0$. It is given in Figure 6.1 whereas the 38 phase portraits are given in Figure 6.2. The symmetric case yields the largest collection of phase portraits among the four cases listed above; the other two cases each generate a few additional phase portraits apart from those in their intersection with the symmetric case. The final critical points are located at $P_0 : (0,0), P_1 : (0, \frac{1}{b}), P_2 : (\sqrt{-\frac{b+c}{ac^2}}, -\frac{1}{c})$ and $P_3 : (-\sqrt{-\frac{b+c}{ac^2}}, -\frac{1}{c})$ and are finite in number, except when $P(x,y)$ and $Q(x,y)$ have a common factor $1 + cy$ if $a = 0, b = -c$, yielding an essentially linear system.

As $\Lambda(P_1) = -\frac{b+c}{b}, P_1$ is a saddle if $b(b+c) > 0$ and a center point if $b(b+c) < 0$. If $b + c \neq 0, b = 0$, then P_1 "has gone to infinity" along the y axis to yield a $M_{1,q}^i$ point as $c \neq 0$ so that P_2 and P_3 will remain on $y \equiv -\frac{1}{c}$

either as finite or infinite critical points; moreover $3< q=2$ as a result of the symmetry around the y axis and $i = \pm 1$, to be determined later. (An exception occurs for $a = c, b = 0$; then all infinite points are critical points and $M^i_{1,q}$ is not isolated.) For $b + c=0$, $abc \neq 0, P_1, P_2$ and P_3 coincide into a nilpotent m^i_3 point on the y axis wherein $\text{div}(P, Q)=0$. The index of this point can be determined from $A \equiv c^2_{46} - c_{45}c_{56} = abc^2$. So if $ab >0$, then $A>0$ and $i_f=0$ so $m^i_3 = m^{-1}_3$, whereas for $ab <0$, then $A<0$ and since $c_{45} + c_{56} = c(a - b) = -ab + b^2 >0$, then $i_f=2$ so $m^i_3 = m^1_3$. When $m^i_3 = m^1_3$, there exists a nilpotent saddle point and when $m^i_3 = m^1_3$ a nilpotent critical point with an elliptic and a hyperbolic sector. For $b + c=0$, $a \neq 0, b = 0$, so $c=0$ and P_1, P_2 and P_3 coincide in an $M^i_{3,2}$ point "at the ends of the y axis ". (Note that $a = b = c=0$ need not be considered since $a^2 + b^2 + c^2=1$.)

The critical points P_2 and P_3 are complex if $a(b + c) >0$ and real and distinct if $a(b+c) <0$, being saddle points (nodes) if $c(b+c) <0(>0)$, as can be seen from $\Lambda(P_2)= \Lambda(P_3) = -2\frac{b+c}{c}$. As mentioned above, for $b + c = 0, abc \neq 0, P_1, P_2$ and P_3 form an m^i_3 point, whereas for $b + c = 0, a \neq 0, b = 0$ thus $c=0$ and $P_1, P_2$and P_3 coincide in an $M^i_{3,2}$ point "at the ends of the y axis", whereas $a = b = c = 0$ need not be considered. Finally if $a = 0, b + c \neq 0, P_2$ and P_3 " have gone to the ends of the x axis" to yield an $M^i_{2,q}$ point, where $q=1$ if $c \neq 0$ and $q=3$ if $c=0$.

The location of the infinite critical points is given by $bu^3 + (a - c)u=0$, so there are a finite number of these points, except when $b = a - c=0$. There exists an E^i or an $M^i_{p,1}$ point in $u=0$ if $a - c \neq 0$ and an $M^i_{p,3}$ point if $a - c = 0, b \neq 0$. There exist two other infinite critical points located in $u = \pm\sqrt{-\frac{a-c}{b}}$, respectively, being complex for $b(a - c) <0$; two saddles if $c(a - c) >0$ and two nodes if $c(a - c) <0$, as follows from $(2.3.15),(2.3.16)$ that $\lambda_z\lambda_u = -2(a - c)c$ at these points. If $b=0, (a - c) \neq 0$ these two points coincide with an $M^i_{p,2}$ point at the ends of the y axis and $p=1$ if $c \neq 0$, $p=3$ if $c=0$. The case $b = (c - a)=0$ was discussed before. Finally, if $b(a - c) < 0$,so there exist two real distinct critical points, and $c=0$, the two points are both $M^0_{1,1}$ points since $\lambda_z = -c =0, \lambda_u = 2(a - c) \neq 0$ at these points, which shows that these points are semi-elementary transversally non-hyperbolic points, and since they occur if $m_f=2$ they are of second order.

All critical points can now be determined, using index arguments, the knowledge of the orbit $1 + cy=0$ for $c \neq 0$, the knowledge of the possible infinite critical points in quadratic systems , if needed supplemented by further analysis using [95,RK].

There exist three types of bifurcation lines related to the coincidence of critical points. Those corresponding to the coincidence of finite and infinite critical points occur at $A \equiv c_{46}^2 - c_{45}c_{56} = abc^2 = 0$. The coincidence of infinite critical points occurs for a−c=0, b≠0, then three points coincide in u=0 and for b=0 if two infinite critical points (and one finite critical point) coincide in u=∞. The third type occurs when the three finite critical points outside of (0,0) coincide if b+c=0, a≠0.

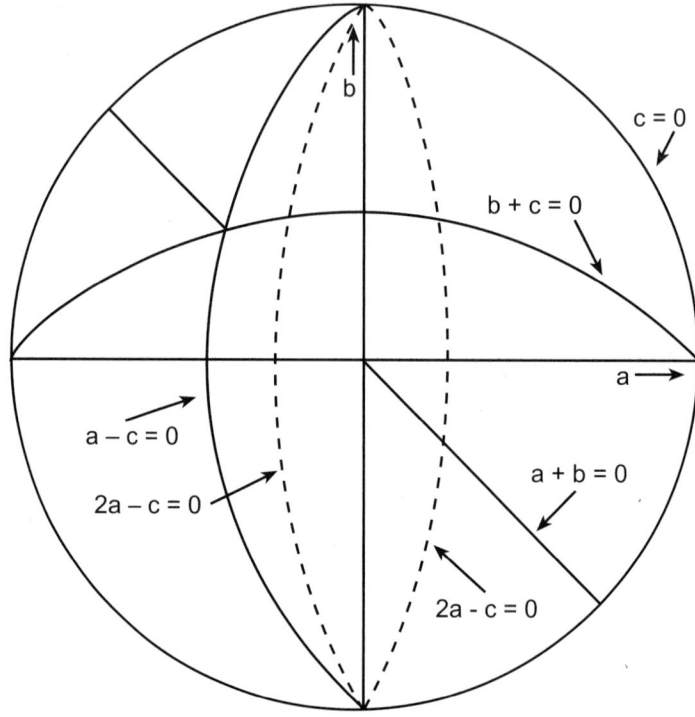

Figure 6.1 Bifurcation diagram for the symmetric case:$\lambda_2 = \lambda_5 = 0$

In most cases the separatrix cycle surrounding the region of closed orbits around a center point changes type at the bifurcation points related to the coincidence of critical points. An exception occurs at the intersection of a−c=0 and b+c=0, so at a$=-\frac{1}{3}$, $b = \frac{2}{3}$, $c = -\frac{2}{3}$, where the type of separatrix cycle changes leaving the critical points of the same type. Other cases can be found on the line a+b=0.

Shown in the bifurcation diagram in Figure 6.1 are also the intersections with the other cases; the condition (ii):$\lambda_3 = \lambda_6$, the case with three orbital lines,yields a+b=0, the condition (iii)$\lambda_4 = \lambda_5$=0, the Hamiltonian case, yields 2a+c=0 and the condition(iv) $\lambda_5 = 5\lambda_3 + \lambda_4 - 5\lambda_6 = \lambda_2^2 - \lambda_3\lambda_6 + 2\lambda_6^2$=0,the lonely case,yields two points: a$=-\frac{1}{10}\sqrt{10}, b = 0, c = -\frac{3}{10}\sqrt{10}$ and $a = -\frac{1}{3}\sqrt{6}, b = \frac{1}{6}\sqrt{6}, c = -\frac{1}{6}\sqrt{6}$.

The first integrals F(x,y)=C of (6.2.1),(6.2.2),where C is constant on a orbit, can be found in several papers of Schlomiuk e.g.[93,S1],[93,S2]. It was pointed out by her that a geometrical characterization of the phase portraits of (6.2.1),(6.2.2) can be obtained by, simultaneously, considering changes in particular algebraic integrals such as conics and straight lines and changes in the phase portraits, as the parameters are varied.

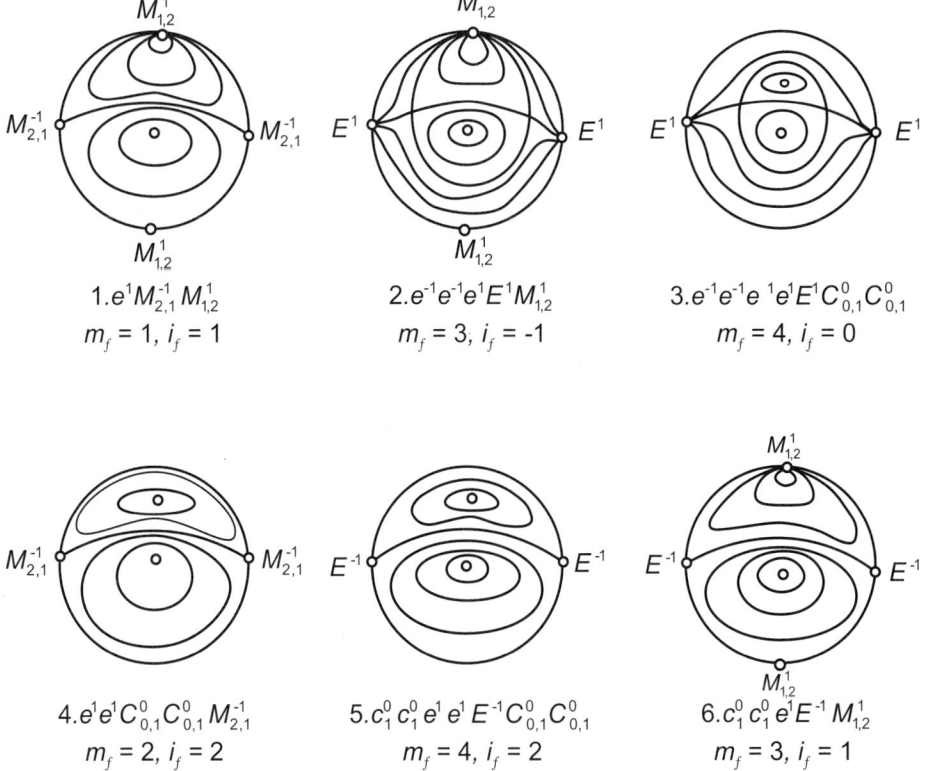

Figure 6.2 Phase portraits for the symmetric case.
(a)Portraits near a=b=0

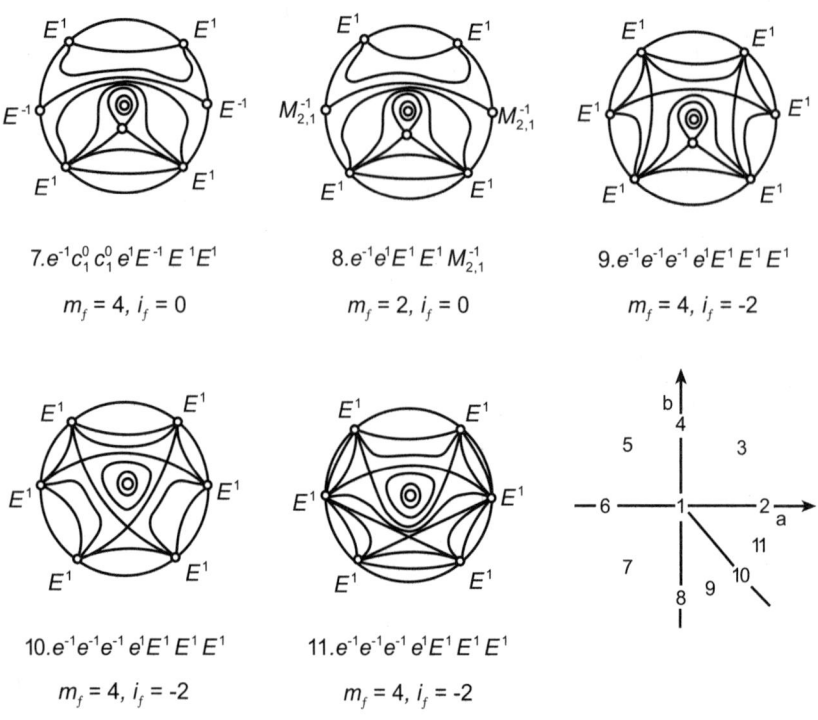

7. $e^{-1}c_1^0 c_1^0 e^1 E^{-1} E^{-1} E^1$

$m_f = 4, i_f = 0$

8. $e^{-1}e^1 E^1 E^1 M_{2,1}^{-1}$

$m_f = 2, i_f = 0$

9. $e^{-1}e^{-1}e^{-1} e^1 E^1 E^1 E^1$

$m_f = 4, i_f = -2$

10. $e^{-1}e^{-1}e^{-1} e^1 E^1 E^1 E^1$

$m_f = 4, i_f = -2$

11. $e^{-1}e^{-1}e^{-1} e^1 E^1 E^1 E^1$

$m_f = 4, i_f = -2$

Figure 6.2 Phase portraits for the symmetric case.
(a) Portraits near a=b=0, continued

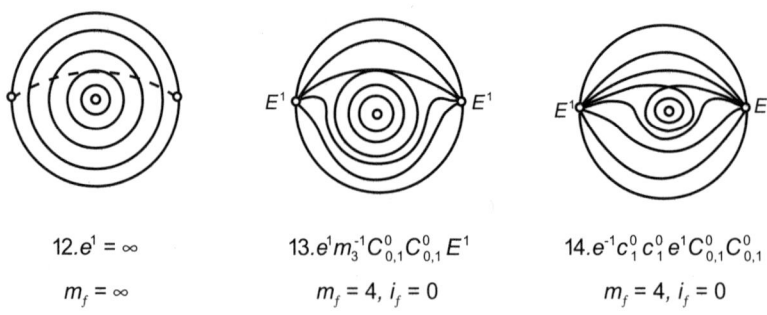

12. $e^1 = \infty$

$m_f = \infty$

13. $e^1 m_3^{-1} C_{0,1}^0 C_{0,1}^0 E^1$

$m_f = 4, i_f = 0$

14. $e^{-1}c_1^0 c_1^0 e^1 C_{0,1}^0 C_{0,1}^0$

$m_f = 4, i_f = 0$

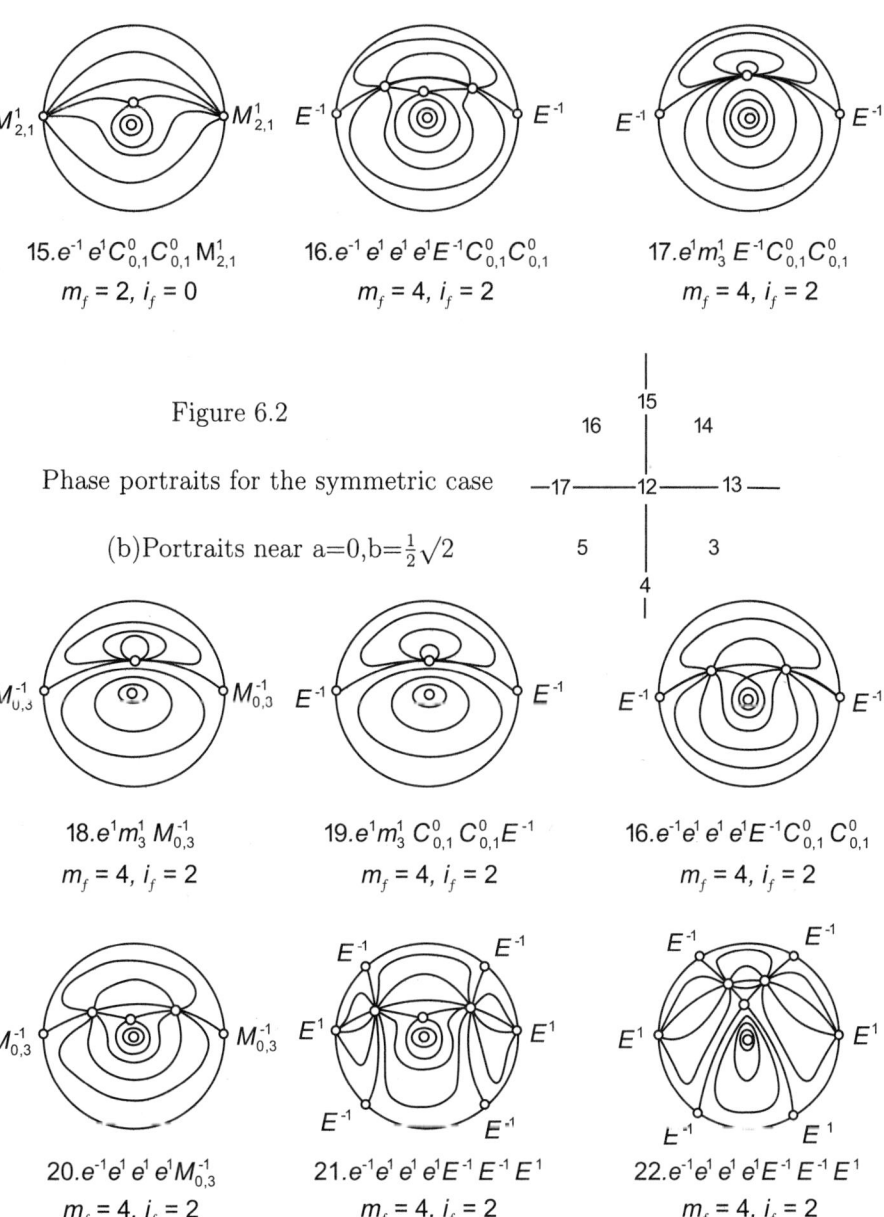

15. $e^{-1} e^{1} C_{0,1}^{0} C_{0,1}^{0} M_{2,1}^{1}$
$m_{f} = 2, i_{f} = 0$

16. $e^{-1} e^{1} e^{1} e^{1} E^{-1} C_{0,1}^{0} C_{0,1}^{0}$
$m_{f} = 4, i_{f} = 2$

17. $e^{1} m_{3}^{1} E^{-1} C_{0,1}^{0} C_{0,1}^{0}$
$m_{f} = 4, i_{f} = 2$

Figure 6.2

Phase portraits for the symmetric case

(b)Portraits near a=0,b=$\frac{1}{2}\sqrt{2}$

18. $e^{1} m_{3}^{1} M_{0,3}^{-1}$
$m_{f} = 4, i_{f} = 2$

19. $e^{1} m_{3}^{1} C_{0,1}^{0} C_{0,1}^{0} E^{-1}$
$m_{f} = 4, i_{f} = 2$

16. $e^{-1} e^{1} e^{1} e^{1} E^{-1} C_{0,1}^{0} C_{0,1}^{0}$
$m_{f} = 4, i_{f} = 2$

20. $e^{-1} e^{1} e^{1} e^{1} M_{0,3}^{-1}$
$m_{f} = 4, i_{f} = 2$

21. $e^{-1} e^{1} e^{1} e^{1} E^{-1} E^{-1} E^{1}$
$m_{f} = 4, i_{f} = 2$

22. $e^{-1} e^{1} e^{1} e^{1} E^{-1} E^{-1} E^{1}$
$m_{f} = 4, i_{f} = 2$

Figure 6.2 Phase portraits for the symmetric case.
(c)Portraits near a=$-\frac{1}{3}\sqrt{3}, b = \frac{1}{3}\sqrt{3}$. (i)

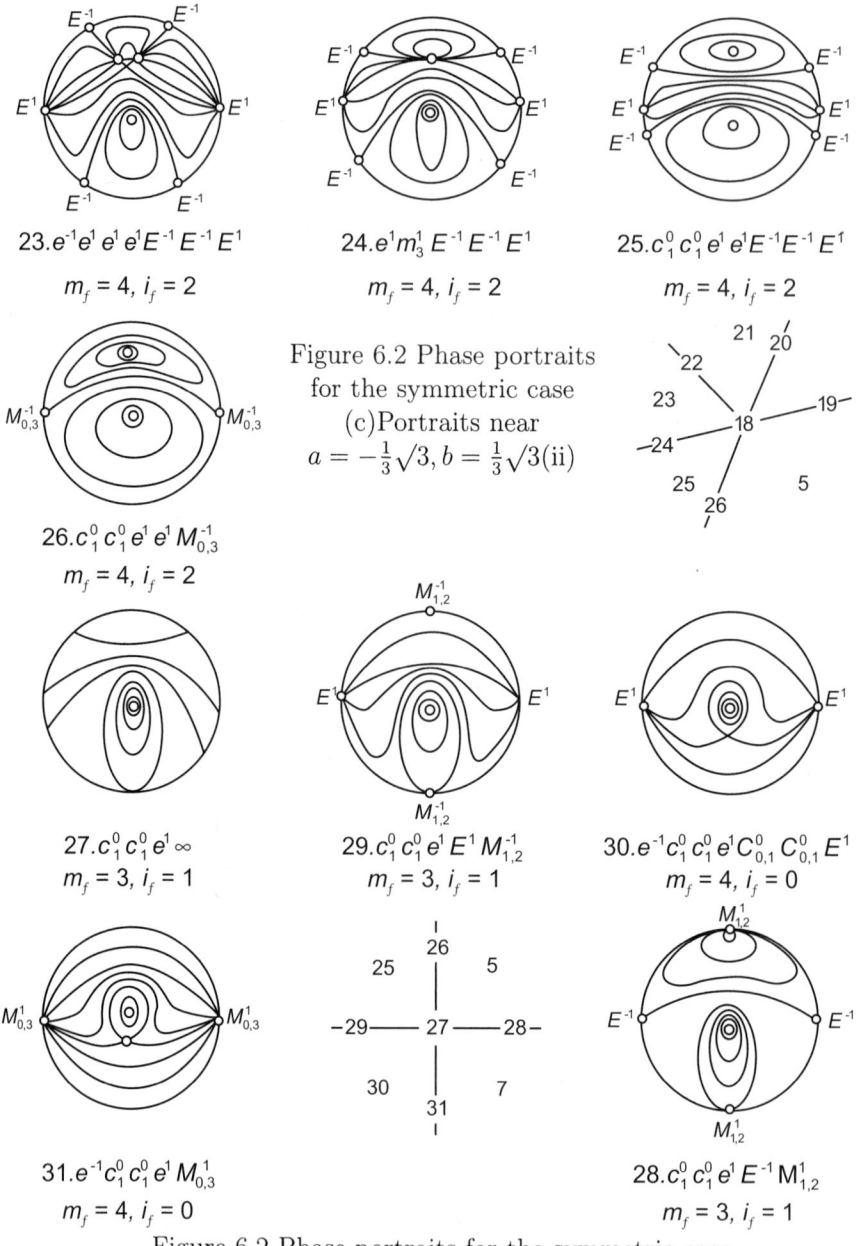

23. $e^{-1} e^1 e^1 e^1 E^{-1} E^{-1} E^1$

$m_f = 4,\ i_f = 2$

24. $e^1 m_3^1 E^{-1} E^{-1} E^1$

$m_f = 4,\ i_f = 2$

25. $c_1^0 c_1^0 e^1 e^1 E^{-1} E^{-1} E^1$

$m_f = 4,\ i_f = 2$

26. $c_1^0 c_1^0 e^1 e^1 M_{0,3}^{-1}$

$m_f = 4,\ i_f = 2$

Figure 6.2 Phase portraits
for the symmetric case
(c) Portraits near
$a = -\frac{1}{3}\sqrt{3}, b = \frac{1}{3}\sqrt{3}$ (ii)

27. $c_1^0 c_1^0 e^1 \infty$

$m_f = 3,\ i_f = 1$

29. $c_1^0 c_1^0 e^1 E^1 M_{1,2}^{-1}$

$m_f = 3,\ i_f = 1$

30. $e^{-1} c_1^0 c_1^0 e^1 C_{0,1}^0\ C_{0,1}^0\ E^1$

$m_f = 4,\ i_f = 0$

31. $e^{-1} c_1^0 c_1^0 e^1 M_{0,3}^1$

$m_f = 4,\ i_f = 0$

28. $c_1^0 c_1^0 e^1 E^{-1} M_{1,2}^1$

$m_f = 3,\ i_f = 1$

Figure 6.2 Phase portraits for the symmetric case
(d) Portraits near $a = -\frac{1}{2}\sqrt{2}, b = 0$

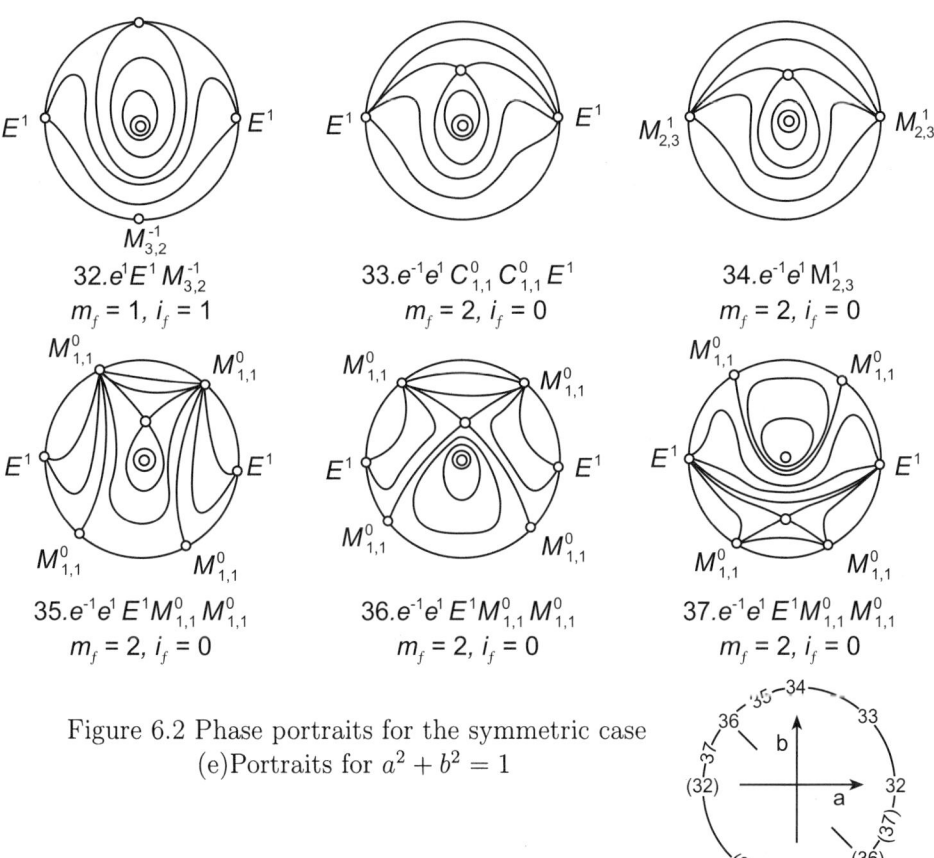

Figure 6.2 Phase portraits for the symmetric case
(e)Portraits for $a^2 + b^2 = 1$

6.2.2   The case with three orbital lines

Applying the condition $\lambda_3 - \lambda_6 = 0$ to (6.1.10),(6.1.11) yields a system for which $a_{20} + a_{02} = b_{20} + b_{02} = 0$. As was shown by Frommer [34,F] this condition is invariant under rotation of the coordinate system around the origin. As a result a rotation may be chosen such that $b_{20} = b_{02}$ (so $\lambda_2 = 0$). We then rewrite (6.1.10),(6.1.11) as

$$\dot{x} = -y - ax^2 + cxy + ay^2, \qquad (6.2.3)$$
$$\dot{y} = x + bxy, \qquad (6.2.4)$$

which again may be analyzed on the unit sphere $a^2 + b^2 + c^2 = 1$, of which the lower half $c \leq 0$ is projected onto the disk $a^2 + b^2 \leq 1$ in the plane $c \equiv 0$ to yield the bifurcation diagram as given in Figure 6.3.

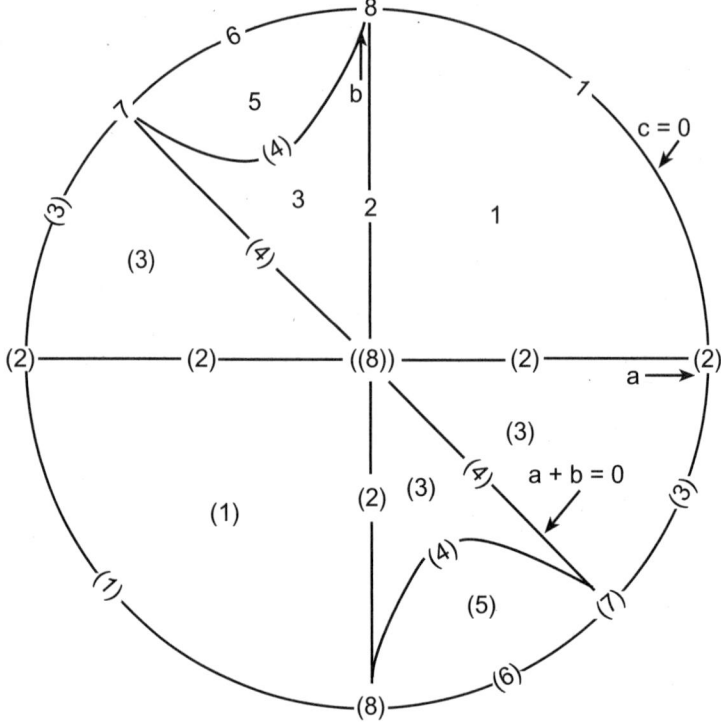

Figure 6.3 Bifurcation diagram for the case with three orbital
lines:$\lambda_3 - \lambda_6 = 0$

Orbital lines are straight lines consisting of one or more straight line orbits, one orbit separated from an adjacent orbit by a critical point. An orbital line connects the diametrically opposite points, that represent an infinite critical point, whereas an orbital line obviously also is an isoclinic line.

The Lamé equation for the degenerate isoclines of (6.2.3),(6.2.4) reads

$$\lambda[-a\lambda^2 + c\lambda\mu + (a+b)\mu^2] = 0. \tag{6.2.5}$$

The solution $\lambda=0$ yields the degenerate isocline $x(1+by)=0$ containing the orbital line $L_b : 1 + by = 0$. The solutions $\lambda_\pm - \frac{c\pm\sqrt{\triangle}}{2a}\mu_\pm = 0$,-with $\triangle = c^2 + 4a(a+b)$-, yield the degenerate isoclines

$$[x - \frac{c \pm \sqrt{\triangle}}{2a}y][1 - \frac{c \pm \sqrt{\triangle}}{2}x - ay] = 0, \tag{6.2.6}$$

containing the orbital lines $L_{\pm} : 1 - \frac{c \pm \sqrt{\triangle}}{2} x - ay = 0$.

The lines $L_{\pm}$ and $L_{\pm}$ intersect at the critical point $P_a = (0, \frac{1}{a})$, the lines L_b and $L_{\pm}$ at the critical point $P_{\mp} = (\frac{c \mp \sqrt{\triangle}}{-2ab}, -\frac{1}{b})$. The location of the infinite critical points is given by $au^3 + cu^2 - (a + b)u = 0$, the roots of this equation corresponding to the direction of L_b and $L_{\pm}$.

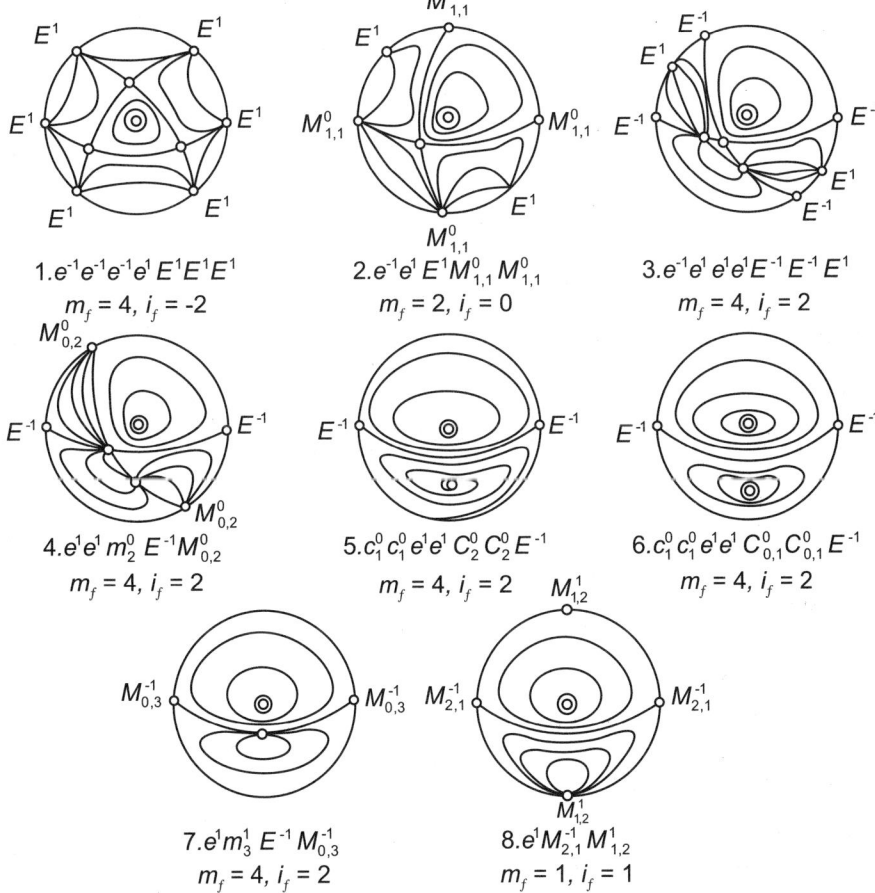

$M^0_{1,1}$

E^1 $\quad$ E^1

E^1 $\quad$ $M^0_{1,1}$ $\quad$ E^1

E^1 $\quad$ E^1

$M^0_{1,1}$

E^1

E^1 $\quad$ $M^0_{1,1}$ $\quad$ E^{-1}

E^1

$M^0_{1,1}$

E^{-1}

E^1

E^{-1} $\quad$ E^{-1}

E^{-1}

1.$e^{-1}e^{-1}e^{-1}e^{1} E^1 E^1 E^1$
$m_f = 4,\ i_f = -2$

2.$e^{-1}e^{1} E^1 M^0_{1,1} M^0_{1,1}$
$m_f = 2,\ i_f = 0$

3.$e^{-1}e^{1} e^{1}e^{1} E^{-1} E^{-1} E^1$
$m_f = 4,\ i_f = 2$

$M^0_{0,2}$

E^{-1} $\quad$ E^{-1}

$M^0_{0,2}$

E^{-1} $\quad$ E^{-1}

E^{-1} $\quad$ E^{-1}

4.$e^{1}e^{1} m^0_2 E^{-1} M^0_{0,2}$
$m_f = 4,\ i_f = 2$

5.$c^0_1 c^0_1 e^{1}e^{1} C^0_2 C^0_2 E^{-1}$
$m_f = 4,\ i_f = 2$

6.$c^0_1 c^0_1 e^{1}e^{1} C^0_{0,1} C^0_{0,1} E^{-1}$
$m_f = 4,\ i_f = 2$

$M^1_{1,2}$

$M^{-1}_{0,3}$ $\quad$ $M^{-1}_{0,3}$ $\quad$ $M^{-1}_{2,1}$ $\quad$ $M^{-1}_{2,1}$

$M^1_{1,2}$

7.$e^{1}m^1_3 E^{-1} M^{-1}_{0,3}$
$m_f = 4,\ i_f = 2$

8.$e^{1}M^{-1}_{2,1} M^1_{1,2}$
$m_f = 1,\ i_f = 1$

Figure 6.4 Phase portraits for the case with three orbital lines:$\lambda_3 - \lambda_6 = 0$.

There exist five phase portraits in addition to the phase portraits in the intersection with the symmetric case.

For $a > 0, b > 0, c < 0$ the phase portrait is given by portrait 1 in Figure 6.4. Since $\Lambda = c^2_{46}$ $c_{45}c_{56} = a^2b^2 < 0$ so $m_f = 4$ and as $c_{45} + c_{56} = -2ab < 0$, then $i_f = -2$. As $\Delta > 0, a + b \neq 0$, all critical points apart from the center point are also elementary and, in fact, saddle points, coinciding with the three intersection points of the three real orbital lines L_b, L_+ and L_-. The three infinite critical points at the ends of these lines are elementary nodes; elementary since there exist three infinite critical points and $m_f = 4$ and nodes since $i_f = -2$ so $i_i = 3$.

If $a = 0, b > 0, c < 0$ the phase portrait is given by portrait 2 in Figure 6.4. There exist two real finite orbital lines: $L_b : 1 + by = 0$ and $L_- : 1 - cx = 0$ intersecting at the saddle point $P_+ = (\frac{1}{c}, -\frac{1}{b})$. The critical point P_-, having gone to the ends of the x axis to yield the point $M_{1,1}^0$, and the critical point P_a, having gone to the ends of the y axis to yield another point $M_{1,1}^0$ there, are located on the orbital line L_b which becomes an orbital line at infinity if a$\rightarrow$ 0. We have $m_f = 2$ and $i_f = 0$. Obviously the infinite critical points are located in $u = 0, u = \frac{b}{c}$ and $u = \infty(v = 0)$.

If $a < 0, 0 < -a < b < -a - \frac{c^2}{4a}, c < 0$ the phase portrait is given in portrait 3 of Figure 6.4. Since A$=-a^2b^2 < 0$ then $m_f = 4$, whereas $c_{45} + c_{56} = -2ab > 0$ then $i_f = 2$. There exist three real different finite orbital lines, yielding elementary critical points at their intersections, which then should be a saddle point plus two nodes. Since $\Lambda(0, \frac{1}{a}) = -\frac{a+b}{a} > 0$, P_a is a node, so on L_b there exist a saddle and a node, their relative location given by the theorem of Berlinski. At infinity there exist two elementary saddles and one node; their character can be determined from the positions of L_b, L_+ and L_- with respect to the center point.

If $a < 0, b > 0, c < 0, \Delta = 0$ the phase portrait is given by portrait 4 in Figure 6.4. Seen as a limit from portrait 3 also follows $m_f = 4, i_f = 2$. The orbital lines L_+ and L_- now coincide and a saddle node is formed in $(\frac{c}{-2ab}, -\frac{1}{b})$, the point where P_+ and P_- coincide; P_a is still a node. The coincidence of L_+ and L_- also leads to the second saddle node $M_{0,2}^0$ at infinity.

If $a < 0, -a - \frac{c^2}{4a} < b, c < 0$ it follows that $\Delta < 0$ and L_+ and L_- are complex orbital lines intersecting at $P_a = (0, \frac{1}{a})$ which is now a focus. The critical points P_+ and P_- are complex as are the two infinite critical points at the ends of L_+ and L_-. The third orbital line is L_b. The phase portrait is given as portrait 5 in Figure 6.4.

For the other values of a and $b, c \neq 0$ apart from $(a, b) = (0, 0)$, similar arguments yield topological equivalent phase portraits. In fact, there exists the relation 1..5 $\leftrightarrow$ (1)..(5).

For $c=0$ the system is symmetric and the phase portraits are as in Figure 6.2; in the bifurcation diagram in Figure 6.1 they are indicated by $a + b=0$.

In phase portrait 7 the three orbital lines L_b, L_+ and L_- coincide as well as the critical points P_a, P_+ and P_- whereas the infinite critical points coincide in $M_{0,3}^{-1}$. In portrait 8, P_a has gone to the ends of the y axis and P_+ and P_- to the ends of the x axis, bringing L_+ and L_- together as an orbital line at infinity. The third orbital line is L_b. In the intermediate phase portrait 6, L_+ and L_- are complex and L_b is real. Finally, the phase portrait ((8)) for $a = b = 0, c = -1$ can be obtained by the transformation $x \rightarrow y, y \rightarrow -x$.

The case with three orbital lines adds five phase portraits to the 38 phase portraits already known from the symmetric case. (In fact only portraits 4 and 5 are topologically new.) The intersection with the symmetric case occurs for $c=0$; that with the Hamiltonian case for $b = 2a, c = 0$.

The solutions of (6.2.3),(6.2.4) are given by Schlomiuk [93,S1],[93,S2].

6.2.3 The Hamiltonian case

System (6.1.10),(6.1.11) can now be written as

$$\dot{x} = -y + ax^2 + 2cxy + by^2, \qquad (6.2.7)$$
$$\dot{y} = x + cx^2 - 2axy - cy^2, \qquad (6.2.8)$$

which again may be analyzed on the unit sphere $a^2 + b^2 + c^2 = 1$ of which the lower half $c \leq 0$ is projected onto the disk $a^2 + b^2 \leq 1$ in the plane $c \equiv 0$ to yield the bifurcation diagram as given in Figure 6.5.

The solutions of (6.2.7),(6.2.8) are given by $H(x, y) = h=$constant, where

$$H(x,y) = -\frac{1}{2}(x^2 + y^2) - \frac{1}{3}cx^3 + ax^2y + cxy^2 + \frac{1}{3}by^3. \qquad (6.2.9)$$

From it follows that the phase portraits are invariant under the transformation $a \rightarrow -a, b \rightarrow -b, y \rightarrow -y$. As a result only the phase portraits for $a + b \geq 0, c \leq 0$ will be considered.

For $c=0$, (6.2.7),(6.2.8) represent a symmetric system and the phase portraits correspond to the curve $2a + c=0$ in the bifurcation diagram as given in Figure 6.1.

For $c \neq 0$ there exist four additional topologically different phase portraits 1...4 as given in Figure 6.6. They can all be regarded as bifurcations from the case $c=0$, in particular from the phase portrait for $a = 0, b = 1, c = 0$.

The case with three orbital lines corresponds to $a+b=0$. If $a+b=0$, we may use a result of Frommer that $a_{20}+b_{02}=b_{20}+b_{02}=0$ is invariant under rotation of the coordinate system around the origin, with the result that $c=0$ may be taken in (6.2.7),(6.2.8) to determine the topological character of the phase portraits on $a+b=0$. The phase portrait is thus given by rotation of the phase portrait 10 for the symmetric case.

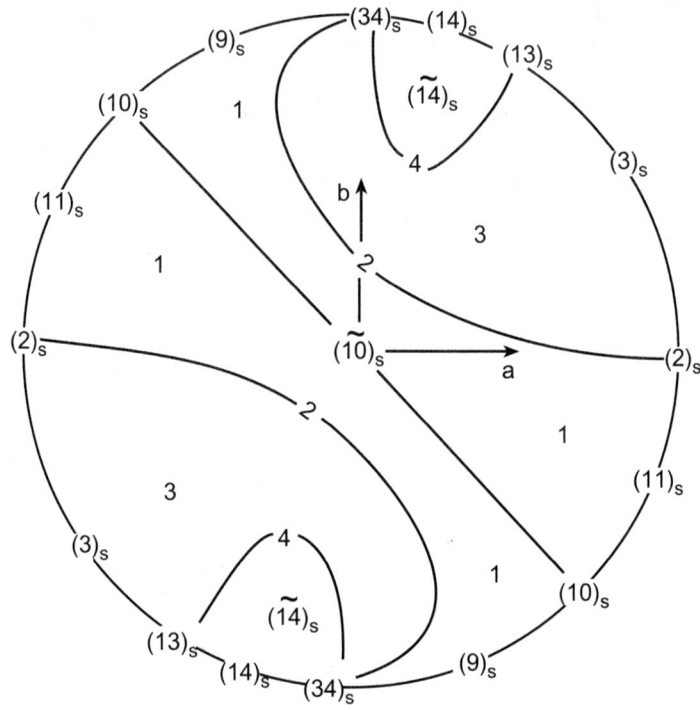

Figure 6.5 Bifurcation diagram for the Hamiltonian case

It follows for (6.2.7),(6.2.8) that

$$A \equiv c_{46}^2 - c_{45}c_{56} = (a+b)^2c^2 + 4(a^2+c^2)(ab-c^2), \tag{6.2.10}$$

so that for $c \neq 0$, there is $A \neq 0$ and $m_f = 4$, except on the curve 2, given by $A=0$. On it $m_f = 3$ since $B_1 = 2c(a+b^2) \neq 0$ and Theorem 2.1 can be applied.

The region in the a,b plane wherein $A < 0$ ($a+b \neq 0$) is indicated by 1. So, within this region $ab - c^2 < 0$ so that $c_{45} + c_{56} = -2(a^2+c^2) + 2(ab-c^2) < 0$

and $i_f = -2$. Phase portrait 1 contains three saddle points and a center point in the finite part of the plane and three nodes at infinity. It may be seen as a bifurcation from either $9_s, 10_s$ or 11_s. Starting from 9_s it may be seen that the orbital line $y \equiv \frac{1}{2a}$ is broken as a result of the increase of $|c|$ away from zero. Moreover from (6.2.9) it may be calculated that the value of h is different at the saddle points, whence saddle connections are not possible. The only boundary to enclose the region of periodic solutions is then a saddle loop as illustrated in phase portrait 1.

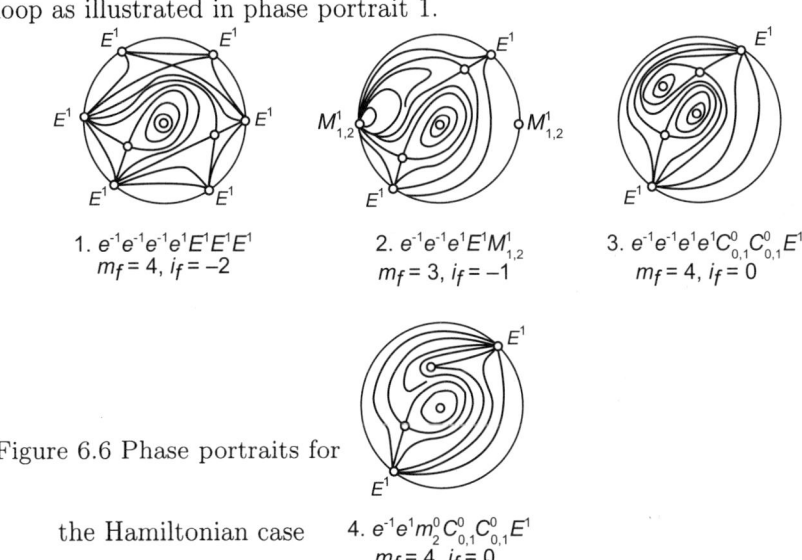

1. $e^{-1}e^{-1}e^{-1}e^1E^1E^1E^1$
   $m_f = 4, i_f = -2$

2. $e^{-1}e^{-1}e^1E^1M_{1,2}^1$
   $m_f = 3, i_f = -1$

3. $e^{-1}e^{-1}e^1e^1C_{0,1}^0 C_{0,1}^0 E^1$
   $m_f = 4, i_f = 0$

Figure 6.6 Phase portraits for

the Hamiltonian case   4. $e^{-1}e^1m_2^0 C_{0,1}^0 C_{0,1}^0 E^1$
   $m_f = 4, i_f = 0$

On the curve 2 there is $m_f=3$, and one of the saddle points in portrait 1 has gone to infinity to yield an point $M_{1,q}^i$ which upon further analysis appears to be a $M_{1,2}^1$ critical point with an elliptic and a hyperbolic sector. Since $i_f = -1$ (as $c_{45} + c_{56} <0$) there exists another infinite critical point being a node E^1. The bifurcation from 2_s to 2 through the break of the straight saddle connection may also be seen to yield phase portrait 2.

In region 3, there is A>0 so $m_f=4$ and $i_f=0$. Regarded as bifurcation from 3_s it may be seen that two saddle points and two antisaddles prevail after bifurcation, the latter remaining center points. There exist two saddle loops in phase portrait 3 as a result of the breaking up of the straight orbit connecting the saddle points in $(3)_s$. Region 3 is bounded; apart from by curve 2 where one saddle point has gone to infinity, and by case $(3)_s$ on c≡0, also by curve 4 where a saddle point and a center point coincide to form a cusp. Curve 4 is given by

$$-4ab^3 - 9abc^2 + 24a^2b^2 + 117a^2c^2 - 48a^3b + 32a^4 + 13b^2c^2 + 108c^4 = 0, \quad (6.2.11)$$

with $a^2 + b^2 + c^2 = 1$. The phase portrait corresponding to this curve is given in portrait 4. Upon crossing curve 4 the cusp disappears and two complex critical points are generated from it yielding a phase portrait topologically equivalent to $(14)_s$.

The statements about the phase portraits are confirmed by the analysis of Vulpe in [83,V]; see also Artes and Llibre [94,AL1],[96,AL].

6.2.4 The lonely case

$$\lambda_5 = 5\lambda_3 + \lambda_4 - 5\lambda_6 = \lambda_2^2 - \lambda_3\lambda_6 + 2\lambda_6^2 = 0.$$

We rewrite (6.1.10),(6.1.11) as

$$\dot{x} = -y - ax^2 + 2cxy - by^2, \quad (6.2.12)$$
$$\dot{y} = x + cx^2 - (3a + 5b)xy - cy^2, \quad (6.2.13)$$

with $c^2 + ab + 2b^2 = 0$. This system will again be analyzed on the unit sphere $a^2 + b^2 + c^2 = 1$ of which the lower half $c \leq 0$ is projected onto the disk $a^2 + b^2 \leq 1$ in the plane $c \equiv 0$. The analysis is then restricted to the hyperbola given by $-a^2 + ab + b^2 + 1 = 0$ on the open disk. It meets $c \equiv 0$ in the points $(a, b) = (\pm\frac{2}{5}\sqrt{5}, \mp\frac{1}{5}\sqrt{5})$ where phase portrait $(24)_s$ applies and the points $(a, b) = (\pm 1, 0)$ where phase portrait $(6)_s$ applies. See Figure 6.7.

If $c \neq 0$, then $b \neq 0$ and since $A = -2\frac{b^2+c^2}{b^2} < 0$ so $m_f = 4$, whereas $c_{45} + c_{56} = \frac{3(b^2+c^2)^2}{b^2} > 0$ yields $i_f = 2$. For the Lamé equation of (6.2.12),(6.2.13) follows

$$a\lambda^3 - 2c\lambda^2\mu + 3(a + 2b)\lambda\mu^2 + c\mu^3 = 0. \quad (6.2.14)$$

Since $a \neq 0$ we may use that $F = [-\frac{1}{216}c^3 + \frac{1}{32}ac(a+2b) + \frac{1}{128}ca^2]^2 + [\frac{1}{16}a(a+2b) - \frac{1}{36}c^2]^3 > 0$, as $a(a+2b) - \frac{4}{9}c^2 = (a+2b)(a+\frac{4}{9}b) > 0$. As a result (6.2.14) has two complex solutions plus one real solution and (6.2.12),(6.2.13) has two complex finite critical points and an antisaddle apart from the center point. The antisaddle appears to be a node as may be concluded from the fact that the parabola $\lambda + 2\bar{y} + (\bar{x} - \bar{y})^2 = 0$ and the cubic curve $\lambda + 3\bar{y} - 3\bar{x}\bar{y} + 3\bar{y}^2 - (\bar{x} - \bar{y})^3 = 0$ ($\lambda = \frac{b^2}{b^2+c^2}, \bar{x} = cx, \bar{y} = by$) intersect and both curves consist of orbits. At infinity there exist one node and two saddle points. A discussion may further be found in [93,S1].

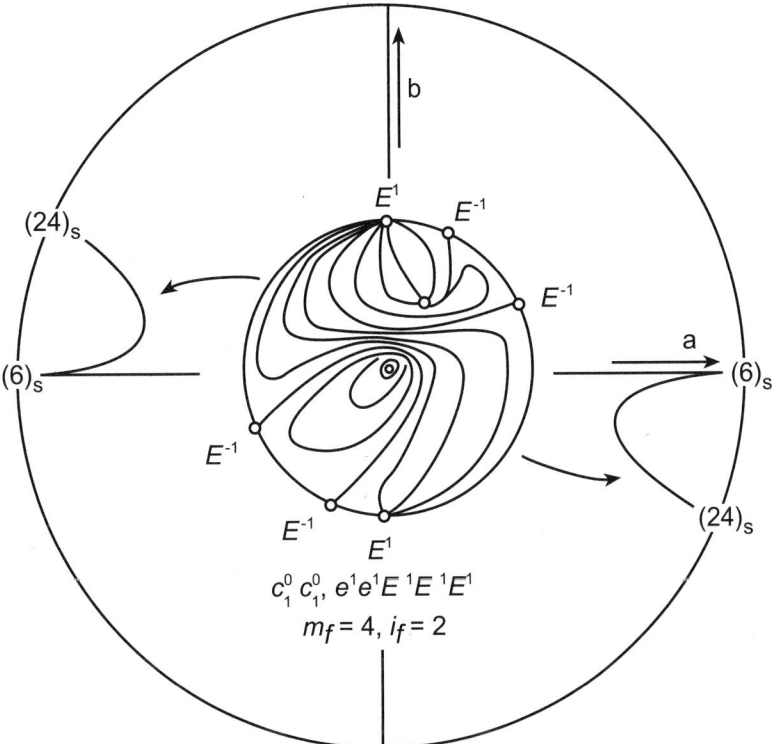

Figure 6.7 Bifurcation diagram and phase portrait for the lonely case

6.2.5 Theorem 6.1

The phase portraits of quadratic systems with a center point can be found by analyzing systems containing two (or one) parameter(s). There exist four cases: (i) the symmetric case, with 38 phase portraits, (ii) the case with three orbital lines, having an additional five phase portraits, (iii) the Hamiltonian case, with four additional phase portraits, and (iv) the lonely case, giving one extra phase portrait. This yields 48 phase portraits all together (among which some are topologically equivalent).

6.3 Invariant theory for quadratic systems with center points

As was mentioned above, the first formulation of the center conditions for real quadratic systems by Kapteyn [11,K] in 1911 contained some inaccuracies. The next paper [34,F] dealing with this subject appeared in 1934 by Frommer, which, unfortunately, also contained some errors. Additional errors in the subsequent literature based on these references suggest that the reader should be warned for these pitfalls [90,SGR]. After that the correct center conditions for (6.1.1),(6.1.2) , satisfying the condition $b_{20} + b_{02}=0$ were obtained by Bautin, this condition was lifted in 1948 by Sakharnikov [48,S], in 1954 by Belustina and by Sibirskii [54,S] who gave the most compact form for these conditions; they can be found in the survey paper on quadratic systems by Coppel in 1966 [66,C]. Further progress in the direction of formulating the center conditions for general quadratic systems was obtained in 1963 by Sibirskii [63,S1] assuming only that the origin is a critical point. Formulation of these conditions in terms of centro-affine invariants were given later by Sibirskii and can be found in his book in 1976 [76,S].

Obviously the center conditions become more complicated if more coefficients are left free as parameters in the system. In fact, as a quadratic system has a center point if it can be transformed by a linear transformation into (6.1.10),(6.1.11) satisfying one of the stated conditions on $\lambda_i(i=1,..6)$, then the corresponding partition of the coefficient space can be found by applying the affine (linear) transformation

$$\bar{x} = \alpha_0 + \alpha_1 x + \alpha_2 y, \qquad (6.3.1)$$
$$\bar{y} = \beta_0 + \beta_1 x + \beta_2 y, \qquad (6.3.2)$$

with arbitrary values of α_i, β_i (i=0,1,2) to (6.1.10),(6.1.11), taking the mentioned conditions on λ_i into account. For the new coefficients a_{ij}, b_{ij} (i,j=0,1,2) then follow 12 nonlinear equations , which upon elimination of the α_i and β_i yield the sought relations between the coefficients giving the wanted partition. That a quadratic system contains a center point is a topological property that is invariant under an affine transformation as, in fact, phase portraits are topologically equivalent under such transformations. For a partition of coefficient space according to topologically different phase portraits a more suitable way to structurize this coefficient space therefore seems to use combinations of these coefficients that in some sense are invariant under these transformations. A distinction is made between affine ($\alpha_0^2 + \beta_0^2 \neq 0$) and centro-affine ($\alpha_0^2 + \beta_0^2 = 0$) transformations ; correspondingly between affine and centro-affine invariants.

In [89,BVS] Bularas, Vulpe and Sibirskii obtained the necessary and sufficient conditions for a general quadratic system to have (one or two) center point(s). These conditions are expressed in affine relations between centro-affine invariants. An improvement of this result was given by Voldman and Vulpe [97,VV] by using affine invariants.

Although phase portraits of quadratic systems with center points already appear in the papers of Büchel[04,B], Frommer [34,F], Bautin [39,B1], Latipov and Shirov [63,LS] the first systematic investigation to obtain all phase portraits appeared in 1964 in the paper by Kukles and Khasanova [64,KK]. They found 32 phase portraits at variance with the 48 phase portraits mentioned in Theorem 6.1 as a result of a slightly different definition of topological equivalence of phase portraits. A similar study was made by Lukashevich [65,L]. For a quadratic system with a critical point in the origin (system (1.1),(1.2)with $a_{00}^2 + b_{00}^2 \neq 0$) by Kalin and Sibirskii [90,KS], [92,K]. The classification of the phase portraits given in the present book in terms of affine invariants has yet to be given.

Expressions for the solutions of quadratic systems with a center point in the origin using centro- invariants and commitants were given by Lunkevich and Sibirskii [82,LS].

Invariant theory is also used to indicate the regions in coefficient space corresponding to quadratic systems with isochronic center points, being points surrounded by periodic solutions all having the same period. The conditions on (6.1.1),(6.1.2) to have an isochronic center point in (0,0) were derived by Loud [64,L]. A subsequent formulation of these conditions in terms of invariants was given by Plechkan and Sibirskii [71,PS] and Sibirskii [76,S] for quadratic systems with a critical point at the origin and for general quadratic systems by Bularas [88,B]. The phase portraits corresponding to isochronic center points can be found in the symmetric case : in the portraits 5 and 6 in Figure 6.2a,portrait 25 in Figure 6.2c and portrait 27 in Figure 6.2d.

6.4 Limit cycles and separatrix cycles

As stated in section 6.1 the analysis of phase portraits for $m_f > 0$ induces the possibility of periodic solutions, because of the presence of critical points in such systems. In the previous sections closed orbits were considered, imbedded in a region completely filled with such orbits, surrounding precisely one common center point. As the phase portraits show, the outer boundary of a region of closed orbits is a separatrix cycle, being a closed curve consisting of a finite number of critical points connected by orbits on this curve. It might be considered as a limit set of the periodic solutions in its interior, the period having gone to infinity.

It is of interest to collect the various types of these cycles and complement them later with those present in quadratic systems without center points, separatrix cycles being important elements in phase portraits.

Another question of interest is concerned with the possibility of a simultaneous occurrence in a quadratic system with a center point having also a limit cycle. The answer appears to be negative, at variance with that for polynomial systems with degree higher than 2.

6.4.1 Separatrix cycles in quadratic systems with center points

Apart from the information obtainable from an inspection of the phase portraits given in section 6.2, the types of regions surrounding a center point and containing all the closed orbits around that point, can also be studied directly, as was done by Conti[87,Co]. It then can be shown that the outer boundary of a region of closed orbits surrounding a center point cannot be another closed orbit but consists of (an) orbit(s) connecting (at least one) critical point(s) either at infinity or in the finite part of the plane. In all cases these orbits are the separatrices that form in all critical points on the cycle the boundary of a hyperbolic sector situated in the interior of the cycle.

The following types of separatrix cycles can be observed in quadratic systems with a center point. See Figure 6.8.

1.Bounded separatrix cycles.

1.1.Loops. They consist of one separatrix connecting a finite critical point with itself, the critical point being either (i) an elementary saddle point e^{-1}, (ii) a nilpotent point with an elliptic and a hyperbolic sector m_3^1 or (iii) a nilpotent saddle point m_3^{-1}. Examples are : (i) portrait 7,8 and 9 in Figure 6.2a, (ii) portrait 17 in Figure 6.2b , and (iii) portrait 13 in Figure 6.2b.

1.2. Twogons. They have two finite critical points, being elementary saddle points e^{-1} connected by a straight line orbit and a curved orbit. Examples are: 2, 3 and 11 in Figure 6.2a.

1.3. Triangles. They have three finite critical points, being elementary saddle points e^{-1}, the straight line orbits connecting the saddle points forming a triangle. Example: Portrait 10 in Figure 6.2a.

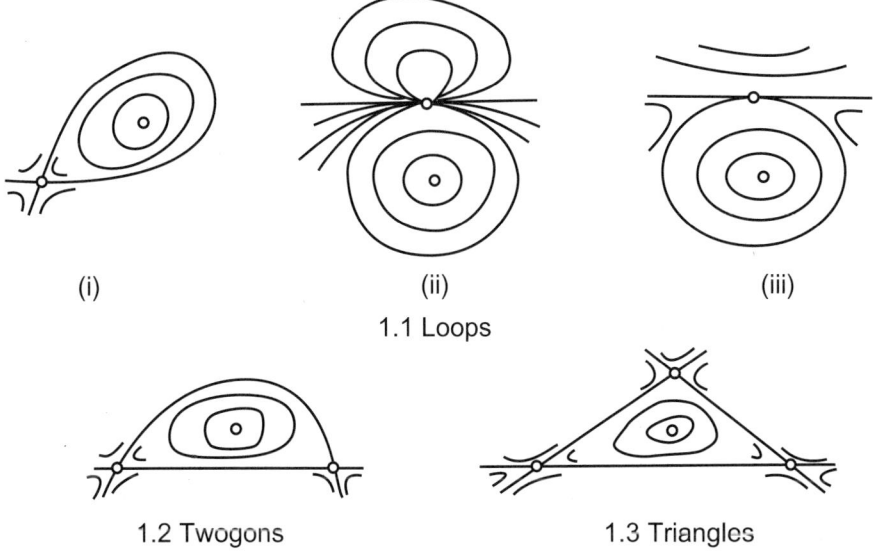

1.1 Loops

1.2 Twogons   1.3 Triangles

Figure 6.8 Separatrix cycles in quadratic systems with a center point
1. Bounded separatrix cycles

2. Unbounded separatrix cycles.

2.1. Without finite critical points.

2.1.1. Loops. They consist of one separatrix connecting an infinite critical point with itself. The infinite critical point is either (i)a nilpotent critical point with an elliptic and a hyperbolic sector $M_{1,2}^1$, (ii) a nilpotent saddle point of third order $M_{1,2}^{-1}$ or (iii) of fifth order $M_{3,2}^{-1}$ or (iv) one of the points on the Poincaré circle filled with critical points. Examples are: (i) phase portrait 28 in Figure 6.2d, (ii) portrait 29 in Figure 6.2d, (iii) portrait 32 in Figure 6.2e, (iv) 27 in Figure 6.2d.

2.1.2 Twogons. They have two corner points on the Poincaré circle, being infinite critical points. One side of the twogon coincides with part of the Poincaré circle. If these corner points are elementary saddle points E^{-1}, they are also connected by a finite (i) straight or (ii) curved orbit. If they are semi-elementary, then there exist the possibilities: (iii) saddle nodes $M_{1,1}^0$

which are also connected by a finite curved orbit , or they are (iv) tangentially or (v) transversally non-hyperbolic saddle points $M_{0,3}^{-1}$ or $M_{2,1}^{-1}$, respectively; they are also connected by a straight finite orbit. Moreover, there exist the possibility (vi) that the corner points are third order saddle points $M_{2,1}^{-1}$, connected in the finite part of the plane by a straight orbit. On the part of the Poincaré circle bordering the cycle there exists a nilpotent critical point $M_{1,2}^{1}$ having an elliptic and a hyperbolic sector; the latter being located in the interior of the cycle. Examples: (i) portrait 5 in Figure 6.2a, (ii) portrait 25 in Figure 6.2c, (iii) portrait 37 in Figure 6.2 e, (iv) portrait 18 in Figure 6.2c and (v) portrait 1 in Figure 6.2a. Moreover for case (vi) an example can be found in phase portrait 8 of Figure 6.4.

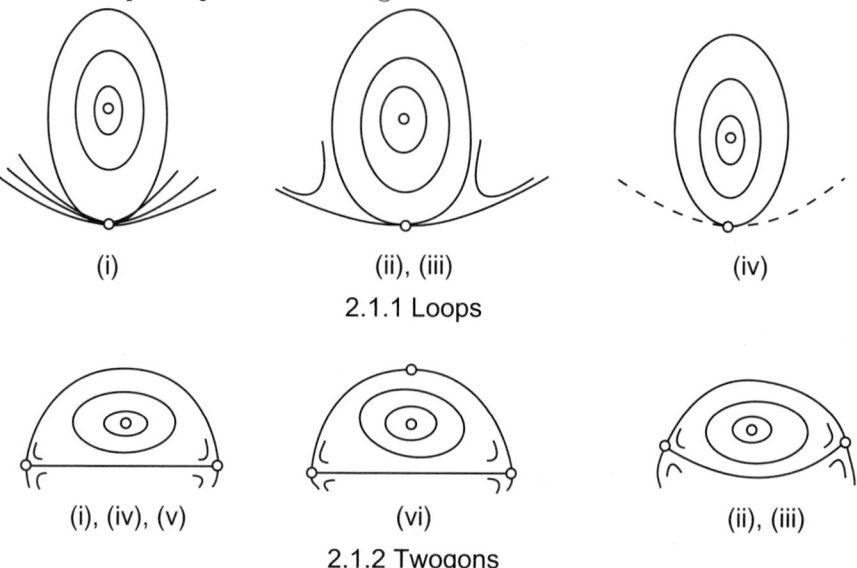

(i)   (ii), (iii)   (iv)

2.1.1 Loops

(i), (iv), (v)   (vi)   (ii), (iii)

2.1.2 Twogons

Figure 6.8 Separatrix cycles in quadratic systems with a center point

2. Unbounded separatrix cycles.

2.1 Cycles without finite critical points.

2.2 Cycles with one finite critical point.

2.2.1 Wedges. They consist of two straight finite orbits making an angle $\alpha < \pi$ each tending to an infinite critical point connected by the intermediate part of the Poincaré circle. There exist three types: (i)$e^{-1}E^{-1}E^{-1}$, (ii)$e^{-1}M_{1,1}^{0}M_{1,1}^{0}$ and (iii)$m_{2}^{0}E^{-1}M_{0,2}^{0}$. Examples are: (i) portrait 3 in Figure 6.4, (ii) portrait 36 in Figure 6.2e, (iii) portrait 4 in Figure 6.4.

2.2.2 Halfcircles. They consist of half of the Poincaré circle plus the straight line finite orbit connecting the corner points at infinity. There exist the types: (i)$m_3^1 E^{-1} E^{-1}$ and (ii)$m_3^1 M_{0,3}^{-1} M_{0,3}^{-1}$. Examples are: portraits 18 and 19 in Figure 6.2c, respectively.

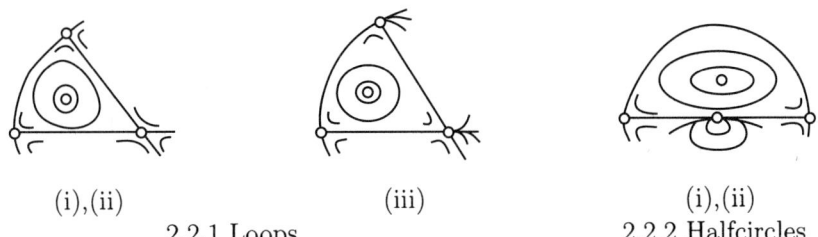

| (i),(ii) | (iii) | (i),(ii) |
| 2.2.1 Loops | | 2.2.2 Halfcircles |

Figure 6.9 Separatrix cycles in quadratic systems with a center point
2.Unbounded separatrix cycles
2.2 Cycles with one finite critical point

6.4.2 Limit cycles and center points

Theorem 6.2 *A quadratic system cannot have both a center point and a limit cycle. (Lukashevich [65,L]*

Proof. According to Theorem 4.5 the interior of a closed orbit contains precisely one critical point and this point is either a center point or a focus. From the previous sections it follows that a center point is surrounded by a region completely filled with closed orbits. This leaves the focus as the only critical point possible in the interior of a limit cycle.

The only phase portrait with a center point also containing a focus is that given in portrait 5 (5) of Figure 6.4 and corresponds to

$$\dot{x} = -y - ax^2 + cxy + ay^2 \equiv P_c(x, y), \qquad (6.4.1)$$
$$\dot{y} = x + bxy \equiv Q_c(x, y), \qquad (6.4.2)$$

where $c < 0, \Delta \equiv c^2 + a(a + b) < 0, a^2 + b^2 + c^2 = 1$.

It has a focus in $P(0, \frac{1}{a})$, stable (unstable) if $\frac{c}{a} < 0 (> 0)$. For $c = 0$, P is a center point surrounded by closed orbits of the system given by the vector field $(P_0(x, y), Q_0(x, y)) = (-y - ax^2 + ay^2, x + bxy)$. Outwards pointing vectors normal to these orbits are given by $(-Q_0(x, y), P_0(x, y))$. In phase portrait 5 there is $a < 0, b > 0$, then the outward flow of (P_c, Q_c) across the closed orbits of $(P_0(x, y), Q_0(x, y))$ is given by $P_c Q_0 - P_0 Q_c = cr^2 y(1 + by) < 0$ in the region below the orbit $y \equiv -\frac{1}{b} < 0$ wherein the focus is located. The orbits of (P_c, Q_c) cannot return upon themselves. So, no closed orbits and in particular no limit cycles are possible.

Chapter 7

Limit cycles in quadratic systems

7.1 Introduction

The other type of periodic orbits apart from the type encountered in the previous chapter on quadratic systems with a center point is the limit cycle; its introduction into mathematics goes back to Poincaré [P]. The limit cycle problem is considerably more difficult to solve than the center problem. A limited number of papers over the twentieth century was sufficient to clarify the center problem. However, by the end of that century, despite a multitude of papers, the limit cycle problem for quadratic systems is still left with quite a number of unsolved questions, some of them with poor hope for answers on the short run. This is due to the limitations of available methods of investigation and, in some questions, the pure absence of them. For that reason, it is difficult to give a well structured presentation of what is known about limit cycles in quadratic systems, illustrating at the same time that the underlying structure of this non linear phenomenon is not yet well understood.

It is the aim of the present chapter to give a sketch of what is known so far about limit cycles in quadratic systems in order to explore that information in subsequent chapters when considering classes with finite multiplicity $m_f \geq 1$. A more detailed discussion will be avoided and postponed if it may be useful in these later chapters and more practical to discuss it there.

Statements about limit cycles in quadratic systems can roughly be divided in those valid for all quadratic systems and those restricted to classes characterized by a particular property.

7.2   General remarks on limit cycles, quadratic systems

A limit cycle is a periodic orbit isolated in the sense that it is embedded in an
annular region, wherein it is the only periodic orbit; the other orbits in the
annulus approach the limit cycle either for $t \to -\infty$ or $\to +\infty$. An example
of a quadratic system containing a limit cycle was already encountered in
the system given by (1.0.15),(1.0.16) of which the phase portrait is given in
Figure 1.6. In that example all orbits in an annular region around the limit
cycle approach it for $t \to \infty$, as a result of which the limit cycle is *stable* both
from the inside and from the outside. Replacing (P, Q) by $(-P, -Q)$ makes
the orbits approach for $t \to -\infty$ and the limit cycle is *unstable* on both sides.
Limit cycles may also have a different stability on, in and outside; these are
semi-stable .

A representation of the behavior of the orbits near a limit cycle can be
obtained from the *Poincaré return map*. In order to illustrate this point we
introduce curvilinear coordinates s and n in a neighborhood of a limit cycle
L, as defined by the relations

$$x \;=\; \varphi(s) + n\psi'(s), \tag{7.2.1}$$
$$y \;=\; \psi(s) - n\varphi'(s), \tag{7.2.2}$$

where the limit cycle L, being given by $x = \varphi(t), y = \psi(t)$ is now given by
$x = \varphi(s), y = \psi(s)$ so corresponds to $n \equiv 0$ in the new coordinate system
[ALGM2]. See Figure 7.1

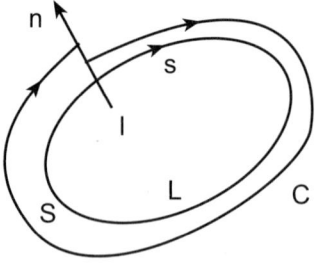

Figure 7.1 Flow near a limit cycle

Introducing these new coordinates in (1.0.1),(1.0.2) yields the equation

$$n'(s) = R(s, n), \qquad (7.2.3)$$

of which the solutions can be written as $n = f(s : s_0, n_0)$ with $n_0 = f(s_0; s_0, n_0)$. Choosing an arbitrary point P on the limit cycle L and letting P correspond to $s=0$, allows these solutions to be written as $n = f(s; 0, n_0)$. For (7.2.3) this means in geometrical terms, that every path of the system passing through a point in a sufficiently small neighborhood of L crosses the normal l (for $s = 0$) and also all other normals to L for all other values of s (on L coinciding with the values of t). Let the (smallest) period of the flow on L be indicated by τ, then $\varphi(0) = \varphi(\tau), \psi(0) = \psi(\tau)$. The Poincaré return map is then the mapping of the point $(s, n) = (0, n_0)$ onto the point $(s, n) = (\tau, f(\tau; 0, n_0)$ both located on the normal l through P on L. If we write $f(\tau; 0, n_0) = f(n_0)$, then this mapping is given by $n_0 \to f(n_0)$ on the normal l. Obviously, any closed orbit satisfies $f(n_0) = n_0$, so $f(0) = 0$. Together with the Poincaré return map consider the distance function $d(n_0) = f(n_0) - n_0$; then $d(n_0)=0$ represents a closed orbit through $(s, n) = (0, n_0)$ and in particular $d(0) = 0$ for the limit cycle L. If $d'(0) < 0$ then the limit cycle is stable, if $d'(0) > 0$ unstable. Moreover if $d'(0) \neq 0$, L is *hyperbolic* implying that small perturbations in the coefficients of the system leave precisely one limit cycle in a neighborhood of L, leaving also the stability unchanged. For $d'(0)$ then may be derived the expression $d'(0) = e^k - 1$, where the *characteristic exponent* k is given by [ALGM2]

$$k = \int_0^\tau [P_x(\varphi(t), \psi(t)) + Q_y(\varphi(t), \psi(t))]dt. \qquad (7.2.4)$$

If $d'(0)=0$, the limit cycle is *non-hyperbolic* or of *higher multiplicity*. In fact a limit cycle is of *multiplicity n* if $d'(0) = d''(0) = \cdots = d^{n-1}(0)=0, d^n(0) \neq 0$, where n is a natural number [ALGM2]. In quadratic systems, limit cycles with multiplicity up to 3 three have been encountered and there seems a general belief among research workers in this field that *the multiplicity of a limit cycle in a quadratic system is at most 3*, although it is difficult to find a statement of this conjecture in the literature [97,G]. It can hardly be expected that a proof of this conjecture will be given soon. What is known, however, is that the *multiplicity of a limit cycle* in a polynomial (thus also a *quadratic)system* is *finite*. The conditions $d'(0) = d''(0) = \cdots = d^{n-1}(0)=0, d^n(0) \neq 0$ with n$\to \infty$ would indeed impose restrictions on the finite number of coefficients in the system that are difficult to realize unless the periodic orbit is imbed ded in a family of periodic orbits around a center point and $d(n_0) \equiv 0$. The situation is analogous to the center problem, being global instead of local.

Through small perturbations of the coefficients a_{ij}, b_{ij} in system (1.0.1), (1.0.2) a multiple limit cycle can split up in a number of limit cycles at most equal to its multiplicity. Also limit cycles may merge into a multiple limit cycle, whereby an even number of them with pairwise opposite stability may disappear upon confluence, in the sense that this lowers the multiplicity of the limit cycle.

Some general properties of limit cycles in quadratic systems may directly be deduced from those valid for all closed orbits in these systems, as discussed in section 4.3.2 of Chapter 4. So a limit cycle is a convex curve enclosing a region, wherein precisely one critical point exists (Theorems 4.4,4.5). According to Theorem 4.5 this point must be a focus, since it cannot be a center point, as follows from Chapter 6. Moreover, the motion along a limit cycle and along the orbits in the interior enclosed by that limit cycle are either both in clockwise or both in anti-clockwise direction (Theorem 4.6). Also, two limit cycles are oppositely oriented if external to each other and similarly oriented if one is inside the other (Theorem 4.7), whereas there exist at most two nests of limit cycles in a quadratic system (Theorem 4.8).

A natural question that poses itself is to ask how many limit cycles a nest of limit cycles can contain, or more generally; what is the maximum number of limit cycles in quadratic systems. This question is dealt with, in a more general form, in the second part of the 16th Hilbert problem, as stated by Hilbert in his address to the International Congress of Mathematicians in Paris in 1900 [H].

Hilbert's 16th problem, second part.

Find the maximum number and relative position of the limit cycles of a polynomial vector field of degree $\leq n$

$$\dot{x} \;=\; \sum_{i+j=0}^{n} a_{ij} x^i y^j, \tag{7.2.5}$$

$$\dot{y} \;=\; \sum_{i+j=0}^{n} b_{ij} x^i y^j, \tag{7.2.6}$$

and let $H(n)$ be that number.

After a century has passed the problem has not been solved yet, even not for the simplest case of quadratic systems (n=2). As the history on the subject reveals, however, many results point in the direction of H(2)=4 being probably the correct answer.

In 1939 Bautin proved that H(2)$\geq$3 by constructing a system with a nest of three limit cycles surrounding a focus [39,B1]. In the same paper Bautin also proved that 3 is the maximum number of limit cycles that can appear from a focus or center point by variation of the coefficients in the system. The full proof of this result appeared in 1952.

A direct attempt to solve the 16th Hilbert problem was undertaken in the second half of the 1950s by Petrovsky and Landis [55,PL], using ideas from algebraic geometry and allowing the variables to be complex. Unfortunately, their proofs contained errors that could not be removed, what lead to withdrawal of them [59,PL]. As follows from the subsequent papers on the subject, however, their result, being H(2)=3, was accepted for quite some time. In fact, it took more than twenty years thereafter until examples were given of quadratic systems with (at least) four limit cycles, which lead to the idea that H(2)$\geq$ 4. These examples involved two nests of limit cycles with (at least) one limit cycle around one focus and (at least) three limit cycles around the other focus. The first examples of these configurations were published in 1979,1980 by Chen Lansun and Wang Mingshu [79,CW] and Shi Songling [80,S1]. There followed several other examples by various authors. As follows from a discussion to be given later in this chapter all the examples discussed there are in the class with finite multiplicity m_f=4. Now, trivially, for m_f=0 there are no limit cycles, whereas for m_f=1 it can be shown that there exists at most one limit cycle [97,R]. There is, moreover, strong evidence that 2 is the maximum number of limit cycles in the class m_f=2 [97,RK] and 3 in the class m_f=3 [97,RH],[96,H]. Obviously there exist no limit cycles in the class $m_f = \infty$. As a result there may be conjectured [97,R]:

Conjecture 1. *The number of limit cycles in a quadratic system with finite multiplicity m_f is at most equal to m_f, and systems exist wherein this number is attained.*

This conjecture includes the general conviction that H(2)=4.

The problem of determining H(2) still being open, yet another question came up. In the paper [23,D] published in 1923, Dulac stated the result that a given polynomial (so also a quadratic) system can only have a finite number of limit cycles. It was realized in the 1970s, however, that the proof of this statement contained a flaw. In the late 1980s the claim of Dulac was finally proved, independently, by Ecalle,Martinct,Moussu and Ramis and by

Il'yashenko [91,I]. Their proofs are very elaborate as they deal with arbitrary n. Using the special properties of quadratic systems, a more elementary proof was given in 1986 by Bamon [86,B], which was later still more simplified by Coppel [89,C].

Knowing that a given quadratic system only has a finite number of limit cycles does not guarantee, however, that H(n) exists, i.e. that H(n) is a finite number. It cannot a priori be excluded that there exist a sequence of points in the space of coefficients such that the number of limit cycles increases indefinitely while approaching the limit point of such a sequence. In that case, the limit point would represent a *limit periodic set*, which, itself not being a limit cycle, upon vanishing small perturbations would generate an infinite number of limit cycles. In 1991 Dumortier, Roussarie and Rousseau presented a program to prove that H(2) exists by investigating all possible limit periodic sets in quadratic systems and determining the number of limit cycles each of them can generate [94,DRR1].

As a result of the availability of methods to prove existence, non-existence or uniqueness of limit cycles, many results on limit cycles in quadratic systems are concerned with these questions. These methods are applied to classes, determined by a specific property, which leads to studying specific vector fields or to a general form that includes all quadratic systems which possibly might contain limit cycles.

Such a form was given, i.e., in 1958 by Ye Yanqian [58,Y2] and studied extensively since, in China [65,Y] and also in the rest of the world. From Chapter3 it follows that there always exists a real degenerate isocline through a real critical point, which if this point is located in the origin can be written in the form

$$\lambda P(x, y) + \mu Q(x, y) = (\alpha x + \beta y)(\gamma + \delta x + \varepsilon y) = 0. \qquad (7.2.7)$$

As we want (0,0) to be a focus, then $\alpha\mu - \beta\lambda \neq 0$, since otherwise $\alpha x + \beta y = 0$ would be an orbit through (0,0). The non-singular transformation $\bar{x} = \alpha x + \beta y, \bar{y} = \lambda x + \mu y$ then yields, after scaling of x, y and t, that a quadratic system with a focus or a center point in the origin can be written in the form

$$\dot{x} = dx - y + lx^2 + mxy + ny^2, \qquad (7.2.8)$$
$$\dot{y} = x(1 + ax + by), \qquad (7.2.9)$$

where $-2 < d < 2$ should be imposed to assure that (0,0) is a focus or a center point.

Three classes are defined: I.a=b=0, II.a≠0,b=0, III.b≠0. In view of the numerous results on limit cycles obtained using (7.2.8),(7.2.9), it is important to establish the relation between these classes and those defined in Chapter 3. This is done in Table 7.1 making use of the theory developed in Chapters 2 and 3.

Class I

$$\dot{x} = dx - y + lx^2 + mxy + ny^2, \tag{7.2.10}$$
$$\dot{y} = x. \tag{7.2.11}$$

$$(1)n \neq 0, l \neq 0 (m_f = 2, i_f = 0)$$
$$m^2 - 4ln < 0 : e^{-1}e^1 C_{0,1}^0 C_{0,1}^0 E^1$$
$$m^2 - 4ln = 0 : e^{-1}e^1 E^1 M_{0,2}^0$$
$$m^2 - 4ln > 0 : e^{-1}e^1 E^1 M_{1,1}^0 M_{1,1}^0$$
$$(2)n \neq 0, l = 0 (m_f = 2, i_f = 0)$$
$$m \neq 0 : e^{-1}e^1 M_{1,1}^0 M_{1,2}^1$$
$$m = 0 : e^{-1}e^1 M_{2,3}^1$$
$$(3)n = 0, l \neq 0 (m_f = 1, i_f = 1)$$
$$m \neq 0 : e^1 E^1 M_{1,1}^0 M_{2,1}^{-1}$$
$$m = 0 : e^1 E^1 M_{3,2}^{-1}$$
$$(4)n = l = 0 (m_f = 1, i_f = 1)$$
$$m \neq 0 : e^1 M_{2,1}^{-1} M_{1,2}^1$$
$$m = 0 : \text{linear system}$$

Class II

$$\dot{x} = dx - y + lx^2 + mxy + ny^2, \tag{7.2.12}$$
$$\dot{y} = x + ax^2, a \neq 0. \tag{7.2.13}$$

$$(1)n \neq 0 (m_f = 4, i_f = 0)$$
$$D < 0 : e^{-1}c_1^0 c_1^0 e^1 E^{-1} E^1 E^1, e^{-1}c_1^0 c_1^0 e^1 C_{0,1}^0 C_{0,1}^0 E^1, e^{-1}c_1^0 c_1^0 e^1 E^1 M_{0,2}^0$$
$$D = 0 : e^{-1}e^1 m_2^0 E^{-1} E^1 E^1, e^{-1}e^1 m_2^0 C_{0,1}^0 C_{0,1}^0 E^1, e^{-1}e^1 m_2^0 E^1 M_{0,2}^0$$
$$D > 0 : e^{-1}e^{-1}e^1 e^1 E^{-1} E^1 E^1, e^{-1}e^{-1}e^1 e^1 C_{0,1}^0 C_{0,1}^0 E^1, e^{-1}e^{-1}e^1 e^1 E^1 M_{0,2}^0$$
$$D = a^2[(a+m)^2 - 4n(l - ad)]$$

$$(2.1)n = 0, a + m \neq 0 (m_f = 2)$$
$$(2.1.1)m \neq 0,$$
$$\tfrac{a}{m} < -1 (i_f = 0)$$
$$l^2 + 4am < 0 : e^{-1}e^1 C_{0,1}^0 C_{0,1}^0 M_{2,1}^{-1}$$
$$l^2 + 4am = 0 : e^{-1}e^1 M_{0,2}^0 M_{2,1}^{-1}$$
$$l^2 + 4am > 0 : e^{-1}e^1 E^{-1} E^1 M_{2,1}^{-1}$$
$$-1 < \tfrac{a}{m} < 0 (i_f = 2):$$
$$l^2 + 4am < 0 : e^1 e^1 C_{0,1}^0 C_{0,1}^0 M_{2,1}^{-1}$$
$$l^2 + 4am = 0 : e^1 e^1 M_{0,2}^0 M_{2,1}^{-1}$$
$$l^2 + 4am < 0 : e^1 e^1 E^{-1} E^1 M_{2,1}^{-1}$$
$$\tfrac{a}{m} > 0 (i_f = 0):$$
$$e^{-1}e^1 E^{-1} E^1 M_{2,1}^{-1}$$
$$(2.1.2)m = 0 (i_f = 0)$$
$$l \neq 0 : e^{-1}e^1 E^1 M_{2,2}^0$$
$$l = 0 : e^{-1}e^1 M_{2,3}^1$$
$$(2.2)\text{n=0,a+m=0.}$$
$$(2.2.1)ad - l \neq 0 (m_f = 1, i_f = 1)$$
$$l^2 - 4a^2 < 0 : e^1 C_{0,1}^0 C_{0,1}^0 M_{3,1}^0$$
$$l^2 - 4a^2 = 0 : e^1 M_{0,2}^0 M_{3,1}^0$$
$$l^2 - 4a^2 > 0 : e^1 E^{-1} E^1 M_{3,1}^0$$
$$(2.2.2)ad - l = 0 (m_f = \infty),$$
$$\text{degeneration}$$

Class III

$$\dot{x} = dx - y + lx^2 + mxy + nt^2, \tag{7.2.14}$$
$$\dot{y} = x + ax^2 + bxy, b \neq 0. \tag{7.2.15}$$

$$A = n(a^2 n - b^2 l - abm), B_1 = n(ab + 2an + b^2 d - bm), C_1 = n(b + n),$$
$$B_2 = -abdn + abm - amn - 2a^2 n + 2bln - b^2 l),$$
$$C_2 = a^2 + am + abd - adn - 2bl + ln, D_2 = ad - l.$$

$$(1)n \neq 0, a^2 n + b^2 l - abm \neq 0 (m_f = 4)$$
$$(1.1)n(a^2 n + b^2 l - abm) < 0, -am + bl - bn < 0 (i_f = -2)$$
$$e^{-1}e^{-1}e^{-1}e^1 E^1 E^1 E^1$$

$$(1.2) n(a^2n + b^2l - abm) < 0, -am + bl - bn > 0 (i_f = 2)$$
$$(1.2.1) b + n \neq 0$$
$$B_1^2 - 4AC_1 < 0 :$$
$$c_1^0 c_1^0 e^1 e^1 E^{-1} E^{-1} E^1, c_1^0 c_1^0 e^1 e^1 C_{0,1}^0 C_{0,1}^0 E^{-1}, c_1^0 c_1^0 e^1 e^1 E^{-1} M_{0,2}^0, c_1^0 c_1^0 e^1 e^1 M_{0,3}^{-1}$$
$$B_1^2 - 4AC_1 = 0 :$$
$$e^1 e^1 m_{0,2}^0 E^{-1} E^{-1} E^1, e^1 e^1 m_{0,2}^0 C_{0,1}^0 C_{0,1}^0 E^{-1}, e^1 e^1 m_2^0 E^{-1} M_{0,2}^0, e^1 e^1 m_{0,2}^0 M_{0,3}^{-1}$$
$$B_1^2 - 4AC_1 > 0 :$$
$$^{-1} e^1 e^1 e^1 E^{-1} E^{-1} E^1, e^{-1} e^1 e^1 e^1 C_{0,1}^0 C_{0,1}^0 E^{-1}, e^{-1} e^1 e^1 e^1 E^{-1} M_{0,2}^0, e^{-1} e^1 e^1 e^1 M_{0,3}^{-1}$$
$$(1.2.2) b + n = 0$$
$$B_1 \neq 0 : e^1 e^1 m_2^0 E^{-1} E^{-1} E^1, e^1 e^1 m_2^0 C_{0,1}^0 C_{0,1}^0 E^{-1}, e^1 e^1 m_2^0 E^{-1} M_{0,2}^0, e^1 e^1 m_2^0 M_{0,3}^{-1}$$
$$B_1 = 0 : e^1 m_3^1 E^{-1} E^{-1} E^1, e^1 m_3^1 C_{0,1}^0 C_{0,1}^0 E^{-1}, e^1 m_3^1 E^{-1} M_{0,2}^0, e^1 m_3^1 M_{0,3}^{-1}$$
$$(1.3) n(a^2n + b^2l - abm) > 0 (i_f = 0, m_f = 4)$$
$$(1.3.1) b + n \neq 0$$
$$B_1^2 - 4AC_1 < 0 :$$
$$e^{-1} c_1^0 c_1^0 e^1 E^{-1} E^1 E^1, e^{-1} c_1^0 c_1^0 C_{0,1}^0 C_{0,1}^0 E^1, e^{-1} c_1^0 c_1^0 e^1 E^1 M_{0,2}^0, e^{-1} c_1^0 c_1^0 e^1 M_{0,3}^1$$
$$B_1^2 - 4AC_1 = 0 :$$
$$e^{-1} e^1 m_2^0 E^{-1} E^1 E^1, e^{-1} e^1 m_2^0 C_{0,1}^0 C_{0,1}^0 E^1, e^{-1} e^1 m_2^0 E^1 M_{0,2}^0, e^{-1} e^1 m_2^0 M_{0,3}^1$$
$$B_1^2 - 4AC_1 > 0 :$$
$$^{-1} e^{-1} e^1 e^1 E^{-1} E^1 E^1, e^{-1} e^{-1} e^1 e^1 C_{0,1}^0 C_{-1}^0 E^1, e^{-1} e^{-1} e^1 e^1 E^1 M_{0,2}^0, e^{-1} e^{-1} e^1 e^1 M_{0,3}^1$$
$$(1.3.2) b + n = 0, B_1 \neq 0 :$$
$$e^{-1} e^1 m_2^0 E^{-1} E^1 E^1, e^{-1} e^1 m_2^0 C_{0,1}^0 C_{0,1}^0 E^1, e^{-1} e^1 m_2^0 E^1 M_{0,2}^0, e^{-1} e^1 m_2^0 M_{0,3}^0$$
$$(1.3.3) b + n = 0, B_1 = 0 :$$
$$e^1 m_3^{-1} E^{-1} E^1 E^1, e^1 m_3^{-1} C_{0,1}^0 C_{0,1}^0 E^1, e^1 m_3^{-1} E^1 M_{0,2}^0, e^1 m_3^{-1} M_{0,3}^1$$
$$(2) n \neq 0, a^2n + b^2l - abm = 0$$
$$(2.1) B_1 \neq 0 (m_f = 3)$$
$$(2.1.1) bl - am - bn < 0 (i_f = -1)$$
$$b + n \neq 0 : e^{-1} e^{-1} e^1 E^1 E^1 M_{1,1}^0, e^{-1} e^{-1} e^1 E^1 M_{1,2}^1$$
$$(2.1.2) bl - am - bn > 0 (i_f = 1)$$
$$b + n \neq 0 : e^{-1} e^1 e^1 E^1 M_{1,2}^{-1}, e^{-1} e^1 e^1 M_{0,2}^0 M_{1,1}^1, e^{-1} e^1 e^1 M_{1,3}^0, e^{-1} e^1 e^1 \infty$$
$$b + n = 0 : e^1 m_2^0 E^1 M_{1,2}^{-1}, e^1 m_2^0 M_{0,2}^0 M_{1,1}^1, e^1 m_2^0 M_{1,3}^0, e^1 m_2^0 \infty$$
$$(2.2) B_1 = 0$$
$$(2.2.1) b + n \neq 0 (m_f = 2)$$
$$0 < bn < 1 (i_f = 2) :$$
$$^1 e^1 E^{-1} E^1 M_{2,1}^{-1}, e^1 e^1 C_{0,1}^0 C_{1,0}^0 M_{2,1}^{-1}, e^1 e^1 E^1 M_{2,2}^{-2}, e^1 e^1 E^{-1} M_{2,2}^0, e^1 e^1 M_{0,2}^0 M_{2,1}^{-1}, e^1 e^1 M_{2,3}^{-1}, e^1 e^1 \infty$$
$$bn < 0 \text{ or } bn > 1 (i_f = 0) :$$
$$^{-1} e^1 E^1 E^1 M_{2,1}^{-1}, e^{-1} e^1 E^{-1} E^1 M_{2,1}^1, e^{-1} e^1 C_{0,1}^0 C_{0,1}^0 M_{2,1}^1, e^{-1} e^1 E^1 M_{2,2}^0, e^{-1} e^1 E^{-1} M_{2,2}^2, e^{-1} e^1 M_{0,2}^0 M_{2,1}^1$$
$$\text{or } e^{-1} e^1 M_{2,3}^1, e^{-1} e^1 \infty$$

$$(2.2.2)b + n = 0(m_f = \infty): \text{ essentially linear}$$
$$(3)n = 0, a^2n + b^2l - abm \neq 0(m_f = 3)$$
$$(3.1)bl - am < 0(i_f = -1):$$
$$e^{-1}e^{-1}e^1E^1E^1M_{1,1}^0, e^{-1}e^{-1}e^1E^1M_{1,2}^1$$
$$(3.2)bl - am > 0(i_f = 1)$$
$$C_2^2 - 4B_2D_2 < 0:$$
$$c_1^0c_1^0e^1E^{-1}E^1M_{1,1}^0, c_1^0c_1^0e^1C_{0,1}^0C_{0,1}^0M_{1,1}^0, c_1^0c_1^0e^1E^{-1}M_{1,2}^1, c_1^0c_1^0e^1E^1M_{2,1}^{-1},$$
$$c_1^0c_1^0e^1M_{0,2}^0M_{1,1}^0, c_1^0c_1^0e^1M_{1,3}^0, c_1^0c_1^0e^1\infty$$
$$C_2^2 - 4B_2D_2 = 0:$$
$$e^1m_2^0E^{-1}E^1M_{1,1}^0, e^1m_2^0C_{0,1}^0C_{0,1}^0M_{1,1}^0, e^1m_2^0E^{-1}M_{1,2}^1, e^1m_2^0E^1M_{2,1}^{-1},$$
$$e^1m_2^0M_{0,2}^0M_{1,1}^0, e^1m_2^0M_{1,3}^0, e^1m_2^0\infty$$
$$C_2^2 - 4B_2D_2 > 0: e^{-1}e^1e^1E^{-1}E^{-1}M_{1,1}^0, e^{-1}e^1e^1C_{0,1}^0C_{0,1}^0M_{1,1}^0, e^{-1}e^1e^1E^{-1}M_{1,2}^1,$$
$$e^{-1}e^1e^1E^1M_{2,1}^{-1}, e^{-1}e^1e^1M_{0,2}^0M_{1,1}^0, e^{-1}e^1e^1M_{1,3}^0, e^{-1}e^1e^1\infty$$

$$(4)n = 0, a^2n + b^2l - abm = 0$$
$$(4.1)a^2 + l^2 \neq 0$$
$$C_2 \neq 0(m_f = 2, i_f = 0): e^{-1}e^1E^1M_{1,1}^0M_{1,1}^0, e^{-1}e^1C_{0,1}^0C_{0,1}^0E^1, e^{-1}e^1M_{1,1}^0M_{1,2}^1$$
$$C_2 = 0, D_2 \neq 0(m_f = 1, i_f = 1): e^1E^1M_{1,1}^0M_{2,1}^{-1}, e^1M_{1,1}^0M_{2,2}^0, e^1M_{2,1}^{-1}M_{1,2}^1$$
$$(4.2)a^2 + l^2 = 0$$
$$m - bd \neq 0(m_f = 2, i_f = 0):$$
$$e^{-1}e^1E^1M_{1,1}^0M_{1,1}^0, e^{-1}e^1C_{0,1}^0C_{0,1}^0E^1e^{-1}e^1M_{1,1}^0M_{1,2}^1$$
$$m - bd = 0(m_f = 1, i_f = 1):$$
$$e^1E^1M_{1,1}^0M_{2,1}^{-1}, e^1M_{1,1}^0M_{2,2}^0, e^1M_{2,1}^{-1}M_{1,2}^1$$

Table 7.1 Relations between the classes in the Ye normal form and the classes defined in Chapter 3.

7.3 Limit cycles in particular systems

7.3.1 Quadratic systems with a weak critical point

A weak critical point is a finite or infinite critical point wherein the divergence of the vector field vanishes. If a finite critical point is weak it can either be elementary or degenerate. If elementary it is either a center point, a weak focus or a weak saddle point. If degenerate it is a nilpotent critical point, so a cusp, a third order saddle, a third order point with an elliptic and a hyperbolic sector or a fourth order saddle point, or it is a critical point in an essentially homogeneous system. If an infinite critical point is weak it is also either elementary or degenerate. If elementary it is a weak saddle point. If degenerate it is a nilpotent critical point or a highly degenerate point as listed in section 2.3.2 of Chapter 2.

As the presence of weak critical points appears to affect the behavior of limit cycles, which, in turn, is related to the singular behavior of integrating factors, it seems that in a, not yet fully understood, way the occurrence of weak critical points affects the integrability of a quadratic system. As was seen in Chapter 6 the presence of a center point in a quadratic system makes the system integrable so that singular behavior of integrating factors does not occur in such a way that limit cycles can be present. For Hamiltonian systems there is $\text{div}(P(x,y), Q(x,y)) \equiv 0$ and all critical points are weak. Such systems have no limit cycles, as can be shown using Green's theorem. For consider, as sketched in Figure 7.1 an annular region S bounded on one side by the limit cycle L and on the other side by an orbit C extending from a point on a transversal l to L to the next intersection point with l when following C in clockwise direction, plus the intermediate segment of l cut out by C. Then follows the contradiction

$$0 = \int \int div(P(x,y), Q(x,y))dxdy = \int (P(x,y), Q(x,y)).\underline{n}ds \neq 0, \quad (7.3.1)$$

where the surface integral is taken over S and the line integral over the intermediate segment of l cut out by C. Moreover $\underline{n}$ is the outwards pointing unit vector normal to l.

Apart from in Hamiltonian systems, the other quadratic systems with points wherein $\text{div}(P(x,y), Q(x,y)) \equiv a_{10}+b_{01}+(2a_{20}+b_{11})x+(a_{11}+2b_{02})y=0$ occur are those where these points form a line. Such a line contains at most two weak critical points as two is the maximum number of critical points on a line. It can then be shown that a system with two weak critical points does not contain limit cycles; if these critical points are elementary [72,CZ1],[74,CZ], or, without imposing the latter condition. The possible combinations of weak critical points will be discussed later.

We will now discuss quadratic systems with a single weak critical point.

Quadratic systems with a single weak critical point: a weak focus

The order of a focus may be defined by considering again the Poincaré return map and the distance function, as introduced in section 7.2, now for the flow around a focus. Let the intersection of an orbit near the critical point with the positive x axis be indicated by $x(0)$ and the next intersection point when following the orbit in anti-clockwise direction by $x(2\pi)$, and indicate the distance function by $d(x) = x(2\pi) - x(0)$. Then, when using Bautin's normal form (6.1.10),(6.1.11)

$$\dot{x} = \lambda_1 x - y - \lambda_3 x^2 + (2\lambda_2 + \lambda_5)xy + \lambda_6 y^2, \qquad (7.3.2)$$

$$\dot{y} = x + \lambda_1 y + \lambda_2 x^2 + (2\lambda_3 + \lambda_4)xy - \lambda_2 y^2, \qquad (7.3.3)$$

it was shown by him [39,B1] that near $(0,0)$ the distance function d(n) may be written as

$$d(x) = \lambda_1 \sum_{k=1}^{\infty} \delta_k(\lambda)x^k + \bar{v}_3(\lambda)x^3[1 + \sum_{k=1}^{\infty} \alpha_k(\lambda)x^k]$$

$$+\bar{v}_5(\lambda)x^5[1 + \sum_{k=1}^{\infty} \beta_k(\lambda)x^k] + \bar{v}_7(\lambda)x^7[1 + \sum_{k=1}^{\infty} \gamma_k(\lambda)x^k], \qquad (7.3.4)$$

where $\delta_1(\lambda) = \frac{e^{2\pi\lambda_1}-1}{\lambda_1}$, the δ_k, k=2,3... are analytic functions of λ, while α_k, β_k and γ_k are homogeneous polynomials of degree k in the variables $\lambda_2, ..., \lambda_6$ and the *focal values* $\bar{v}_k(\lambda)$ are given by

$$\bar{v}_3(\lambda) = -\frac{\pi}{4}\lambda_5(\lambda_3 - \lambda_6), \qquad (7.3.5)$$

$$\bar{v}_5(\lambda) = \frac{\pi}{24}\lambda_2\lambda_4(\lambda_3 - \lambda_6)(\lambda_4 + 5\lambda_3 - 5\lambda_6), \qquad (7.3.6)$$

$$\bar{v}_7(\lambda) = -\frac{5\pi}{32}\lambda_2\lambda_4(\lambda_3 - \lambda_6)^2(\lambda_3\lambda_6 - 2\lambda_6^2 - \lambda_2^2). \qquad (7.3.7)$$

(It appeared later that both sign and scaling of $\bar{v}_7(\lambda)$ are incorrect in Bautin's paper; by now they have been checked by several authors to be correct in the form given above.)

Obviously if $\lambda_1 < 0(> 0)$ the origin is a rough stable (unstable)focus. If $\lambda_1=0$ the origin is a *weak focus of order 1* if $\bar{v}_3 \neq 0$, of *order 2* if $\bar{v}_3 = \bar{v}_5 = 0, \bar{v}_7 \neq 0$. If all focal values $\bar{v}_3, \bar{v}_5$ and $\bar{v}_7$ vanish the origin is a center point. A weak focus of order k is stable(unstable) if $\bar{v}_{2k+1} < 0(> 0)$.

With regard to the *maximum number of limit cycles in one nest*, at several places in the literature there may be found the following

Conjecture 2. *There exist at most $3-k$ limit cycles in a nest around a weak focus of order k (k=0,1,2,3).*

In fact for k=3 the conjecture has been shown to be true; *around a third order focus there exists no limit cycle.* By investigating particular classes with a third order focus the proof was developed by Cherkas in 1976[76,C1],Cai Suilin in 1981 [81,C]Wang Mingshu and Lin Yingju in 1982[82,WL], and Du Xingfu in 1982 [82,D]. In 1986 the proof for all quadratic systems with a third order focus was completed by Li Chengzhi [86,L] and Cherkas [86,C], whereas in 1989 a proof was also given by Zhang Pingguang [89,Z1],[90,Z1].

The phase portraits of the quadratic systems with a third order focus were determined by Artés [90,A] and can be found in [97,AL]. It appears that only the classes with finite multiplicity $m_f=3$ or 4 can contain a third order focus and such systems in $m_f=3$ do not contain limit cycles. In those within class $m_f=4$ having two complex and two real finite critical points (a) limit cycle(s) occur(s) around a second focus. The question of uniqueness, hyperbolicity and extension in parameter space of these limit cycle(s) was considered by Kooy and Zegeling [98,KZ]. They showed that a unique, hyperbolic limit cycle around a second focus exists in a region of the parameter space given in [98,KZ],[97,AL]. The same question still remains open for the region a>0, $125a^4 + a^2(25l^2 + 170l + 262) + (2l+5)^3 < 0$ in the parameter space of system

$$\dot{x} = -y + lx^2 + 5axy + y^2, \qquad (7.3.8)$$
$$\dot{y} = x + ax^2 + (3l + 5)xy, \qquad (7.3.9)$$

which in this region is of the class $c_1^0 c_1^0 e^1 e^1 C_{0,1}^0 C_{0,1}^0 E^{-1}$.

The form of this statement reflects that weak foci are often studied using Ye's normal form (7.2.8),(7.2.9)

$$\dot{x} = dx - y + lx^2 + mxy + ny^2, \qquad (7.3.10)$$
$$\dot{y} = x(1 + ax + by). \qquad (7.3.11)$$

Then, with

$$W_1 = m(l + n) - ab + 2l, \qquad (7.3.12)$$
$$W_2 = am(5a - m)[(l + n)^2(b + n) - a^2(b + 2l + n)], \qquad (7.3.13)$$
$$W_3 = a^2m[2a^2 + n(l + 2n)][(l + n)^2(b + n) - a^2(b + 2l + n)], \qquad (7.3.14)$$

it follows that, if $d \neq 0$ the origin is a rough focus, if $d=0, W_1 \neq 0$ a weak focus of order 1, if $d=W_1 = 0, W_2 \neq 0$ a weak focus of order 2, if $d=W_1 = W_2 = 0$, $W_3 \neq 0$ a weak focus of order 3 and if $d=W_1 = W_2 = W_3=0$ a center point. A weak focus of order k is stable(unstable) if $W_k < 0(> 0)$.

Also for k=2 the conjecture has been shown to be true: *around a second order weak focus there exists at most one limit cycle* and it is *hyperbolic* if it exists. By investigating various classes with a second order weak focus the proof was developed over the years 1984 through 1999 by Cai Suilin and Wang Zhongwei [84,CW], Chen Weifen [85,C],[86,Ch1],[88,Ch1],[88,Ch2], Han Mao an [85,H], Zhang Pingguang and Cai Suilin [89,ZC],[90,ZC],[91,ZC], Zhang Pingguang [89,Z1],[90,Z1], Chen Wencheng and Zhang Pingguang [90,CZ], [91,ZC], Zhang Pingguang and Li Wenhua [92,ZL], Gao Suzhi and Wang Xuejin [94,GW], Wang Xuejin [98,W], Huang Xianhua and Reyn [97,HR], whereas the final proof was given by Zhang Pingguang in 1999 [99,Z1].

It may be seen that a second order weak focus can only occur in quadratic systems with a finite multiplicity such that $2 \leq m_f \leq 4$. From a classification of the phase portraits in the class $m_f=2$ [97,RK] it follows that systems in this class with a second order weak focus do not contain limit cycles, the second order weak focus being the only focus in such a system. In the class $m_f=3$ there exists at most one limit cycle around a second order weak focus and if it exists it is hyperbolic [97,HR]; there exists no limit cycle around a possible second focus. For class $m_f=4$ the situation is more involved. Obviously a system with an essentially homogeneous system does not contain a second order weak focus and from a classification of the phase portraits of quadratic systems with a nilpotent critical point the same follows for this class [90,J]. Also systems with a third or fourth order semi-elementary critical point do not contain a second order weak focus [89,J]. With regard to systems with a second order semi-elementary critical point the result obtained by Zhang Pingguang and Cai Suilin [89,ZC],[90,ZC] may be used that states that there exists at most one limit cycle around a second order weak focus and none around a possible second focus in quadratic systems with a semi-elementary critical point. This leaves the systems with four elementary critical points. If there exist four real (elementary) critical points, a result by Zegeling and Kooij [99,ZK] may be used that states that if there are two nests of limit cycles there exists precisely one limit cycle in each nest and this limit cycle is hyperbolic. This would also apply if one nest is centered around a second weak focus. Examples wherein only around a second order weak focus there does exist a (unique and hyperbolic) limit cycle may be constructed by bifurcating a second order weak focus with a limit cycle around it from a third order weak focus. In [97,AL] phase portraits with a third order weak focus may be selected in the regions R_{10}, R_6 and R_4, which have 3,2 or 1 antisaddles (1,2 or 3 saddles), respectively and which upon bifurcation yield the various possibilities with four real critical points. In case there exist two real and two complex critical points, from the phase portraits in the regions R_{11} or R_1, which have 2 or 1 antisadddles (0 or 1 saddle), respectively, the cases with a weak second order weak focus surrounded by a single (hyperbolic) limit cycle and another real critical point may be constructed.

The conjecture that there exist *at most two limit cycles around a weak focus of order 1* is difficult to prove since essentially no method of proof is available. Results on non-existence and uniqueness of limit cycles can, however, be obtained for particular cases. It may be seen that a first order weak focus occurs in quadratic systems with a finite multiplicity such that $1 \leq m_f \leq 4$. From a classification of phase portraits in the class $m_f=1$ [97,RK], it follows that if a system in this class contains a first order weak critical point it does not contain a limit cycle. From a classification of the phase portraits in the class $m_f=2$ [97,RK], it follows that if a system in this class contains a first order focus, there exists at most one limit cycle and if it exists it is hyperbolic. The limit cycle can be either around the first order weak focus or around a possible second focus. As far as the class $m_f=3$ is concerned the following can be remarked. Obviously a fourth order critical point cannot occur in class $m_f=3$ and if a third order critical point occurs there is no focus so no limit cycle. A classification of de Jager [90,J] shows that if a system in $m_f=3$ contains a nilpotent critical point of the second order (so a cusp) it can contain a first order weak focus; such a system does not contain a limit cycle either. If a semi-elementary saddle node is present there seems that no clear conclusion can be found in the literature. If there exist three elementary critical points, a classification by Reyn and Huang [97,RH] shows that if there exists a degenerate infinite critical point (an $M_{1,2}^i$ or an $M_{1,3}^0$ point) in addition to a first order weak critical point, then no limit cycle occurs. For the remaining case, that with a transversally non-hyperbolic infinite critical point $M_{1,1}^0$, no conclusion seems to have been given yet, except that in case of two nests of limit cycles, there exists precisely one limit cycle in each nest, which then is a hyperbolic limit cycle [99,ZK]. For the class $m_f=4$ similar remarks can be made as for $m_f=3$. Obviously a system with a fourth order critical point does not contain a (first order weak) focus and so no limit cycle. If a system in $m_f=4$ contains a third order critical point it does not contain a first order focus if the critical point is nilpotent [90,J], and can contain such a point if the critical point is semi-elementary [89,J]. In the latter case no limit cycles exist. If a system in $m_f=4$ contains a second order nilpotent critical point (cusp) it can contain a first order weak focus; in that case there exist no limit cycles in the system [90,J]. If the second order point is semi-elementary (saddle node) no conclusion seems possible at the moment. If there exist four elementary critical points, an inspection of the literature indicates that a proof of the conjecture cannot yet be given.

Although not all limit cycle configurations possible in quadratic systems with a weak focus are known, the available information still yields interesting conclusions. A weak focus of order k (k=1,2,3) can only occur in a system with finite multiplicity such that k$\leq m_f \leq$4. This property is in accordance with Conjecture 1 in the sense that if this property would also be valid for m_f <k, k limit cycles could be bifurcated from the weak focus of order k according to the results of Bautin [39,B1]. This is in contradiction with Conjecture 1 which states that the number of limit cycles is at most equal to m_f. Another conclusion is that for m_f=k there exist no limit cycles around the weak focus and the same remark as above can be made. For m_f=k+1 it was found that a system in this class contains at most one limit cycle and if it exists is hyperbolic, admitting the same remark.

As mentioned earlier in Chapter 6 it can be concluded from the existence and uniqueness statements for the initial value problem given in Chapter 1 that quadratic systems belong to the wide class of systems for which there exists a first integral F(x,y)=c, where c is constant on an orbit. Finding this first integral for a given system amounts to finding an integrating factor I(x,y) such that div$(P(x,y)$I$(x,y),Q(x,y)$I$(x,y))\equiv$0, thus by replacing dt by $\frac{1}{I(x,y)}$dt in $\dot{x} = P(x,y), \dot{y} = Q(x,y)$ in order to make the system Hamiltonian. Since it was shown above, using Green's theorem, that limit cycles cannot occur in Hamiltonian systems, it must be concluded that the application of this theorem near a limit cycle and to the vector field $(P(x,y)$I$(x,y),Q(x,y)$I$(x,y))$ is not allowed, apparently since I(x,y) has a singular behavior at the limit cycle. It may also be seen that if $(P(x,y),Q(x,y))$ is already given to be a Hamiltonian quadratic system, I(x,y) can be taken to be an arbitrary constant, whereas F(x,y) then is a third degree polynomial in x and y. If the term integrable is reserved to those cases in which I(x,y) is defined as an analytic function on the whole plane, we might say that a system is less integrable if I(x,y) exhibits singular behavior somewhere on this plane. In a rather verbal and not precisely defined way we may also wish to speak of degree of integrability and negative or positive integrability effects due to certain elements in a phase portrait. In that sense a weak focus has a positive integrability effect, the more so the higher the order of the focus. This effect becomes noticeable in the limit cycle behavior as discussed above. (It should be noted that in the discussion given above, for the limit cycle behavior use is made of Cartesian coordinates.)

Quadratic systems with a single weak critical point: a weak saddle point

As was mentioned in Chapter 6, the center problem for quadratic systems was first discussed in 1908; in the paper by Dulac [08,D]. He considered the system

$$\dot{x} = x + a_{20}x^2 + a_{11}xy + a_{02}y^2, \qquad (7.3.15)$$
$$\dot{y} = -y + b_{20}x^2 + b_{11}xy + b_{02}y^2, \qquad (7.3.16)$$

allowing x, y, t and the coefficients a_{ij}, b_{ij} (i,j=0,1,2) to take complex values and using a wider definition of center point. For real values of these quantities this system represents a normal form for a quadratic system with a weak saddle point at the origin. Taking in this system for x and y the complex variables $u + iv$ and $u - iv$, respectively, leads to the system

$$u' = -v + \alpha_{20}u^2 + \alpha_{11}uv + \alpha_{02}v^2, \qquad (7.3.17)$$
$$v' = u + \beta_{20}u^2 + \beta_{11}uv + \beta_{02}v^2, \qquad (7.3.18)$$

where $' = \frac{d}{d\tau} = \frac{d}{dt}$. If variables and coefficients are restricted to real values this system represents a normal form for a quadratic system with a weak focus (or a center point) at the origin.

This shows that there exists a correspondence between systems with a weak focus and those with a weak saddle point. To the *focal values* governing the process of constructing an analytic first integral $F(x,y)$ near a center point correspond *saddle values* governing the same process near a weak saddle point. For the normal form (7.3.15),(7.3.16) saddle values were given by Zhu Deming [87,Z] defined to be given by the expressions

$$R_1 = -a_{20}a_{11} + b_{20}b_{11}, \quad (7.3.19)$$
$$R_2 = a_{11}b_{02}(2a_{11} - b_{20})(a_{11} + 2b_{20}) + a_{02}b_{11}(a_{20} - 2b_{11})(2a_{20} + b_{11}), \quad (7.3.20)$$
$$R_3 = (a_{20}b_{20} + a_{11}b_{11} - 5a_{02}b_{02})[a_{11}b_{02}(4b_{20}^2 - b_{11}^2) - a_{02}b_{11}(4a_{20}^2 - b_{11}^2)]. \quad (7.3.21)$$

If the coefficients in this expression are taken such that Bautin's normal form (6,1.10),(6.1.11) results, then the saddle values of the corresponding system (7.3.15),(7.3.16) and the focal values (7.3.5),(7.3.6) and (7.3.7) of the Bautin

form are given by the relations $\bar{v}_3 = i\pi R_1$, $\bar{v}_5 = \frac{1}{3}i\pi R_2$ and $\bar{v}_7 = \frac{1}{24}i\pi R_3$
[87,Z]. The origin in system (7.3.15),(7.3.16) is a *weak saddle* of *order 1* if
$R_1 \neq 0$ of *order 2* if $R_1 = 0, R_2 \neq 0$ and of *order 3* if $R_1 = R_2 = 0, R_3 \neq 0$.
If $R_1 = R_2 = R_3 = 0$ the system is integrable; there exists (globally) an
analytic first integral $F(x,y)$ [87,Z],[87,C] and there would not exist a limit
cycle.

We concentrate on the case that the weak saddle point is the only weak
critical point in the system, since otherwise the result is clear that there
would not exist a limit cycle. Then, as was shown by Zhang Pingguang
and Cai Suilin [87,ZC] and Zhu Demin [89,Z2], in a quadratic system with a
second or third order weak saddle point there does not exist a limit cycle (or
other periodic or singular closed orbit).This leaves the weak saddle of order 1
($R_1 \neq 0$). If limit cycles exist for this case they are nested around one focus.
This may readily be seen using Lemma 1 of Tung, mentioned in section 4.2 of
Chapter 4. In fact, the presence of two nests of limit cycles would imply that
the number of critical points and contact points with the line div(P, Q)=0
would be more than 2, the weak saddle being located on this line, which in
turn must intersect all the limit cycles yielding two points of contact [87,ZC].
Conditions for the non-existence and uniqueness of limit cycles in quadratic
systems with a first order weak saddle point were given by Zhang Pingguang
[90,Z2]. From them follows, for instance, that a quadratic system with two
antisaddles and two saddles, one of them being weak, has at most one limit
cycle and if it exists is hyperbolic.

As for a weak focus, a weak saddle may be said to have a positive integrability effect, the more so the higher the order of the weak saddle, this
being expressed by the limit cycle behavior in systems with a weak saddle.

Although quadratic systems with (an) infinite weak saddle(s) were studied
in the literature, no complete evaluation of their presence on the limit cycle
behavior seems to have been made so far.

Quadratic systems with a degenerate critical point; further remarks on systems with a weak critical point

As was stated in section 7.3.1 a finite critical point that is weak may be
either elementary or degenerate. If degenerate it is either a nilpotent critical
point or a critical point in an essentially homogeneous system. If an infinite
critical point is weak it is also either elementary or degenerate; a saddle
point if elementary. If degenerate it is a nilpotent critical point or a highly

degenerate point, as listed in section 2.3.2 of Chapter2.

A quadratic system with a finite or infinite degenerate critical point has at most one limit cycle and if it exists is hyperbolic. Although part of this statement may already be found in a paper in 1977 by Chen Lansun [77,C], a conclusive treatment was given in 1988 by Coppel [88,C].

It was already remarked above that if there exist two weak critical points in a non-Hamiltonian quadratic system, there exists no limit cycles in such a system. In addition, it can be shown that if two finite elementary critical points exist that are weak, so either are a saddle or a focus, then they are both of first order [83,YW],[86,Z2].

Apart from those indicated in the discussion above various other results on the effects of the presence of (a) weak critical point(s) on the behavior of limit cycles may be found in the literature [99,Z2],[99,Z3],[95,LLZ1],[93,Z].

In relation to the 16th Hilbert problem and the search for quadratic systems with a maximum number of limit cycles the following conjecture, as formulated by Coppel [98,KZ], is of interest

Conjecture 3. *If a quadratic system contains a finite weak critical point, then it has at most one limit cycle not surrounding this critical point. Moreover, if such a limit cycle does exist it is hyperbolic.*

There are plausibility arguments to believe this conjecture to be true. In fact, for systems with a degenerate weak critical point the statement in the conjecture already follows from the result that there exists at most one (hyperbolic) limit cycle in such systems and because a limit cycle cannot encircle a degenerate critical point. This leaves the weak focus and the weak saddle. In both cases there can exist at most one nest of limit cycles not surrounding the critical point. Suppose now there exist more than one limit cycle in such a nest and assume a path in the space of coefficients is possible such that the weak critical point remains weak and the innermost limit cycle in the nest shrinks until it disappears in a weak focus leaving at least one limit cycle in the nest. This would contradict the result that there exist no limit cycles in a quadratic system with two weak critical points. Contributions to the proof of the conjecture can be found in [91,ZC],[92,ZL],[99,ZK],[98,KZ] [90,Z2].

Another interesting point is raised by the result that a quadratic system with a straight line solution and a weak critical point not on this line does not have a limit cycle [70,CZ]. This leads to the idea of indicating the existence of an algebraic solution as a positive integrability effect in addition to that of the weak critical point. This idea may also be illustrated by quadratic systems with center points, where algebraic (conic or cubic) solutions and weak critical points occur simultaneously and the systems are integrable and without limit cycles.

As a result it is of interest to consider quadratic systems with algebraic solutions and the limit cycle behavior therein.

7.3.2   Quadratic systems with algebraic solutions

The simplest algebraic solution is a straight line solution and in the early 1970s it was shown by the work of Cherkas and Zhilevich [70,CZ],[72,CZ2] and finally by Ryckov [72,R] that quadratic systems with a straight line solution have at most one limit cycle, which is hyperbolic if it exists [88,Co].

An early result concerning the effect on limit cycle behavior in quadratic systems of the presence of an orbit coinciding with a conic was given in 1954 by Bautin. A degenerate hyperbola consists of two intersecting straight lines and, by a linear transformation, a quadratic system having such lines as solutions may be brought into the form

$$\dot{x} = x(a_0 + a_1 x + a_2 y) \equiv P(x, y), \tag{7.3.22}$$
$$\dot{y} = y(b_0 + b_1 x + b_2 y) \equiv Q(x, y), \tag{7.3.23}$$

where $a_i, b_i (i=0,1$ or $2)$ are real constants. This equation is known as the Volterra−Lotka equation, and, among quadratic systems, it is most frequently used as a model in various fields of applications [87,R]. Obviously $x \equiv 0$ and $y \equiv 0$ are two straight line solutions which cannot be crossed by a periodic solution, which as a result must then be located wholly in one quadrant, having a critical point in its interior determined by the intersection point of $a_0 + a_1 x + a_2 y = 0$, and $b_0 + b_1 x + b_2 y = 0$ with D=$a_1 b_2 - a_2 b_1 \neq 0$.

A method to demonstrate the non-existence of a (periodic solution) limit cycle consists of finding a Dulac function and applying Green's theorem. For the Volterra−Lotka system this can be shown by using the Dulac function $B(x, y) = x^{k-1} y^{h-1}$, where $k = b_2(b_1 - a_1)D^{-1}$, $h = a_1(a_2 - b_2)D^{-1}$, as given by Bautin [54,B].

For (7.3.22),(7.3.23) it follows that

$$div(B(x, y)P(x, y), B(x, y)Q(x, y)) = \frac{\sigma}{D} B(x, y), \tag{7.3.24}$$

with $\sigma = a_1 b_0(a_2 - b_2) + a_0 b_2(b_1 - a_1)$.

Let R be the bounded region enclosed by a periodic orbit γ, then Green's theorem yields

$$0 = \oint_\gamma (\dot{x}\dot{y} - \dot{x}\dot{y})B(x,y)dt = \oint_\gamma B(x,y)Q(x,y)dx - B(x,y)P(x,y)dy$$

$$= -\int\int_R div(B(x,y)P(x,y), B(x,y)Q(x,y))dxdy$$

$$= -\frac{\sigma}{D}\int\int_R B(x,y)dxdy. \qquad (7.3.25)$$

If $\sigma \neq 0$, this leads to a contradiction since $B(x,y) \equiv 0$ is not true in a quadrant; there exist no periodic orbits in particular no limit cycles. If $\sigma = 0$, it follows from (7.3.24) that $div(BP, BQ)=0$ and no limit cycles are possible, as was shown in section 7.3.1. Periodic orbits can occur, however. The phase portraits for the Volterra–Lotka system are given in [87,R]. It is an interesting aspect of the used Dulac function that, for specific values of the parameters ($\sigma=0$), it becomes an integrating factor. This phenomenon can sometimes help to find a Dulac function for other systems.

The result for the Volterra–Lotka system may also be seen as a result of the presence of two algebraic solutions stengthening the result for quadratic systems with a single algebraic (straight line) solution stating only the uniqueness of a possible limit cycle.

If there exists a solution, coinciding with a non-degenerate hyperbola, there cannot exist a limit cycle either [77,Ce],[89,Ch].

We may work with the system

$$\dot{x} = \alpha(xy - \gamma) + x(ax + by + c), \qquad (7.3.26)$$
$$\dot{y} = \beta(xy - \gamma) - y(ax + by + c), \qquad (7.3.27)$$

for which $xy - \gamma=0$ is the hyperbolic orbit. Now, let the Dulac function be given by $B(x,y) \equiv | xy - \gamma |^n$, then $div(B(x,y)P(x,y), B(x,y)Q(x,y)) =B(x,y)((a + (n+1)\beta)x + (-b + (n+1)\alpha)y)$ and if $\alpha \neq 0(\beta \neq 0)$ n may be chosen such that $-b + (n+1)\alpha=0$ $(a + (n+1)\beta = 0)$ and the y axis (x axis) is the zero divergence line which, however, cannot be crossed by a limit cycle since it is a line without contact.(Note that $\alpha^2 + \beta^2 \neq 0$.) If, moreover, also $a + (n+1)\beta=0$ $(-b + (n+1)\alpha = 0)$, $B(x,y)$ again works as an integrating factor.

Orbits of elliptic type may either be situated on a real ellipse, a complex ellipse or on two complex straight lines intersecting in a real point. Moreover, the ellipse may consist of one or several orbits and no or some critical points.

The question whether an elliptic limit cycle is possible in a quadratic system was answered in the affirmative, as early as 1957, by Qin Yuanxun [57,Q],[58,Q]. It may be seen that a quadratic system containing a curve coinciding with a real ellipse and which consists of one or more orbits can be transformed by a linear transformation into the form

$$\dot{x} = x^2 + y^2 - 1 + y(ax + by + c), \qquad (7.3.28)$$
$$\dot{y} = -x(ax + by + c). \qquad (7.3.29)$$

If $a=0$, the system has a center point and no limit cycles are possible; a family of elliptic solution curves exist if $(b+1)(b+2)=0$ and a single elliptic solution curve if $(b+1)(b+2) \neq 0$. If $a \neq 0$ there exists a single elliptic solution curve which becomes a limit cycle if $a^2 + b^2 < c^2$; it is unique and hyperbolic [58,Q],[75,D2].

An elliptic limit cycle is possible in the classes $e^{-1}c_1^0 c_1^0 e^1 C_{0,1}^0 C_{0,1}^0 E^1$ and $c_1^0 c_1^0 e^1 E^{-1} C_{0,1}^0 C_{0,1}^0$ in $m_f=4$ and the class $c_1^0 c_1^0 e^1 C_{0,1}^0 C_{0,1}^0 M_{1,1}^0$ in $m_f=3$.

If the elliptic solution curve is degenerate it consists of two conjugate complex straight line orbits intersecting in a real point. It was shown by Suo Guangjiang and Chen Yongshao [86,SC] that then there exists at most one limit cycle. If, in addition, the system also has a real straight line orbit or two conjugate complex straight line orbits, then Ye Weiyin showed [88,Y] that the system has no limit cycles.

Affine invariant conditions on the coefficients of (1.0.1),(1.0.2) to possess two conjugate complex straight line orbits as well as for two pairs of such orbits were established by Voldman and Vulpe [99,VV]. For some classes with two complex straight line orbits the neccesary and sufficient affine invariant conditions for possessing just one limit cycle were also determined in [99,VV].

Finally, if a quadratic system has an orbit coinciding with a complex ellipse it does not contain a limit cycle; this was shown by Gasull [93,G].

Orbits of parabolic type may either be situated on a non-degenerate parabola, two parallel straight lines either real or complex, two coinciding straight lines or on two lines, one or both lying at infinity.

Limit cycles of quadratic systems with a non-degenerate parabola as orbit were first studied in 1985 by Chen Shuping [85,Ch], thereupon followed by several other papers , by Chen Guangqing [86,Cg], by Christopher [89,Ch], by Gasull [93,G], by Shen Boqian and Song Yan [92,SS], by Xie Xiangdong and Cai Cuilin [93,XC] and by Zhuang Yu [91,Zy],[93,Zy].

A proof that at most one limit cycle can exist in a quadratic system with a non-degenerate parabolic curve consisting of (an)orbit(s) was given

by Zegeling and Kooij [94,ZK]. In case two real different or coinciding parallel
lines are orbits, no limit cycles can exist since they must encircle a critical
point on such a line yet cannot cross this line. If the parallel lines are complex
no real critical points exist and no limit cycles are possible as well.

It was shown by Cherkas in 1963 that a quadratic system cannot have a
limit cycle which lies on an algebraic curve of degree 3 [63,C]. Of course , this
result can already directly seen to be true for Hamiltonian systems, being
systems with only cubic orbits and without limit cycles (although periodic
orbits may exist). In a series of papers Evdokimenko extensively studied
quadratic systems with cubic orbits and no limit cycles were found by him
[70,E],[71,E],[76,E],[79,E]. Two later papers, by Shen Boqian [91,S1] and Xu
Changjiang [92,X], seem to indicate , however, that a unique hyperbolic limit
cycle may still be possible in these systems. A new proof of the theorem of
Cherkas was given, however, by Chavarriga and Llibre [98,CL].

The question of the existence in quadratic systems of algebraic limit cycles
of degree four or more seems still imperfectly understood. Attention should
be given to the results of Jablonskii [70,J],[71,J],Filipcov [73,F1],[73,F2] and
Chavarriga and Llibre [98,CL].

Algebraic solution curves of any degree exist, however, whether isolated
or in a family of such solutions. Examples can be found in a paper by
Druzhkova [75,D2], which, in fact, is concerned with polynomial systems.

7.4   Limit cycle distribution over two nests

As follows from Theorem 4.8 in a quadratic system there exist at most two
nests of limit cycles, the motion in one nest being clockwise and in the other
counterclockwise. Moreover, in relation to the second part of the 16th Hilbert
problem some remarks on the possible distribution of limit cycles over two
nests were already made in section 7.2.

We will use the notation (m,n), meaning that there are precisely m limit
cycles in one nest and n in the other.

Obviously, in the classes $m_f=0$ and 1, two nests of limit cycles cannot
occur. With regard to class $m_f=2$ only the combination (1,1) can occur
if two nests are present; in that case the limit cycles are hyperbolic. This
follows from the classification of the phase portraits in $m_f=2$ if the number of
infinite critical points is finite given by Reyn and Kooij [97,RK]. Those with
an infinite number of infinite critical points is given by Gasull and Prohens

[96,GP1]; they contain no limit cycles at all.

For systems in $m_f=3$ two nests of limit cycles are not possible if there exists a degenerate infinite critical point, i.e. a point $M_{1,2}^i$ or $M_{1,3}^i$, since in this case there exists at most one limit cycle (which is then hyperbolic). The phase portraits of this class are given by Reyn and Huang [97,RH]. This leaves the class with an $M_{1,1}^0$ saddle node at infinity. It was shown by Huang and Reyn [95,HR2] and Coppel [96,C] that if two nests of limit cycles occur, then in at least one nest there exists precisely one limit cycle, which is hyperbolic if it occurs. This result can be improved using the uniqueness results obtained by Zegeling and Kooij [99,ZK] to show that only the hyperbolic (1,1) combination is possible.

When considering $m_f=4$, first, the class with four real finite critical points is mentioned since a definite result for this class was obtained. In fact, also for this class only a (1,1) combination is possible; with hyperbolic limit cycles. This result was obtained by Zegeling and Kooij [99,ZK]. If two of the four finite critical points are made to coincide such that two foci and a second order critical point results, the class $e^1e^1m_2^0$ is obtained. If m_2^0 is a second order degenerate critical point, so a cusp, there exists at most one limit cycle in the system [88,C], and the distribution problem does not exist. For the other case, when m_f is a second order saddle node, the problem does not seem to have been attacked yet. Seen as a limiting case of that with four real critical points it seems likely, however, that only the (1,1) combinations are possible.

Since two foci are necessary to obtain two nests of limit cycles, the only class that remains as a candidate for a different combination than (1,1) is that with two real and two complex finite critical points $c_1^0c_1^0e^1e^1$. In fact, various examples in this class were given that show that also (1,2) and (1,3) are possible and all evidence so far points in the direction of these being the only remaining possibilities; the (1,3) distribution corresponding to the conjecture that H(2)=4.

In the late fifties of the twentieth century an early attempt to investigate the possible relative positions of limit cycles in quadratic systems can be found in the interesting work of Dong Jinzhu [59,D]. Apart from the basic properties of closed orbits in quadratic systems as mentioned earlier in section 4.3.2, also a final answer was given by him to this question, however, in the conviction that Petrovsky and Landis obtained the correct result in that H(2)=3. A first example of a (1,2) combination was given; however, it is in class $c_1^0c_1^0e^1e^1$.

The question of the possible distributions of limit cycles over two nests, seemingly being settled, it took nearly twenty years to realize this was not the case, which subsequently led to the conjecture that H(2)≥4. Looking back, this appears somewhat surprising since the results of Bautin [39,B2] showed that up till three limit cycles could be generated in one nest; could there be no other limit cycle(s) around a possible other focus?

The first example, given by Chen Lansun and Wang Mingshu [79,CW] starts with examining the system

$$\dot{x} = -y - 3x^2 + xy + y^2, \tag{7.4.1}$$

$$\dot{y} = x(1 + \frac{2}{9}x - 3y), \tag{7.4.2}$$

which belongs to the class $c_1^0 c_1^0 e^1 e^1 C_{0,1}^0 C_{0,1}^0 E^{-1}$. In (0,0) the antisaddle is a stable second order weak focus, whereas in (0,1) the antisaddle is an unstable rough focus. It follows that on $1-3y=0$ there is $\dot{y} = \frac{2}{9}x^2 > 0$ for $x \neq 0$. As a result $y < \frac{1}{3}$ and $y > \frac{1}{3}$ are Poincaré–Bendixson regions, so around both (0,0) and (1,0) there exist an odd number (counting multiplicity) of limit cycles, so at least one limit cycle. Around (0,0) the innermost limit cycle is unstable from the inside as is the outmost from the outside. Around (1,0) the innermost limit cycle is stable from the inside as is the outermost from the outside.

Next is considered the system

$$\dot{x} = -\delta_2 x - y - 3x^2 + (1 - \delta_1)xy + y^2, \tag{7.4.3}$$

$$\dot{y} = x(1 + \frac{2}{9}x - 3y), \tag{7.4.4}$$

which is related to (7.4.1),(7.4.2), and where $0 < \delta_2 \ll \delta_1 \ll 1$. For the relevant focal values follows, with (7.3.12),(7.3.13),(7.3.14), $W_1 = 2\delta_1, W_2 = W_2(\delta_1) < 0$ for small $|\delta_1|$. As a result it may be seen that if $\delta_1 = \delta_2 = 0$, the critical point in (0,0) is a stable second order weak focus, which for $\delta_1 > 0, \delta_2 = 0$ becomes an unstable first order weak focus, and for $\delta_1 > 0, \delta_2 > 0$ a stable rough focus. Related to these subsequent changes in stability is the bifurcation from (0,0) of zeros in the distance function (7.3.4), each zero representing a small limit cycle coming out of the origin. If $0 < \delta_2 \ll \delta_1 \ll 1$ two small hyperbolic limit cycles exist near (0,0), the innermost being much smaller than the outermost. Since the already for $\delta_1 = \delta_2 = 0$ existing (large) limit cycles around (0,0) and (1,0) each are odd in number (counting multiplicity), from the behavior of the Poincaré return map near a limit cycle it may be derived that small changes of δ_1 and δ_2 do not change their oddness and around each critical point at least one large limit cycle exists. As a result there exist at least three limit cycles around (0,0) and at least one limit cycle around (1,0).

The example given by Shi Songling reads [80,S1]

$$\dot{x} = \lambda x - y - 10x^2 + (5 + \delta)y + y^2, \qquad (7.4.5)$$
$$\dot{y} = x + x^2 + (-25 + 8\epsilon - 9\delta)xy, \qquad (7.4.6)$$

where $0 < -\lambda \ll -\varepsilon \ll -\delta \ll 1$. The analysis runs as for the example given by Chen Lansun and Wang Mingshu, except that for $\lambda = \varepsilon = \delta = 0$ the origin is a third order weak focus. Again there exists at least one limit cycle around (0,1) and at least three limit cycles around (0,0). A concrete example is given by taking for λ, ε and δ small numerical values [84,P]. System (7.4.5),(7.4.6) with $\lambda = \varepsilon = \delta = 0$ falls within the class of quadratic systems having a third order focus and (at least) one limit cycle around a rough focus. The phase portraits of this class can be found under W16,W19 and W20 in the classification of phase portraits of quadratic systems with a third order weak focus by Artés and Llibre [97,AL]. Phase portraits W16,W19 and W20 fall into the classes $c_1^0 c_1^0 e^1 e^1 C_{0,1}^0 C_{0,1}^0 E^{-1}$, $c_1^0 c_1^0 e^1 e^1 E^{-1} M_{0,2}^0$ and $c_1^0 c_1^0 e^1 e^1 E^{-1} E^{-1} E^1$, respectively. At least four limit cycles can be realized by small perturbations of the coefficients of a system in the classes represented by W16,W19 and W20. Since for the systems in W16 and W19 uniqueness and hyperbolicity of the limit cycle around the rough focus was shown by Kooij and Zegeling [98,KZ], for these cases small perturbations generate systems with precisely four limit cycles; they have the distributions (1,3).

After the examples given by Chen Lansun and Wang Mingshu, and Shi Songling there followed several other examples by various other authors. They all contained two finite foci and two complex critical points.

In an interesting note in 1983 by Andronova [83,A] the starting point to create a quadratic system with (at least) four limit cycles is the system with two center points in the class $c_1^0 c_1^0 e^1 e^1 E^{-1} C_{0,1}^0 C_{0,1}^0$ as discussed in section 6.2.1 of which the phase portrait is given in Figure 6.2a, portrait 5 and which is given by

$$\dot{x} = -y + ax^2 + by^2 \equiv P(x, y), \qquad (7.4.7)$$
$$\dot{y} = x(1 + cy) \equiv Q(x, y), \qquad (7.4.8)$$

where c<a<0,0<b< −c. Adding to Q(x,y) the perturbation term εx^2 and to P(x,y) the term $\frac{2a+c}{a+b}\varepsilon xy$, where $0< \varepsilon \ll 1$ yields for c−3a−5b $\neq 0$ (=0) a system of the type with a second(third) order weak focus proposed by Chen Lansun (Shi Songling). At least four limit cycles can then be generated following the procedure described above. The bifurcation for c−3a−5b=0 used by Andronova is represented in [97,AL] by going from the line A15 into the region R15. Similarly, in class $c_1^0 c_1^0 e^1 e^1 E^{-1} M_{0,2}^0$ one could go from P15 to C16, so representing case W19 and in class $c_1^0 c_1^0 e^1 e^1 E^{-1} E^{-1} E^1$ from A14 to R14, representing case W16.

Another way to generate a system with a third order weak focus and (at least) one limit cycle around the other focus was shown by Wanner [83,W] who remarked that such a system, in fact, can already be found in the work of Frommer[34,F] and Kapteyn [11,K],[12,K]. They can be bifurcated from homogeneous quadratic systems. Using this approach, the phase portraits indicated by W16,W19 and W20 in [97,AL] and belonging to the classes $c_1^0 c_1^0 e^1 e^1 C_{0,1}^0 C_{0,1}^0 E^{-1}, c_1^0 c_1^0 e^1 e^1 E^{-1} M_{0,2}^0$ and $c_1^0 c_1^0 e^1 e^1 E^{-1} E^{-1} E^1$, respectively, can be bifurcated from systems in $m_4^2 C_{0,1}^0 C_{0,1}^0 E^{-1}, m_4^2 E^{-1} M_{0,2}^0$ and $m_4^2 E^{-1} E^{-1} E^1$, respectively, the phase portraits of which are given in Figure 3.3. This can be done by using the normal form given in [97,AL]

$$\dot{x} = -y + lx^2 + 5axy + y^2, \qquad (7.4.9)$$
$$\dot{y} = x + ax^2 + (3l + 5)xy, \qquad (7.4.10)$$

in the appropriate region of parameter space and letting $\bar{x} = \varepsilon x, \bar{y} = \varepsilon y$, and $\tau = \varepsilon^{-1}t$. Then follows

$$\acute{\bar{x}} = -\varepsilon\bar{y} + l\bar{x}^2 + 5a\bar{x}\bar{y} + \bar{y}^2, \qquad (7.4.11)$$
$$\acute{\bar{y}} = \varepsilon\bar{x} + a\bar{x}^2 + (3l + 5)\bar{x}\bar{y}, \qquad (7.4.12)$$

where $' = \frac{d}{d\tau}$, and $\varepsilon \to 0$ then leads to the result.

All examples with (at least) four limit cycles discussed above involve small parameters and very small limit cycles. Numerical investigations to construct systems with (at least) four limit cycles of "normal size" were made by Perko[84,P]. Numerical calculations to produce (1,1) and (1,2) distributions were also made by Perko [84,P] and by Perko and Shu Shih [84,PS].

Further results on the problem of the (1,3) distribution were obtained by Qin Yuanxun, Shi Songling and Cai Suilin [82,QSC], [83,QSC] and several other Chinese authors.

A method to construct a (1,3) distribution by successive (degenerate) Hopf bifurcations was studied by Rousseau and Schlomiuk [88,RS].

The possibility of an answer to the question whether apart from a (1,1) distribution the only other possible distributions are the (1,2) and the (1,3) distributions rests on the possibility to give a thorough analysis of the class of quadratic systems with two foci and two complex critical points in the finite plane. This remains a difficult problem as long as it is difficult to prove that there exist at most two or three limit cycles in one nest.

However, some progress may be obtained by producing uniqueness results.

7.5   Bifurcation of limit cycles

Results obtained in bifurcation theory yield another source of information about limit cycles in quadratic systems. This subject was already slightly touched upon in previous sections. In particular, the basic result of Bautin [39,B1] was mentioned, who showed that by small perturbations of the co-efficients up to three limit cycles can be generated out of a weak focus or a center point. The use of this property was illustrated in the examples of (at least) four limit cycles as given by Chen Lansun and Wang Mingshu [79,CW] and Shi Songling [80,S1]. The generation of a limit cycle from a weak focus of the first order is often indicated in the literature as an Andronov–Hopf bifurcation or Hopf bifurcation; a degenerate or generalized Hopf bifurcation if a higher order weak focus or possibly a center point is involved. One may look at these bifurcations in reversed order and consider e.g. the contraction of a hyperbolic limit cycle upon a point. By definition, a hyperbolic limit cycle has a characteristic exponent unequal to zero (7.2.4), so is structurally stable, meaning that small changes in the coefficients of the system leaves precisely one hyperbolic limit cycle [ALGM2]. Suppose the limit cycle remains hyperbolic in the contracting process upon a point, then, since in the interior of the limit cycle there exists precisely one elementary critical point (a focus), then in the limit, the point of contraction is also a critical point, not necessarily a focus, as the example of the center point shows and also not necessarily an elementary critical point as other critical points outside of the limit cycle may join in the limiting process towards the point of contraction. A necessary condition to be imposed on this point is that it should be divergence free. This follows directly from the property that the divergence integral of the vector field $(P(x,y),Q(x,y))$ over the interior enclosed by a

limit cycle equals zero, also in the limit and part of the line of zero divergence remains in this interior, thereby shrinking to the point of contraction
in this limit. For elementary critical points this leaves the weak focus and
the center point, as already encountered in the (degenerate) Hopf bifurcation. Furthermore, nilpotent critical points and critical points in essentially
homogeneous systems should be considered as candidates.

Apart from contraction on a point, the other possibility for a limit cycle to
lose its character is to end up in a convex closed curve, convex since a limit
cycle encloses a convex region; this property is preserved into its limiting
position. An example of this phenomenon was mentioned in section 7.2;
in the discussion on multiple limit cycles it was revealed that upon small
changes in the coefficients of the system a multiple limit cycle can split up
in a number of limit cycles, in fact at most equal to its multiplicity. Another
example was indicated in section 7.4 in relation to system (7.4.7),(7.4.8),
studied by Andronova [83,A], wherein limit cycles are bifurcated from closed
orbits, contained in an annulus around a center point.

Other types of convex closed curves from which limit cycles can be bifurcated should be considered as well.

Also, combinations of this type of bifurcation with that from a critical
point can occur.

7.5.1   Bifurcation of limit cycles from critical points

As was pointed out in section 2.3.1 a quadratic system with a nilpotent
critical point in the origin can be written as

$$\dot{x} = y + a_{20}x^2 + a_{11}xy + a_{02}y^2, \qquad (7.5.1)$$
$$\dot{y} = b_{20}x^2 + b_{11}xy + b_{02}y^2. \qquad (7.5.2)$$

If $b_{20} \neq 0$, there exists a second order cusp in the origin. In the 1970s of
the twentieth century bifurcation of the cusp critical point was studied by
Bogdanov [81,B1] and Takens [74,T], which led to the Bogdanov– Takens
bifurcation [GH,p.364]. It is concerned with a class in (7.5.1),(7.5.2) that
can, for bifurcational purposes, be considered as being represented by the
system

$$\dot{x} = y \equiv P(x, y), \qquad (7.5.3)$$
$$\dot{y} = x^2 \pm xy = Q(x, y). \qquad (7.5.4)$$

It was shown that addition of the perturbation terms $\varepsilon_1 + \varepsilon_2 y$, where ε_1 and ε_2 are small parameters , to $Q(x,y)$ yields the complete set of qualitatively different phase portraits that can be bifurcated from (7.5.3),(7.5.4). It appears that at most one (hyperbolic) limit cycle comes out of the cusp point due to these perturbation terms.

Another class within (7.5.1),(7.5.2) and for which the cusp bifurcation can be represented by a three term perturbation, yielding at most two (hyperbolic) limit cycles is studied by Dumortier and Fiddelaers [91,DF].

Not all cusp point bifurcations have been studied yet. For instance, bifurcations of limit cycles from the cusp point in phase portrait 4 of Figure 6.6, representing a Hamiltonian system seems not to have been studied so far.

Bifurcations from third and fourth order nilpotent critical points, so from saddles, third order points with an elliptic and a hyperbolic sector, or fourth order saddle nodes are increasingly more complicated and no complete answer exists as yet. Important results in this direction can be found in [87,DRS],[91,DRS],[91,DF]and[92,F].

About bifurcations from critical points in essentially homogeneous systems little seems to be known. The same applies to nilpotent and highly degenerate points at infinity as listed under section 2.3.2.

7.5.2   Bifurcation of limit cycles from quadratic systems with center points

Apart from center points, limit cycles may also bifurcate from the closed orbits in the annular region around a center point and from the separatrix cycle enclosing such a region; the various kinds of separatrix cycles are illustrated in Figure 6.8.

By using an integrating factor, a quadratic system with a center point may be transformed into a Hamiltonian system, as for instance described in [ALGM2]. If the perturbed Hamiltonian system is written in the form

$$\dot{x} = -H_y + \varepsilon p(x, y), \qquad (7.5.5)$$
$$\dot{y} = H_x + \varepsilon q(x, y), \qquad (7.5.6)$$

where ε is a small parameter, then, obviously, for a limit cycle L bifurcating from an orbit of the Hamiltonian system given by $h = H(x, y)$, the divergence integral over the interior S of L,

$$\varepsilon \int\int_S p_x(x, y) + q_y(x, y) dx dy = \oint_L -p(x, y) dy + q(x, y) dx, \qquad (7.5.7)$$

should vanish. Then, it can be shown (theorem 78 in [ALGM2]) that if

$$I(h) = \oint_{h=H(x,y)} -p(x, y) dy + q(x, y) dx \qquad (7.5.8)$$

evaluated over $h = H(x, y)$ vanishes and

$$l \equiv \oint_{h=H(x,y)} [p_x(x(s), y(s)) + q_y(x(s), y(s))] ds \neq 0, \qquad (7.5.9)$$

then there exists a unique and hyperbolic limit cycle L contracting upon $h = H(x, y)$ if $\varepsilon \to 0$. A more refined analysis is needed if $I(h) \equiv 0$ or $l=0$.

The evaluation of the integral $I(h)$, referred to as an Abelian integral if $H(x, y)$ is algebraic, and the determination of its zeros are in general technically difficult. A glance at the literature gives the following result.

The first results seem to have been given in 1973-74 by Averin [73,A],[74,A], who considered quadratic perturbations of a harmonic oscillator, as well as of the system

$$\dot{x} = y(y - 1), \qquad (7.5.10)$$
$$\dot{y} = x + \beta xy, \qquad (7.5.11)$$

where β is a real constant, whereas Bogdanov [81,B2] considered perturbations of the system for $\beta=0$.

In [82,I] Il'yashenko studies the equation

$$\frac{dw}{dz} = \frac{3(1 - z^2)}{2w}, \qquad (7.5.12)$$

where z and w are complex variables, allowing perturbations in the class of equations

$$\frac{dw}{dz} = \frac{P_2}{Q_1}, \qquad (7.5.13)$$

where P_2 and Q_1 are polynomials of degree 2 and 1, respectively.

In 1985 Huang Qiming and Li Jibin [85,HL1] presented a numerical study of the generation of $(0,1),(0,2)$ and $(1,1)$ limit cycle configurations for specific perturbations of the Hamiltonian system

$$\dot{x} = \frac{1}{4c} - cy^2, \tag{7.5.14}$$

$$\dot{y} = -\frac{1}{4c} + ax^2, \tag{7.5.15}$$

and particular values of a and c. In [88,LC] Li Jibin and Chi Yuehua continued the study of this system for all values of a and c. Also Li Cunfu and Xu Yumin [86,LX] determined the generation of limit cycles, in particular of the $(0,1)$ and $(1,1)$ limit cycle configuration, from the system

$$\dot{x} = xy, \tag{7.5.16}$$

$$\dot{y} = \frac{1}{2} - \frac{1}{2}x^2 + \frac{1}{2}y^2, \tag{7.5.17}$$

which has a pair of isochronous centers, whereas Li Jibin and Chen Xiaoqin [86,LC] considered the generation of $(0,1),(0,2)$ and $(1,1)$ limit cycle configuration from the system

$$\dot{x} = xy, \tag{7.5.18}$$

$$\dot{y} = \frac{1}{2} - \frac{1}{2}x^2 + 2y^2, \tag{7.5.19}$$

and Jiang Jifa and Liu Zhangrong [86,JL] studied the generation of the $(0,1),(1,0),(1,1),(0,2)$ and $(2,0)$ configurations of limit cycles from

$$\dot{x} = xy, \tag{7.5.20}$$

$$\dot{y} = \frac{1}{2} - x^2 - y^2. \tag{7.5.21}$$

In 1987 Drachman, van Gils and Zhang Zhifen [87,DGZ] analyzed the Hamiltonian system

$$\dot{x} = -2y + 3\lambda y^2, \tag{7.5.22}$$

$$\dot{y} = -x - \mu_1 y - x^2 + \mu_2 y^2, \tag{7.5.23}$$

where $0 \le \lambda \le \frac{2}{3}\sqrt{2}$ and μ_1 and μ_2 are small parameters.

Chicone and Jacobs [91,CJ] determined, within the class of quadratic systems, the bifurcation of limit cycles from all quadratic systems with an isochronous center. Up to two limit cycles can be bifurcated for genuine quadratic centers, whereas for a linear center three limit cycles can globally be bifurcated in a higher order bifurcation.

Seemingly the first attempt to study the bifurcation of limit cycles from the separatrix cycle enclosing the annular region of closed orbits around a center point was given in 1989 by Guckenheimer, Rand and Schlomiuk [89,GRS]. Numerical support was given for the conjecture that two is the maximum number of limit cycles that can be bifurcated due to quadratic perturbations in a Hamiltonian quadratic system with a loop containing an elementary saddle. It was shown that five is a rigorous upper bound.

In 1990 Gavrilov and Horozov [90,GH] studied perturbations of the Hamiltonian system

$$\dot{x} = -2xy, \tag{7.5.24}$$
$$\dot{y} = -1 + g^2 - 4gx + 3x^2 + y^2, \tag{7.5.25}$$

with g a real constant, whereas later [93,GH] they showed that no more than two limit cycles can appear out of a (generic) quadratic system with three saddle points and one center point; the phase portrait for this case is given by phase portrait 1 of Figure 6.4.

In a paper in 1992 [92,P] Perko gave a global analysis of the phase portraits of the system

$$\dot{x} = y(y-1) + \mu_1 x + \mu_2 xy, \tag{7.5.26}$$
$$\dot{y} = x, \tag{7.5.27}$$

where μ_1 and μ_2 are real constants, which, if they are considered as small parameters, leads to the perturbed Hamiltonian system(7.5.10),(7.5.11) for $\beta=0$ as studied by Bogdanov. In 1994 Żoladek published a large paper [94,Z], first deriving again the center conditions, then studying perturbations of the Volterra−Lotka system. It was shown that up to three limit cycles can be bifurcated. Conjectures were made about the maximum number of limit cycles that could possibly be bifurcated from quadratic systems with a center point.

In the same year results of the bifurcation of limit cycles from the system

$$\dot{x} = -y + ny^2, \tag{7.5.28}$$
$$\dot{y} - x - xy, \tag{7.5.29}$$

where $0 < n < 1$, containing two center points were published by van Horssen and Kooij [94,HK]. It was shown that $(1,0),(2,0)$ and $(1,1)$ combinations of limit cycles can be bifurcated within quadratic systems.

Also in the same year Horozov and Iliev [94,HI2] proved the conjecture mentioned by Guckenheim, Rand and Schlomiuk [89,GRS] that two is the maximum number of limit cycles that can bifurcate within quadratic systems from a loop with an elementary saddle in a Hamiltonian quadratic system.

In 1995 van Horssen and Reyn [95,HR1] considered the Volterra–Lotka system

$$\dot{x} = x(1 - x - ay), \tag{7.5.30}$$
$$\dot{y} = y(-1 + ax + y), \tag{7.5.31}$$

where $1 < a < \infty$, and showed that at most one limit cycle is bifurcated from the triangular annulus around the center point.

In [95,LZ] Li Baoyi and Zhang Zhifeng considered again perturbations of the system with a saddle loop as studied by Bogdanov, by using the form

$$\dot{x} = y + \varepsilon P(x, y), \tag{7.5.32}$$
$$\dot{y} = -x - x^2 + \varepsilon Q(x, y), \tag{7.5.33}$$

where $P(x, y), Q(x, y)$ are polynomials of degree $\leq N$ $(N \geq 2)$ and ε a small parameter. Attention was given to the case that the Abelian integral $I(h) \equiv 0$ so that a higher order bifurcation analysis was needed.

Also in 1995 Li Chengzhi, Jaume Llibre and Zhang Zhifeng [95,LLZ2] proved that the lowest upper bound for the number of isolated zeros of the Abelian integrals associated with quadratic perturbations in Hamiltonian quadratic systems having a center and a straight line orbit is equal to 1.

Furthermore Shafer and Zegeling [95,SZ] studied the bifurcation of limit cycles from quadratic systems with a center for the symmetric case, the phase portraits of which are given in Figure 6.2. A followup through a numerical approach by Shafer,Wu and Zegeling can be found in [96,SWZ].

In 1997 Dumortier, Li Chengzhi and Zhang Zhifeng presented in [97,DLZ] a complete study of the system

$$\dot{x} = -y - 3x^2 + y^2 + \mu_1 x + \mu_2 xy, \tag{7.5.34}$$
$$\dot{y} = x(1 - 2y) + \mu_3 x^2, \tag{7.5.35}$$

where μ_1, μ_2 and μ_3 are small parameters and wherein the undisturbed system belongs to the symmetric case having two centers and two unbounded hetereoclinic loops, the phase portrait of which is given by portrait 25 in Figure 6.2c. Also in 1997, He Yue and Li Chengzhi proved in [97,HL] that at most two limit cycles can appear from a saddle loop in any (not necessarily Hamiltonian) quadratic system with a center point except for one case.

Various of the investigations mentioned above can be considered to be in the line of trying to solve the weak 16th Hilbert problem proposed by V.I.Arnold [A], asking for an upper bound of the number of zeros of the Abelian integral I(h) for the system

$$\dot{x} \;=\; -H_y + \varepsilon f(x,y), \qquad\qquad (7.5.36)$$
$$\dot{y} \;=\; H_x + \varepsilon g(x,y). \qquad\qquad (7.5.37)$$

Here H(x,y) is a polynomial Hamiltonian of degree n+1 and the terms f(x,y) and g(x,y) are polynomials of degree $\leq$n. It was shown by Varchenko and Khovansk that this upper bound exists.

7.5.3 Bifurcation of limit cycles from other convex closed curves

Apart from the closed orbits and separatrix cycles occurring in quadratic systems with a center point, as discussed in the previous section, other types of convex closed curves should be considered as candidates from which limit cycles can be bifurcated.

Therefore attention should also be given to other convex simple closed curves, consisting of a finite or an infinite number of finite and/or infinite critical points and orbits connecting two flowwise consecutive critical points on the curve; the direction of flow on these orbits all being either clockwise or anticlockwise. These curves will be referred to as cycles if they contain a finite number of critical points. Separatrix cycles are then cycles consisting of critical points flowwise connected by orbits being separatrices either at the downstream or the upstream end or at both ends. The separatrix cycles in quadratic systems have been studied by Dong Jinzhu [62,T] ,Chicone and Shafer [83,CS] and de Jager [89,J]; they only considered separatrix cycles with finite critical points. In relation to the 16th Hilbert problem, Dumortier, Roussarie and Rousseau defined a wider class of 121 convex closed curves, referred to as graphs from which limit cycles in quadratic systems can possibly be bifurcated [94,DRR2]. Results on the bifurcational properties of these graphics were reported on in [94,DRR1],[94,DRR2],[96,DER],[97,R2]. Here

we restrict ourselves to bounded cycles. There appear to be seven such cycles
with one critical point, five with two critical points and one cycle with three
critical points.

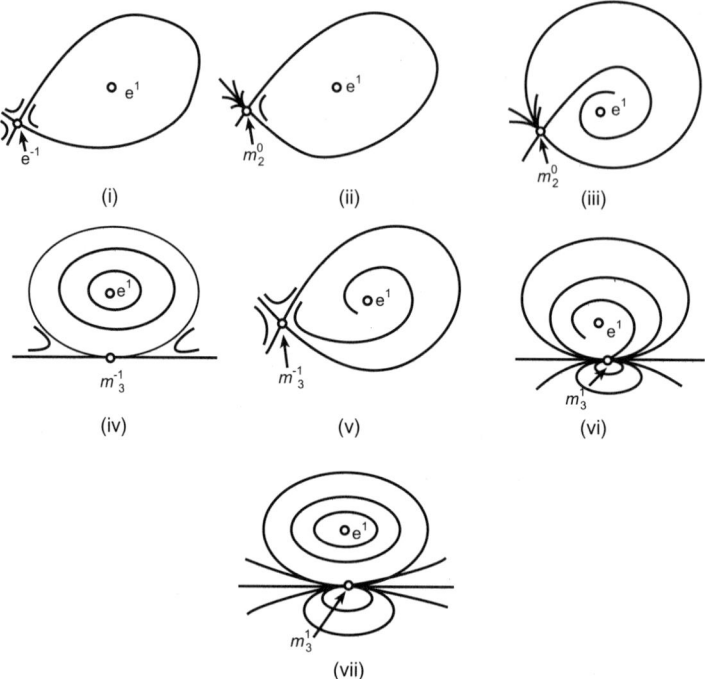

Figure 7.2 Types of bounded cycles with one critical point

The seven types with one finite critical point are illustrated in Figure 7.2,
and can be described as follows:

(i) Loops with an elementary saddle.

The case that the critical point e^1 in the interior of the loop is a center
point was already discussed in the previous section. If not, e^1 is a focus and
at least two limit cycles may exist in the interior of the loop. The focus may
be either rough or weak of the first or second order [89,J]. If the divergence
in the saddle point is unequal to zero, its sign determines the stability of
the loop and precisely one limit cycle can bifurcate, which then is hyperbolic
and has the same stability as the loop [ALGM 2, Theorem 44]. If the saddle
is weak (so the divergence is equals zero and the ratio of the eigenvalues
$\mu < 0 < \lambda$ equals r$=-\frac{\mu}{\lambda}=1$) there exists no loop if the weak saddle is of order
2 or 3 [89,Z].

(ii)Saddle node (m_2^0) loop with a Poincaré return map.

The critical point in the interior of the loop is a strong focus or a first order weak focus, probably not a second order focus and certainly not a third order focus. There may exist at least one limit cycle in the interior of the loop [89,J]. The stability of the loop is determined by the sign of the non-vanishing eigenvalue in m_2^0. Precisely one limit cycle may bifurcate from the loop (using arguments given for Theorem 44 of [ALGM 2], which is then hyperbolic and has the same stability as the loop.

(iii)Saddle node (m_2^0) loop without a Poincaré return map.

The critical point in the interior of the loop is a focus as can be shown using Theorem 4.5. Precisely one (hyperbolic) limit cycle can be bifurcated out of the loop; its stability being determined by the sign of the non-zero eigenvalue in the saddle node m_2^0 [ALGM 2,theorem 52].

An example can be given by the system

$$\dot{x} = x^2 + y^2 - 1 + y(x-1), \qquad (7.5.38)$$
$$\dot{y} = -x(x-1), \qquad (7.5.39)$$

which has a stable focus in $(0, \frac{1}{2}(1-\sqrt{5}))$, a repelling saddle node m_2^0 in $(1,0)$ and a saddle node loop coinciding with the unit circle (Figure 7.3).

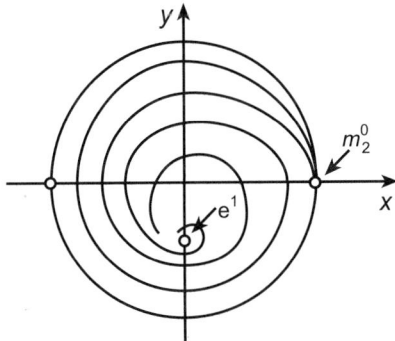

Figure 7.3 Example of a bounded cycle of type (iii)

Adding linear perturbations such that (7.5.38),(7.5.39) becomes

$$\dot{x} = x^2 + y^2 - 1 + y(x-1-\varepsilon), \qquad (7.5.40)$$
$$\dot{y} = -x(x-1-\varepsilon), \qquad (7.5.41)$$

where $0< \varepsilon \ll 1$, makes the saddle node disappear and create an unstable limit cycle coinciding with the unit circle.

(iv)Loop through a nilpotent saddle m_3^{-1}.

It can be shown that the antisaddle in the interior of the loop can only be a center point [89,J]; the loop is mentioned already in section 6.4.1.

(v)Loop through a semi-elementary saddle m_3^{-1}.

It can be shown that the antisaddle in the interior of the loop can only be a rough focus, no limit cycle can exist in the interior of the loop of which the stability is determined by the sign of the non-zero eigenvalue in the saddle m_3^{-1} [89,J]. Precisely one (hyperbolic) limit cycle can bifurcate from the loop (using arguments given in Theorem 44 of [ALGM 2]), which then has the same stability as the loop.

(vi),(vii)Cycles in the elliptic region of a nilpotent critical point with an elliptic and a hyperbolic sector m_3^1 and the finite boundary cycles of these regions of cycles. These boundary cycles have in their interior either a center point (vii) or a focus (vi), then the flow is spiraling out into the nilpotent point. Type (vii) is illustrated by phase portrait 17 in Figure 6.2b. An example of type (vi) can be found in [90,J].

There appear to be five types of cycles with two finite critical points; they are illustrated in Figure 7.4 and can be described as follows:

(i)A polycycle with two elementary saddles and $\Sigma \neq 0$. Here $\Sigma \equiv 1 - \frac{\mu_1 \mu_2}{\lambda_1 \lambda_2}$, where $\mu_i(\lambda_i)$ is the negative (positive) eigenvalue in saddle point i(i=1,2). It can be shown that inside the polycycle no limit cycle can exist and the antisaddle is a rough focus [89,J]. The sign of Σ determines the stability of the polycycle and up to two (hyperbolic)limit cycles can be bifurcated from the cycle [94,DDR1],[79,R].

(ii)A polycycle with two elementary saddles and $\Sigma=0$.

Then it can be shown that the antisaddle in the interior of the cycle is a center point [89,J]. It was shown by Żoladek [95,Z] that at most two limit cycles can be bifurcated from such a polycycle.

(iii)A polycycle with an elementary saddle and an m_2^0 saddle node with a Poincaré return map.

It can be shown that in the interior of the cycle no limit cycle can exist and that the antisaddle is a rough focus. The sign of the non-vanishing eigenvalue in m_2^0 determines the stability of the loop. Up to two limit cycles can be bifurcated from the polycycle [94,DDR].

Remark. Since the polycycles under (i),(ii) and (iii) have a Poincaré return map, it can be shown using Lemma 1 of Tung Chinchu (section 4.2) that the straight line connecting the two critical points is an orbit on the polycycle.

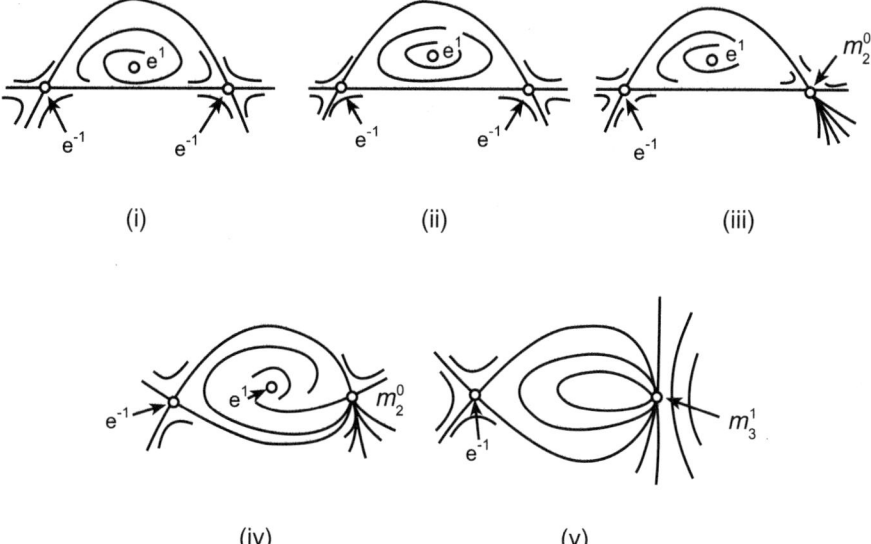

(i)   (ii)   (iii)

(iv)   (v)

Figure 7.4 Types of bounded cycles with two critical points.

(iv)A polycycle with an elementary saddle and an m_2^0 saddle node without a Poincaré return map.

At most two (hyperbolic) limit cycles can be bifurcated from the polycycle [94,DDR].

(v) A polycycle with an elementary saddle and a nilpotent point with an elliptic and a hyperbolic sector.

A classification of the phase portraits wherein these polycycles occur was given by de Jager [90,J].

Quadratic systems having a nilpotent critical point with an elliptic and a hyperbolic sector can be represented by

$$\dot{x} = y + \lambda_1 x^2 + \lambda_2 xy + \lambda_3 y^2, \tag{7.5.42}$$
$$\dot{y} = xy, \tag{7.5.43}$$

with $\lambda_1 > 0, \lambda_2 = 0$ or 1, λ_3 real.

The polycycles occur for $\lambda_3 > 0$. Bifurcation of these cycles do not seem to have been studied although for $\lambda_2 = 0$ integrals are known.

There exists precisely one type of cycle with three critical points. It has a center point in the interior of a triangle the sides of which are straight line orbits (Figure 7.5). It was shown by Żoladek [95,Z] that at most three (hyperbolic) limit cycles can be bifurcated from such a polycycle. If in addition the system is Hamiltonian this maximum equals two [94,HI2].

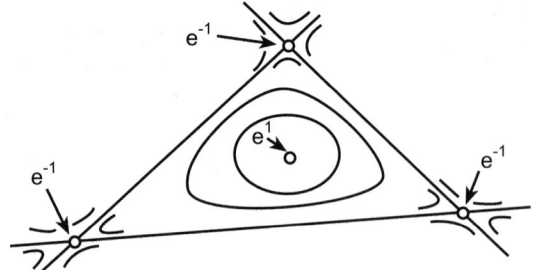

Figure 7.5 Bounded cycle with three critical points

Chapter 8

Phase portraits of quadratic systems in the class $m_f{=}1$

8.1 Introduction

In this chapter the phase portraits are presented of the class of quadratic systems with precisely one (elementary and real) critical point (and no complex critical points), the class with finite multiplicity 1.

A first classification involving this class was given in an investigation in 1988 by Coll, Gasull and Llibre [88,CGL], who considered a wider class, allowing the occurrence of additional complex critical points and allowing the unique finite critical point to be multiple. Their method consists of deriving ten normal forms to present all quadratic systems, then imposing the condition of a unique critical point, and determining the phase portraits. A later investigation by Reyn [97,R] was restricted to the class $m_f{=}1$ which led to a different approach. This approach starts with the observation that, as for class $m_f{=}0$, the dominant influence of the presence of higher order infinite critical points suggests an ordering of the various possible combinations of critical points at infinity as a framework for further classification.

Based on these results the affine invariant partition of the coefficient space corresponding to the topologically different phase portraits was given in [98,VV] by Voldman and Vulpe.

As a result of the classification there follows:

Theorem 8.1 *The number of limit cycles in a quadratic system with finite multiplicity $m_f{=}1$ is at most equal to 1 and if it exists, this limit cycle is* *hyperbolic.*

In the present account we will follow the approach of ordering the various possible combinations of critical points at infinity for $m_f=1$, using the equations developed in section 3.7.

8.2 One transversally non-hyperbolic critical point

For these systems we use the system

$$\dot{x} \;=\; a_{10}x + a_{01}y + a_{11}xy + a_{02}y^2, \qquad (8.2.1)$$
$$\dot{y} \;=\; b_{10}x + b_{01}y + b_{11}xy + b_{02}y^2, \qquad (8.2.2)$$

where $c_{23} \equiv a_{10}b_{01} - a_{01}b_{10} \neq 0$ since the origin is an elementary critical point, and $c_{56} \equiv a_{11}b_{02} - a_{02}b_{11} \neq 0$ since there should be one transversally non-hyperbolic infinite critical point, it being located at "the ends of the x axis". The requirement that the system belongs to $m_f=1$ yields $c_{25} = c_{26} + c_{35} = 0, c_{23}c_{36} \neq 0$. From Lemma 3.4 in section 3.7.1 it follows that the origin is a saddle (antisaddle) if $c_{56} < 0(c_{56} > 0)$.

In order to study the critical points at infinity we use the system

$$z' = -z(a_{10}z + a_{11}u + a_{01}zu + a_{02}u^2), \qquad (8.2.3)$$
$$u' = b_{10}z + b_{11}u + (-a_{10} + b_{01})zu + (-a_{11} + b_{02})u^2 - a_{01}zu^2 - a_{02}u^3, \qquad (8.2.4)$$

so that the infinite critical points are located in $u = \bar{u}$, where $\bar{u}$ is the solution of $f(u) \equiv a_{02}u^3 + (a_{11} - b_{02})u^2 - b_{11}u=0$. The point at the " ends of the x axis" is thus a fourth order point $M_{3,1}^i$ if $b_{11} \neq 0$, a fifth order point $M_{3,2}^i$ if $b_{11}=0, a_{11} - b_{02} \neq 0$, and a sixth order point $M_{3,3}^i$ if $b_{11} = a_{11} - b_{02}=0, a_{02} \neq 0$.

8.2.1 Systems with a fourth order infinite critical point

Using the classification of critical points at infinity [95,RK], it follows that $M_{3,1}^i$ is a fourth order semi-elementary saddle node $M_{3,1}^0$ with center manifold(s) transversal to the Poincaré circle. In fact, from $c_{25} = c_{26} + c_{35}=0$ it follows that $b_{11}^4\gamma_2 \equiv -b_{11}^2c_{15} + b_{10}b_{11}c_{35} + b_{10}^2c_{56}=0$ so $\gamma_2=0$; knowing this and with $c_{23}c_{36} \neq 0$, it also follows that $\alpha \equiv b_{00}b_{11}^2 + b_{10}^2b_{20} - b_{10}b_{01}b_{11} \neq 0$ and $b_{11}^6\delta_2 \equiv \alpha(a_{11}c_{25} + b_{11}c_{35} + 2b_{10}c_{56}) \neq 0$ so $\delta_2 \neq 0$. From figure 2 in [95,RK] the result follows.

Apart from the saddle node $M_{3,1}^0$ there exist two other critical points at infinity located in the non-zero roots of $f(u) = 0$. For $D \equiv (a_{11} - b_{02}) + 4a_{02}b_{11} = (a_{11} + b_{02})^2 - 4c_{56} > 0$ these points are both real and different, for $D=0$ coinciding and for $D<0$ complex. For $D>0$ these points are nodes (saddles) if $c_{56} < 0(>0)$, since the finite critical point is a saddle (antisaddle) and $i_f + i_i = i_t = 1$. If $D=0$ the two critical points coincide to form a tangentially (to the Poincaré circle) semi-elementary second order saddle node $M_{0,2}^0$ with center manifold(s) tangent to the Poincaré circle.

Systems with a finite saddle point

Since $c_{56} < 0$, then $D>0$ and there exist two nodes at infinity. These systems form class $e^{-1}E^1E^1M_{3,1}^0$ in class 1a as mentioned in Table 4 of section 3.7.4 of Chapter 3. The phase portrait is given in Figure 8.1a. No saddle connections are possible.

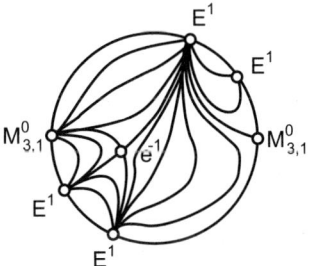

Figure 8.1a Phase portrait of class $e^{-1}E^1E^1M_{3,1}^0$.

Systems with a finite antisaddle

By using the stated conditions for this case and a linear transformation it may be seen that (8.2.1),(8.2.2) can also be written as

$$\dot{x} = \lambda_1 + \lambda_2 y + \lambda_1 xy - y^2, \qquad (8.2.5)$$
$$\dot{y} = 1 + xy, \qquad (8.2.6)$$

where $-\infty < \lambda_1 < \infty$, $0 < \lambda_2$.

The form of this system allows us to apply the theory of rotated vector fields as introduced in 1953 in the theory of limit cycles by Duff and since then further developed and widely used in quadratic systems [65,Y],[75,P].

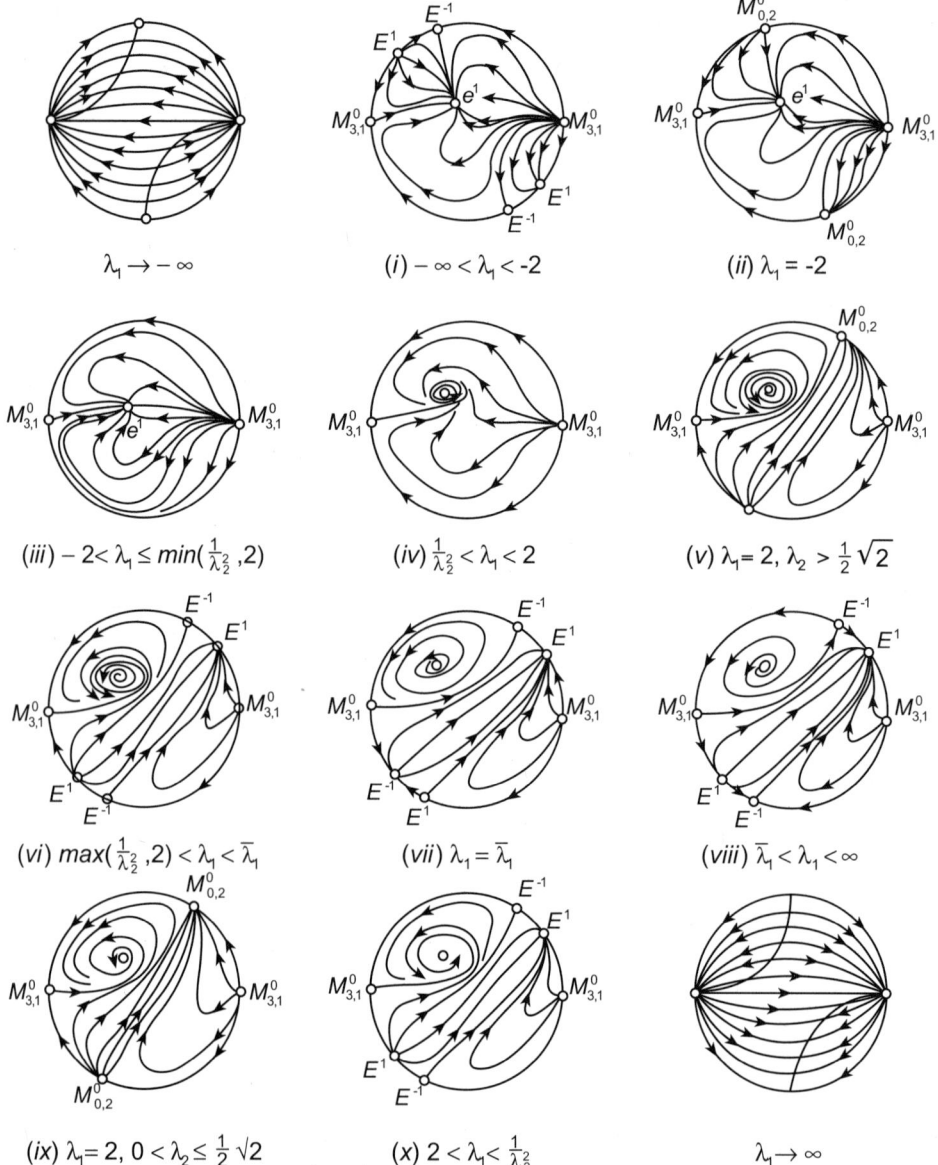

$\lambda_1 \to -\infty$
$(i) -\infty < \lambda_1 < -2$
$(ii)\ \lambda_1 = -2$

$(iii) -2 < \lambda_1 \le min(\frac{1}{\lambda_2^2},2)$
$(iv)\ \frac{1}{\lambda_2^2} < \lambda_1 < 2$
$(v)\ \lambda_1 = 2,\ \lambda_2 > \frac{1}{2}\sqrt{2}$

$(vi)\ max(\frac{1}{\lambda_2^2},2) < \lambda_1 < \overline{\lambda}_1$
$(vii)\ \lambda_1 = \overline{\lambda}_1$
$(viii)\ \overline{\lambda}_1 < \lambda_1 < \infty$

$(ix)\ \lambda_1 = 2,\ 0 < \lambda_2 \le \frac{1}{2}\sqrt{2}$
$(x)\ 2 < \lambda_1 < \frac{1}{\lambda_2^2}$
$\lambda_1 \to \infty$

Figure 8.1b Phase portraits for the classes $e^1 C_{0,1}^0 C_{0,1}^0 M_{3,1}^0$, $e^1 M_{0,2}^0 M_{3,1}^0$ and $e^1 E^{-1} E^1 M_{3,1}^0 (\lambda_2 > 0)$

The finite critical point is an antisaddle located in $(-\frac{1}{\lambda_2}, \lambda_2)$. It is a stable node if $\lambda_1 \le \frac{1}{\lambda_2^2} - 2$, a stable rough focus if $\frac{1}{\lambda_2^2} - 2 < \lambda_1 < \frac{1}{\lambda_2^2}$, a stable first order fine focus if $\lambda_1 = \frac{1}{\lambda_2^2}$, an unstable rough focus if $\frac{1}{\lambda_2^2} < \lambda_1 < \frac{1}{\lambda_2^2} + 2$ and an unstable node if $\frac{1}{\lambda_2^2} + 2 \le \lambda_1$. The stability and the order of the fine focus for $\lambda_1 = \frac{1}{\lambda_2^2}$ may be determined by transforming (8.2,5),(8.2.6) into the Bautin normal form (6.1.10),(6.1.11) or the form of Ye Yanqian (7.2.8),(7.2.9).

Apart from the saddle node $M_{3,1}^0$ there exist infinite critical points in $u_\pm = \frac{1}{2}(-\lambda_1 \pm \sqrt{\lambda_1^2 - 4})$, a saddle E^{-1} and a node E^1 if $|\lambda_1| > 2$, a second order transversally semi-elementary saddle node $M_{0,2}^0$ if $|\lambda_1| = 2$ and two complex infinite critical points $C_{0,1}^0$ if $|\lambda_1| < 2$.

It can further be shown that saddle connections are not possible between diametrically opposite saddle points that are located at infinity. Using similar arguments involving contact points, the relative positions of the separatrices of the saddle node $M_{0,2}^0$ at $\lambda_1 = 2$ can be found.

The phase portraits and the corresponding (λ_1, λ_2) diagram are given in Figure 8.1b and Figure 8.1c, respectively.

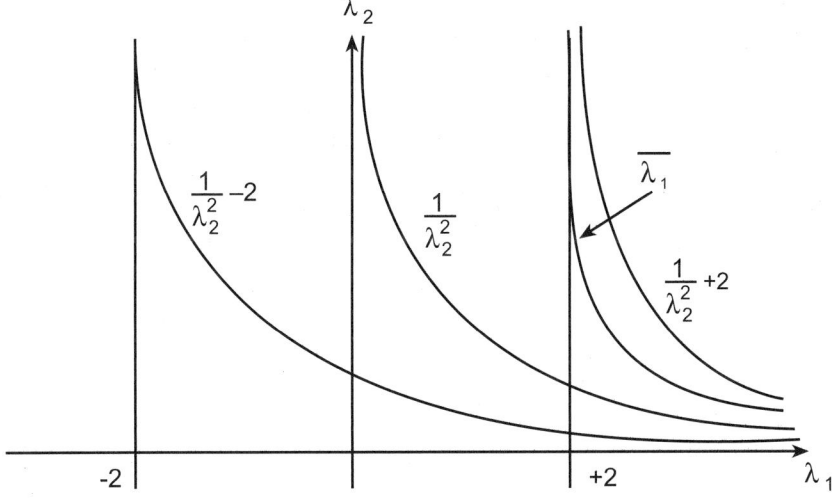

Figure 8.1c The (λ_1, λ_2) diagram for $\lambda_2 > 0$

The phase portraits given in (i),(vi)-(viii) and (x) represent the systems forming the class $e^1 E^{-1} E^1 M_{3,1}^0$, those in (ii),(v) and (ix) the class $e^1 M_{0,2}^0 M_{3,1}^0$, whereas those in (iii) and (iv) the class $e^1 C_{0,1}^0 C_{0,1}^0 M_{3,1}^0$, all in class 2a as mentioned in Table 4 of Chapter 3.

The system (8.2.5),(8.2.6) can be transformed to

$$x' = -y + (-\lambda_1 + \frac{1}{\lambda_2^2})x - \lambda_1 x^2 - xy, \qquad (8.2.7)$$

$$y' = x(1+x), \qquad (8.2.8)$$

which is a system in the class $(II)_{n=0}$ of the normal form presented by Ye Yanqian (Table 5 of Chapter 7), for which a uniqueness theorem is available. In [71,Z] or [65,Y] it was shown that there exists at most one limit cycle around the origin, so around a focus of (8.2.7), (8.2.8), which if it exists is hyperbolic.

When increasing λ_1, at $\lambda_1 = \frac{1}{\lambda_2^2}$ a stable hyperbolic limit cycle is bifurcated from the first order weak critical point, the flow direction on the limit cycle being counterclockwise. From the unstable focus in the region enclosed by the limit cycle, orbits spiral out in counterclockwise direction towards the limit cycle. From (8.2.5),(8.2.6) there follows

$$\frac{dx}{dy} = \lambda_1 + \frac{y(\lambda_2 - y)}{1 + xy},$$

which shows that increasing λ_1 rotates the vector field (P, Q) in clockwise direction. As a result, the outward motion on the spiraling orbits is strengthened and cross flow occurs over the curve on which the limit cycle was located before increasing λ_1. Due to its structural stability the (hyperbolic) limit cycle remains present in the phase portrait. In fact the result of increasing the rotation parameter λ_1 is an expansion of the limit cycle until it is absorbed by a stable unbounded separatrix cycle, as sketched in phase portrait (vii) of Figure 8.1b. The values of $\bar{\lambda}_1 = \bar{\lambda}_1(\lambda_2)$ where this occurs has been obtained numerically and are given in the following table:

$\bar{\lambda}_1$....2.25.....2.50.....2.75.....3.00.....3.25.....3.50.....3.75.....4.00.....4.25
λ_2...0.743....0.684...0.641...0.606...0.578...0.553....0.532...0.514...0.497

8.2.2   Systems with a fifth order infinite critical point

Now $b_{11} = 0$, $a_{11} - b_{02} \neq 0$ and using the conditions stated below (8.2.1),(8.2.2), that system may be rewritten by stretching the y axis and rescaling the parameter to yield

$$\dot{x} = \lambda_1 x + \lambda_2 y + \lambda_1 xy + \lambda_3 y^2, \qquad (8.2.9)$$

$$\dot{y} = y + y^2, \qquad (8.2.10)$$

where $\lambda_1 \neq 0, -\infty < \lambda_2 < \infty, -\infty < \lambda_3 < \infty, \lambda_2 \neq \lambda_3$. Indeed, from $c_{56} \equiv a_{11}b_{02} - a_{02}b_{11} = a_{11}b_{02} \neq 0$ there follows with $c_{25} \equiv a_{10}b_{11} - a_{11}b_{10} = -a_{11}b_{10} = 0$ that $b_{10}=0$, whereas from $c_{23} \equiv a_{10}b_{01} - a_{01}b_{10} \neq 0$ then follows $a_{10}b_{01} \neq 0$. From $c_{26} + c_{35}=0$ it follows that $a_{10}b_{02} - a_{11}b_{01}=0$ which shows that $a_{10} = a_{11}$ and, by scaling we may put $b_{01} = b_{02}=1$. The condition $c_{36} \neq 0$ leads to $\lambda_2 = \lambda_3$. It may further be seen that $(0,0)$ is a saddle (node) if $\lambda_1 <0 (\lambda_1 >0)$ and that $y \equiv 0$ and $y \equiv -1$ are orbits of the system.

In order to determine the character of the infinite critical point $M_{3,2}^i$ we consult [95,RK], where it is shown that $M_{3,2}^i$ is a highly degenerate critical point. Following the classification in [95,RK,section 4] it can be seen that since $\Delta=1, \gamma=0, \delta \neq 0$, $M_{3,2}^i$ is an $M_{3,2}^1$ point with one elliptic, one hyperbolic and three parabolic sectors if $\lambda_1 >1$, an $M_{3,2}^{-1}$ point with four hyperbolic sectors and one parabolic sector if $0< \lambda_1 <1$ and an $M_{3,2}^1$ point with one elliptic sector, one hyperbolic sector and three parabolic sectors if $\lambda_1 <0$.

The other critical point at infinity can now easily be determined to be a node E^1 if $\lambda_1 <1$ and a saddle E^{-1} if $\lambda_1 >1$. As a result it may be seen that class $e^{-1}E^1M_{3,2}^1$, mentioned under class 1a in Table 4, Chapter 3, occurs for $\lambda_1 <0$, class $e^1E^1M_{3,2}^{-1}$ occurs for $0< \lambda_1 <1$ and class $e^1E^{-1}M_{3,2}^1$ for $1< \lambda_1$, both being mentioned under class 2a in Table 4, Chapter3.

The phase portraits are given in Figure 8.1d.

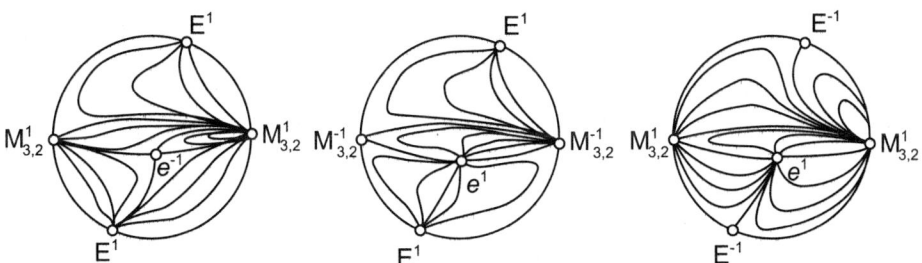

Figure 8.1d Phase portraits for the classes $e^{-1}E^1M_{3,2}^1$, $e^1E^1M_{3,2}^{-1}$ and $e^1E^{-1}M_{3,2}^1$

8.2.3 Systems with a sixth order infinite critical point

Now $b_{11} = a_{11} - b_{02} = 0, a_{02} \neq 0$ and (8.2.7),(8.2.8) may be used again with $\lambda_1=1, \lambda_3 \neq 0$ as additional conditions. The point $M_{3,3}^i$ may again be seen to be a highly degenerate critical point. Following the classification in [95,RK,section4] it can be seen from Figure 13 therein, since $\Delta=1$, (a_{10} |

$\frac{1}{2}b_{01})^2 - \frac{1}{4}\Delta=0$, that the $M_{3,3}^i$ point is an $M_{3,3}^0$ point with two hyperbolic and two parabolic sectors if $\lambda_3(\lambda_2 - \lambda_3) < 0$ and an $M_{3,3}^0$ point with one elliptic, three hyperbolic and two parabolic sectors if $\lambda_3(\lambda_2 - \lambda_3) > 0$. The two phase portraits are given in Figure 8.1d. They belong to the class $e^1 M_{3,3}^0$ of classs 2a in Table 4 of Chapter 3.

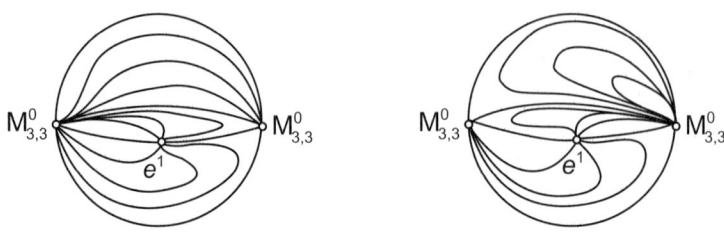

Figure 8.1e Phase portraits for the class $e^1 M_{3,3}^0$

8.2.4   Systems with infinitely many infinite critical points

In this case $b_{11} = a_{11} - b_{02} = a_{02} = 0$ and (3.7.15),(3.7.16) may be used,

$$\dot{x} = a_{10}x + a_{01}y + xy, \qquad (8.2.11)$$
$$\dot{y} = b_{10}x + b_{01}y + y^2, \qquad (8.2.12)$$

where $a_{10}a_{01} \neq 0, b_{10}=0$, $b_{01} = a_{10} \neq 0$. From $c_{23} = a_{10}^2 > 0$ it follows that $(0,0)$ is an antisaddle, in fact, a one tangential node tangent to the orbits on $y \equiv 0$.

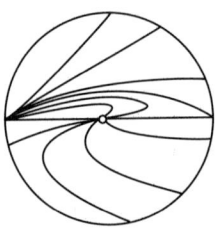

Figure 8.1f Phase portrait for the class $e^1(\infty)$

The behavior at infinity can be studied using

$$z' = a_{10}z + u - a_{01}zu, \qquad (8.2.13)$$
$$u' = a_{01}u^2, \qquad (8.2.14)$$

which has a transversally second order saddle node $M_{1,1}^0$ "at the ends of the x axis". The phase portrait is given in Figure 8.1e. It confirms that given in the paper by Gasull and Prohens on quadratic and cubic systems with degenerate infinity [96,GP1].

The systems belong to the class $e^1(\infty)$ of class 2a in Table 4, Chapter 3.

8.3 Two transversally non-hyperbolic critical points

8.3.1 Two real, different transversally non-hyperbolic critical points

For these systems we use (3.7.6),(3.7.7)

$$\dot{x} = a_{10}x + a_{01}y + a_{11}xy, \tag{8.3.1}$$
$$\dot{y} = b_{10}x + b_{01}y + b_{11}xy, \tag{8.3.2}$$

as a result of which the two transversally non-hyperbolic critical points at infinity are " at the ends of the x axis and the y axis ". Furthermore $a_{11}^2 + b_{11}^2 \neq 0$ and $c_{23} \neq 0$ since the origin is an elementary critical point. In accordance with Theorem 2.3, system (8.3.1),(8.3.2) is of class $m_f=1$ if either $c_{25} = 0, c_{35} \neq 0, c_{15} + c_{23} \neq 0$ or $c_{25} \neq 0, c_{35}=0, c_{15} - c_{23} \neq 0$; the latter condition is obtainable from the first by the transformation $x \leftrightarrow y, a_{ij} \leftrightarrow b_{ji}$. We will work with the first condition.

In order to study the critical points at infinity we use the system

$$z' = -z(a_{10}z + a_{11}u + a_{01}zu), \tag{8.3.3}$$
$$u' = b_{10}z + b_{11}u + (-a_{10} + b_{01})zu - a_{11}u^2 - a_{01}zu^2. \tag{8.3.4}$$

Systems with three infinite critical points

Since the locations of these three points are given by $f(u) \equiv u(a_{11}u - b_{11})=0$ it follows that $a_{11} \neq 0, b_{11} \neq 0$.

At the ends of the x axis there exists a semi-elementary $M_{p,1}^i$ point, which may be analyzed using the classification in [95,RK]. In fact, it follows with $c_{25}=0$, so $a_{10} = \frac{a_{11}b_{10}}{b_{11}}$, that $c_{23} = b_{10}(\frac{a_{11}}{b_{11}}b_{01}-a_{01}) \neq 0$, thus $b_{10} \neq 0$ (and $a_{10} \neq 0$) and $b_{11}^4 \gamma_2 = b_{11}^2 c_{15} + b_{10}b_{11}c_{35} + b_{10}^2 c_{56} \neq 0$ since $c_{15} - c_{56} - 0, c_{35} \neq 0$, so

$\gamma_2 \neq 0$. From figure 2 in [95,RK] it then follows that for $c_{23} < 0$, $M_{p,1}^i$ is a third order semi-linear transversally non-hyperbolic node $M_{2,1}^1$ and for $c_{23} > 0$ a third order semi-linear transversally non-hyperbolic saddle $M_{2,1}^{-1}$.

At the ends of the y axis then there exists an $M_{1,1}^i$ point which according to figure 2 in [95,RK] can only be a semi-linear second order saddle node $M_{1,1}^0$ with center manifold(s) transversal to the Poincaré circle.

As for $c_{23} < 0 (> 0)$ the finite critical point is a saddle(antisaddle) and it follows from index theory that the third infinite critical point is an elementary node E^1. Thus for $c_{23} < 0$ (8.3.1),(8.3.2) belongs to class $e^{-1} E^1 M_{1,1}^0 M_{2,1}^1$ and for $c_{23} > 0$ to class $e^1 E^1 M_{1,1}^0 M_{2,1}^{-1}$; thus they belong to class 1b and class 2b, respectively, as mentioned in Table 4, Chapter3.

Since $a_{10}a_{11}b_{10}b_{11} \neq 0$, $c_{25}=0$, (8.3.1),(8.3.2) can be linearly transformed to

$$\dot{x} = x + \alpha y + xy, \qquad (8.3.5)$$
$$\dot{y} = x + \beta y + xy, \qquad (8.3.6)$$

where $\alpha = \frac{a_{01}b_{11}}{a_{10}a_{11}}, \beta = \frac{b_{01}}{a_{10}}$.

(i)For class $e^{-1} E^1 M_{1,1}^0 M_{2,1}^1$ there is $c_{23} = \beta - \alpha < 0$. It can easily be shown that for $\alpha < 0$ the phase portrait is as shown in Figure 8.2a(i), for $\alpha=0$ in Figure 8.2a(ii) and for $\alpha > 0$ in Figure 8.2a(iii).

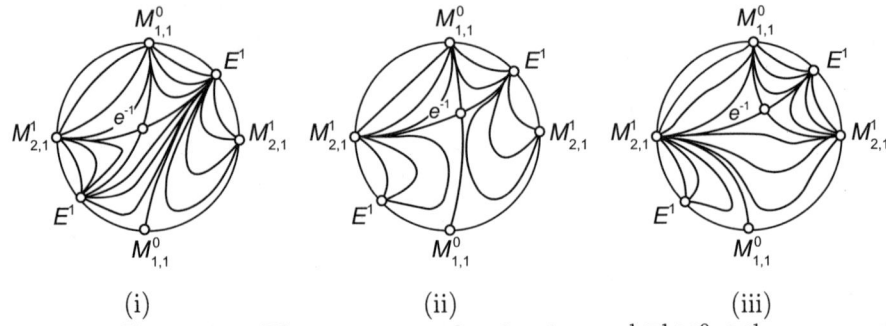

(i)   (ii)   (iii)

Figure 8.2a Phase portraits for the class $e^{-1} E^1 M_{1,1}^0 M_{2,1}^1$

(ii)For class $e^1 E^1 M_{1,1}^0 M_{2,1}^{-1}$ there is $c_{23} = \beta - \alpha > 0$. The origin is a node if $\Delta \geq 0$, a focus if $\Delta < 0$ and a stable first order fine focus if $\Delta < 0, \beta = -1$; here $\Delta = (1 - \beta)^2 + 4\alpha$. The origin is stable for $\beta \leq -1$ and unstable for $\beta > -1$. In Figure 8.2b the α, β plane for (8.3.5),(8.3.6) is shown. A stable limit cycle bifurcates out of the stable fine focus with increasing β from the line $\beta \equiv -1$. The uniqueness and hyperbolicity of this limit cycle can be demonstrated by transforming (8.3.5),(8.3.6) to the system

$$\dot{x} = dx - y + lx^2 + xy, \qquad (8.3.7)$$

$$\dot{y} = x, \qquad (8.3.8)$$

where $\alpha = \frac{1+\beta}{\sqrt{\beta-\alpha}}$, $l = -\frac{1}{\sqrt{\beta-\alpha}}$, and which is of class $(I)_{n=0}$ in the classification of Ye Yanqian [65,Y,p.415]. (From (7.3.12) it is easily seen that $W_1 = l < 0$ so that the fine focus is stable and of first order.)

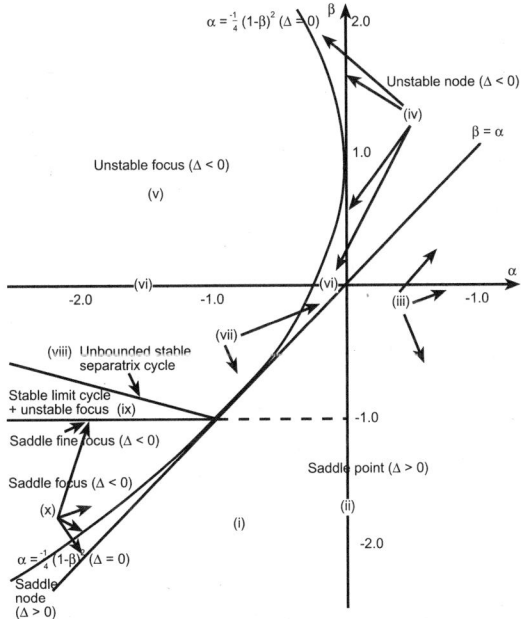

Figure 8.2b Parameter plane for (8.3.5),(8.3.6)

From (8.3.7),(8.3.8) it follows that

$$\frac{dx}{dy} = d - \frac{y}{x} + lx + y, \qquad (8.3.9)$$

so increasing β on a line $\beta - \alpha = \lambda > 0$ rotates the vector (P,Q) of the anticlockwise motion on the (stable) limit cycle in the clockwise direction, as a result of which the limit cycle expands until it is absorbed into an unbounded stable separatrix cycle (viii in Figure 8.2c).The phase portraits for class $e^1 E^1 M_{1,1}^0 M_{2,1}^{-1}$ are given in Figure 8.2c. The trace of the unbounded separatrix in the α, β plane can only be obtained numerically. In [97,R] the following table is given:

$-\alpha$ ..16.01..11.14..6.42..4.33..3.25..2.39..2.22..2.09..1.77..1.19..1.08..1.05

$-\beta$ ....0.01....0.03..0.17..0.33..0.47..0.61..0.66..0.70..0.77..0.94..0.97..0.98

A perturbation analysis, also given in [97,R], shows that for the trace $\beta = \beta(\alpha)$ of this cycle there is $\beta(\alpha) \to 0$ as $\alpha \to -\infty$.

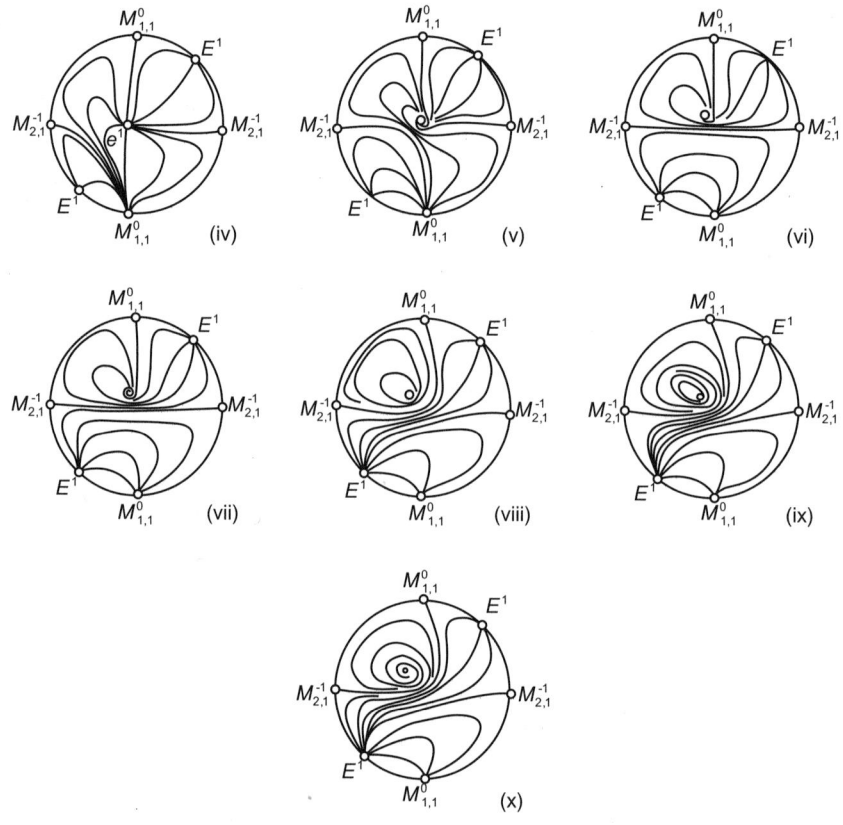

Figure 8.2c Phase portraits for the class $e^1 E^1 M_{1,1}^0 M_{2,1}^{-1}$

Systems with two infinite critical points

We consider again the system

$$\dot{x} = a_{10}x + a_{01}y + a_{11}xy, \tag{8.3.10}$$
$$\dot{y} = b_{10}x + b_{01}y + b_{11}xy, \tag{8.3.11}$$

with $a_{11}^2 + b_{11}^2 \neq 0$, $c_{23} \neq 0$ and the conditions $c_{25} = 0, c_{35} \neq 0, c_{15} + c_{23} \neq 0$.

The location of an infinite critical point is given by $f(u) \equiv u(a_{11}u - b_{11}) = 0$, which shows that in order to yield two infinite critical points either there should be $a_{11} \neq 0, b_{11} = 0$ or $a_{11} = 0$, $b_{11} \neq 0$.

(i)$a_{11} \neq 0, b_{11} = 0$.

At the ends of the x axis there exists an $M_{p,2}^i$ point, which may be analyzed using the classification in [95,RK]. In fact, it follows from $c_{25} \equiv a_{10}b_{11} - a_{11}b_{10} = 0$ that $b_{10} = 0$, so $M_{p,2}^i$ is a highly degenerate critical point and figure 11 in [95,RK] should be consulted. Since $c_{35} \equiv a_{01}b_{11} - a_{11}b_{01} \neq 0$ there follows $b_{01} \neq 0$. Then if $c_{23} \equiv a_{10}b_{01} - a_{01}b_{10} = a_{10}b_{01} < 0$ there follows that $M_{p,2}^i$ is an $M_{2,2}^2$ point with two elliptic and two parabolic sectors and the finite critical point is a saddle, whereas for $c_{23} > 0$ it may be seen that $M_{p,q}^i = M_{2,2}^0$ with two hyperbolic and two parabolic sectors and the finite critical point is an antisaddle.

In both cases this leaves for the point at the ends of the y axis only to be an $M_{1,1}^i$ point, which can only be a second order semi-linear saddle node $M_{1,1}^0$. Thus for $c_{23} < 0$, (8.3.10),(8.3.11) represents the class $e^{-1}M_{1,1}^0 M_{2,2}^2$ and for $c_{23} > 0$ the class $e^1 M_{1,1}^0 M_{2,2}^2$, occurring in the classes 1b and 2b, respectively, both being mentioned in Table 4, Chapter 3.

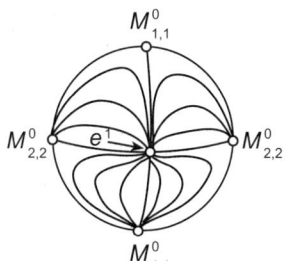

Figure 8.2d Phase portrait for class $e^1 M_{1,1}^0 M_{2,2}^0$

For both cases (8.3.10),(8.3.11) can be linearly transformed to

$$\dot{x} = \alpha x + \beta y + xy, \tag{8.3.12}$$
$$\dot{y} = y, \tag{8.3.13}$$

where $\alpha = \frac{a_{10}}{b_{01}} \neq 0, \beta = \frac{a_{11}}{b_{11}}$.

For the class $e^1 M_{1,1}^0 M_{2,2}^0 (c_{23} > 0)$ the origin is a node and the only phase portrait is given in Figure 8.2d.

For the class $e^{-1} M_{1,1}^0 M_{2,2}^2$ ($c_{23} < 0$) there exist two phase portraits, for $\beta=0$ and for $\beta \neq 0$; they are given in Figure 8.2e.

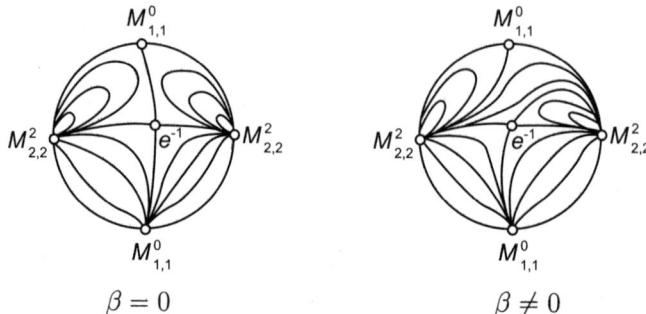

$$\beta = 0 \qquad\qquad\qquad \beta \neq 0$$

Figure 8.2e Phase portraits for class $e^{-1} M_{1,1}^0 M_{2,2}^2$.

(ii)$a_{11} = 0, b_{11} \neq 0$.

At the ends of the x axis there exists a semi-elementary $M_{p,1}^i$ point and the same analysis applies as under the previous section. So for $c_{23} < 0$, $M_{p,1}^i$ is a third order semi-linear node $M_{2,1}^1$ and for $c_{23} > 0$ a third order semi-linear saddle $M_{2,1}^{-1}$. Since for $c_{23} < 0(c_{23} > 0)$ the finite critical point is a saddle (antisaddle) this leaves it for the critical point "at the ends of the y axis" to be an $M_{1,2}^1$ point. As this cannot be a semi-linear point $M_{p,1}^i$ or $M_{0,q}^i$ or a highly degenerate critical point, as such a point must be at least of fourth order, it is a nilpotent critical point. Figure 4 of [95,RK] then shows that $M_{1,2}^1$ is a critical point with an elliptic and a hyperbolic sector. Thus for $c_{23} < 0$, (8.3.10),(8.3.11) represents the class $e^{-1} M_{2,1}^1 M_{1,2}^1$ and for $c_{23} > 0$ the class $e^1 M_{2,1}^{-1} M_{1,2}^1$, occurring in the classes 1b and 2b, respectively, both being mentioned in Table 4, Chapter 3.

In both cases system (8.3.10),(8.3.11) may be transformed to

$$\dot{x} = y, \tag{8.3.14}$$
$$\dot{y} = -(sgnc_{23})x + \beta y + xy, \tag{8.3.15}$$

where $\beta = \frac{b_{01}}{\sqrt{|c_{23}|}}, -\infty < \beta < \infty$.

For the class $e^{-1} M_{2,1}^1 M_{1,2}^1 (c_{23} < 0)$, there exists one phase portrait; it is given in Figure 8.2f. For the class $e^1 M_{2,1}^{-1} M_{1,2}^1 (c_{23} > 0)$, the origin is a (rough) focus for $0 < |\beta| < 2$ and a node for $|\beta| \geq 2$. For $\beta=0$ the origin is a center, as may be concluded from the oval shape of the orbits near it, and their symmetry around the y axis. There are no limit cycles. This may be concluded from the observation that the topographic system of curves [ALGM p.210] formed by the oval orbits for $\beta=0$, along the orbits around the focus for $\beta > 0(\beta < 0)$ is, overflown in outward (inward) direction due to the term βy. The two phase portraits are given in Figure 8.2g.

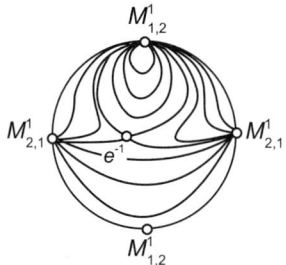

Figure 8.2f Phase portrait for class $e^{-1}M_{2,1}^1 M_{1,2}^1$

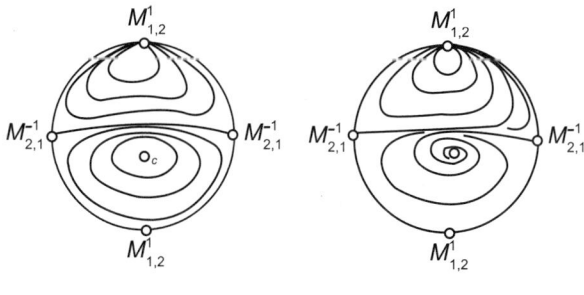

$$\beta = 0 \qquad\qquad \beta \neq 0$$

Figure 8.2g Phase portraits for class $e^1 M_{2,1}^{-1} M_{1,2}^1$

8.3.2   Two coinciding transversally non-hyperbolic infinite critical points

For these systems we use (3.7.10),(3.7.11)

$$\dot{x} = a_{10}x + a_{01}y + a_{02}y^2, \tag{8.3.16}$$
$$\dot{y} = b_{10}x + b_{01}y + b_{02}y^2, \tag{8.3.17}$$

with $a_{02}^2 + b_{02}^2 \neq 0$. There exist two coinciding critical points "at the ends of the x axis". Furthermore $c_{23} \neq 0$ and Theorem 2.1 says that (8.3.16),(8.3.17) belongs to $m_f=1$ iff $c_{26} = 0, c_{23}c_{36} \neq 0$.

In order to study the critical points at infinity we use the system

$$z' = -z(a_{10}z + a_{11}zu + a_{02}u^2), \tag{8.3.18}$$
$$u' = b_{10}z + (-a_{10} + b_{01})zu + b_{02}u^2 - a_{01}zu^2 - a_{02}u^3. \tag{8.3.19}$$

The locations of the infinite critical points are given by $f(u) = (a_{02}u - b_{02})u^2$, from which it follows that "at the ends of the x axis" there exists an $M_{3,2}^i$ point $(b_{02} \neq 0)$ and then there are two infinite critical points or an $M_{3,3}^i$ point $(b_{02}=0)$ and there exists only one such point.

Systems with two infinite critical points($b_{02} \neq 0$)

If $b_{02} \neq 0$, $c_{26}=0$ yields $a_{10} = \frac{a_{02}b_{10}}{b_{02}}$ thus $c_{23} \equiv a_{10}b_{01} - a_{01}b_{10} = -\frac{b_{10}}{b_{02}}c_{36} \neq 0$ yields $b_{10} \neq 0$. With $a_{20} = b_{20} = b_{11} = 0, b_{10} \neq 0$ it may be seen from [95,RK] that $M_{3,2}^i$ is a nilpotent critical point and we consult figure 5 in [95,RK]. It then follows that for $c_{23} < 0$, $M_{3,2}^i$ is a nilpotent fifth order point $M_{3,2}^i$ with an elliptic and a hyperbolic sector and for $c_{23} > 0$ a nilpotent fifth order point $M_{3,2}^{-1}$. Since for $c_{23} < 0(c_{23} > 0)$ the finite critical point is a saddle (antisaddle), index theory tells that the other infinite critical point is an elementary node E^1. Thus for $c_{23} < 0$ system (8.3.16),(8.3.17) represents the class $e^{-1}E^1 M_{3,2}^1$ and for $c_{23} > 0$ the class $e^1 E^1 M_{3,2}^{-1}$ belonging to classes 1c and class 2c, respectively, as mentioned in Table 4, Chapter 3. In both cases (8.3.16), (8.3.17) can be linearly transformed to

$$\dot{x} = -(sgnc_{23})y, \tag{8.3.20}$$
$$\dot{y} = x + \beta y + y^2, \tag{8.3.21}$$

where $\beta = \frac{b_{01}}{\sqrt{|c_{23}|}}, -\infty < \beta < \infty$.

For the class $e^{-1}E^1 M_{3,2}^1(c_{23} < 0)$ there exists one phase portrait; it is given in Figure 8.3a. For the class $e^1 E^1 M_{3,2}^{-1}(c_{23} > 0)$, the origin is a (rough) focus for $0 < |\beta| < 2$ and a node for $|\beta| \geq 2$. For $\beta=0$, the origin is a center and for $|\beta| \geq 0$ there exist no limit cycles. These conclusions may be drawn using the same arguments as for class $e^1 M_{2,1}^{-1} M_{1,2}^1$. The two possible phase portraits are given in Figure 8.3b.

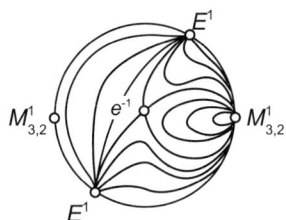

Figure 8.3a Phase portrait for the class $e^{-1}E^1M_{3,2}^1$

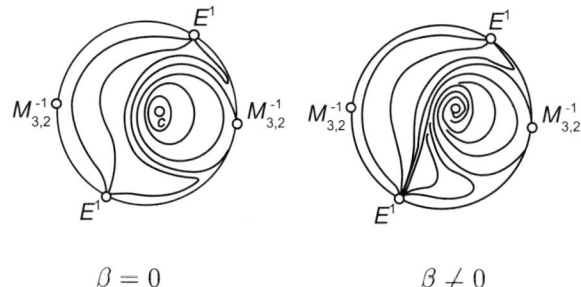

$$\beta = 0 \qquad\qquad\qquad \beta \neq 0$$
Figure 8.3b Phase portraits for the class $e^1E^1M_{3,2}^{-1}$

Systems with one infinite critical point ($b_{02}=0$)

If $b_{20}=0$ then $a_{20} \neq 0$ and $c_{26} = a_{10}b_{02} - a_{02}b_{10} = -a_{02}b_{10} =0$ yields $b_{10}=0$. Furthermore from $c_{36} \equiv a_{01}b_{02} - a_{02}b_{01} = -a_{02}b_{11} \neq 0$ it follows that $a_{10} \neq 0$. With $a_{10} = b_{20} = b_{11} = b_{10}=0$ it may be seen [95,RK] that $M_{3,3}^1$ is a highly degenerate critical point "at the ends of the x axis". From figure 15 of [95,RK] then follows that $M_{3,3}^i$ is an $M_{3,3}^2$ point with two elliptic and one parabolic sector if $c_{23} <0$, an $M_{3,3}^0$ point with two hyperbolic and two parabolic sectors for $0< c_{23} \leq 2b_{01}^2$, and an $M_{3,3}^0$ point with two hyperbolic and four parabolic sectors for $c_{23} > 2b_{01}^2$. For $c_{23} < 0(c_{23} > 0)$ the finite critical point is a saddle(node). Thus for $c_{23} <0$ system (8.3.16),(8.3.17) represents the classes 1c and 2c, respectively, as mentioned in Table 4, Chapter3.

In both cases (8.3.16),(8.3.17) can be linearly transformed to

$$\dot{x} = \alpha x + \beta y + y^2, \qquad\qquad (8.3.22)$$
$$\dot{y} - y, \qquad\qquad (8.3.23)$$

where $\alpha = \frac{c_{23}}{b_{01}^2}, \beta = \frac{a_{01}}{a_{02}}, -\infty < \alpha, \beta < \infty$.

For the class $e^{-1}M_{3,3}^2 (c_{23} < 0)$ the phase portrait is given in Figure 8.3c. For the class $e^1 M_{3,3}^0$ the phase portrait 8.3d(a) corresponds to $0 < c_{23} \le 2b_{01}^2$ and phase portrait 8.3d(b) to $c_{23} > 2b_{01}^2$.

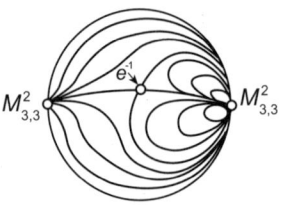

Figure 8.3c Phase portrait for class $e^{-1}M_{3,3}^2$

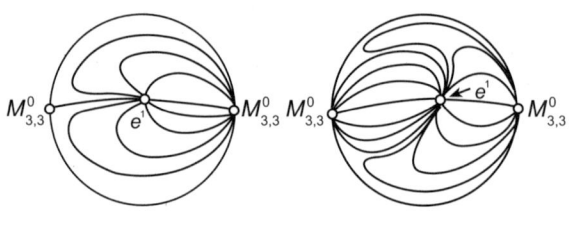

(a)   (b)
Figure 8.3d Phase portraits for class $e^1 M_{3,3}^0$

8.4   Conclusions

Theorem 8.2 *There exist 38 phase portraits in the class of quadratic systems with finite multiplicity $m_f=1$. They are given in Figures 8.1-8.3.*

Remark. If the antisaddle or the limit cycle (separatrix cycle) encircling the antisaddle in the phase portraits given in Figures 8.1b(iii) and (iv), Figure 8.2d and Figure 8.3d are stable, these portraits represent systems for which all orbits remain in the finite part of the plane for increasing values of t. The study of the class of bounded quadratic systems was initiated in 1970 by Dickson and Perko [70,DP] and studied since by various authors.

Chapter 9

Phase portraits of quadratic systems in the class $m_f{=}2$

9.1 Introduction

Compared with the number of phase portraits in the classes with finite multiplicity 0 and 1, there is a drastic increase of this number for $m_f{=}2$. Relaxing of the restrictions on the coefficients when increasing the finite multiplicity leads to more free parameters in the system yielding more combinations of the elements structuring a phase portrait. This also effects the limit cycle problem to the extent that for $m_f{=}2$ this problem is not completely solved yet. Nevertheless, the results of the classification justifies the following statement.

Statement. A proof that the number of limit cycles in a quadratic system with finite multiplicity $m_f{=}2$ is at most equal to 2 has been given for almost all values of the coefficients a_{ij}, b_{ij} yielding $m_f{=}2$, except that the following problem, wherein only affirmative numerical evidence exists, still prevails.

Problem. Show that there exist at most two (hyperbolic) limit cycles around (0,0) for the system

$$\dot{x} = dx - y + lx^2 + mxy, \qquad (9.1.1)$$
$$\dot{y} = x + x^2, \qquad (9.1.2)$$

where $-2 < d < 0$, $m \geq 2$, $l > 0$.

The system belongs to the class $(II)_{n-0}$ in the classification of Ye Yanqian.

For the present discussion we follow the classificaton given by Reyn and Kooij [97,RK], who also determined the partition of the $\mathbb{R}^{12}$ coefficient space corresponding to the phase portraits. An ordering of the various possible combinations of infinite critical points will again be used to structure the classification; also, the equations developed in section 6 of Chapter 3 will be used.

9.2   One transversally non-hyperbolic critical point

For these systems we use (3.6.1),(3.6.2)

$$\dot{x} = a_{00} + a_{10}x + a_{01}y + a_{11}xy + a_{02}y^2, \qquad (9.2.1)$$
$$\dot{y} = b_{00} + b_{10}x + b_{01}y + b_{11}xy + b_{02}y^2, \qquad (9.2.2)$$

,with $a_{11}^2 + b_{11}^2 \neq 0$, $a_{02}^2 + b_{02}^2 \neq 0$, whereas $c_{56} \neq 0$ yields that the transversally non-hyperbolic infinite critical point is "at the ends of the x axis". System (9.2.1), (9.2.2) belongs to the class $m_f=2$ iff $c_{25}=0$ and $c_{26}^2+c_{26}c_{35}-c_{15}c_{56} \neq 0$.

In order to study the infinite critical points we use the system

$$z' = -z(a_{10}z + a_{11}u + a_{00}z^2 + a_{01}zu + a_{02}u^2) \equiv \bar{P}(z,u), \qquad (9.2.3)$$
$$u' = b_{10}z + b_{11}u + b_{00}z^2 + (-a_{10} + b_{01})zu + (-a_{11} + b_{00})u^2$$
$$-a_{01}zu^2 - a_{00}z^2u - a_{02}u^3 \equiv \bar{Q}(z,u), \qquad (9.2.4)$$

so that the infinite critical points are located in $u=\bar{u}$, where $\bar{u}$ is a solution of $f(u) \equiv a_{02}u^3 + (a_{11} - b_{02})u^2 - b_{11}u=0$. The point "at the ends of the x axis" is thus a third order point $M_{2,1}^i$ if $b_{11} \neq 0$, a fourth order point $M_{2,2}^i$ if $b_{11}=0$, $a_{11} - b_{02} \neq 0$ and a fifth order point if $b_{11} = a_{11} - b_{02}=0$, $a_{02} \neq 0$.

9.2.1   Systems with a third order infinite critical point

Using the classification of infinite critical points by Reyn and Kooij [95,RK], it follows that $M_{2,1}^i$ is a third order semi-linear critical point with center manifold(s) transversal to the Poincaré circle. In fact from $c_{25}=0$ it follows with $b_{11} \neq 0$, that $a_{10} = a_{11}\frac{b_{10}}{b_{11}}$ and $c_{26} = \frac{b_{10}}{b_{11}}c_{56}$ whereas $b_{11}^2(c_{26}^2 + c_{26}c_{35} - c_{15}c_{56}) \equiv c_{56}b_{11}^4\gamma_2$ which defines γ_2 as in Figure 2 of [95,RK]. From this figure it then follows that if $\gamma_2 < 0$, $M_{2,1}^i$ is a third order saddle $M_{2,1}^{-1}$ and if $\gamma_2 > 0$ a third order node $M_{2,1}^1$.

Using the stated conditions, by a linear transformation, (9.2.1),(9.2.2) may be transformed to

$$\dot{x} = \alpha b_{11}^2 (a_{11} + b_{02}) - \gamma_2 b_{11}^6$$
$$+ b_{11}^2 (b_{11} c_{35} + 2 b_{10} c_{56}) y + b_{11}(a_{11} + b_{02}) xy - b_{11}^2 c_{56} y^2, \qquad (9.2.5)$$
$$\dot{y} = \alpha b_{11} + xy, \qquad (9.2.6)$$

where $\alpha \equiv -b_{10} b_{01} b_{11} + b_{10}^2 b_{02} + b_{00} b_{11}^2, -\infty < \alpha < \infty, -\infty < a_{11} + b_{02} < \infty, -\infty < b_{11} c_{35} + 2 b_{10} c_{56} < \infty$ and $\gamma_2 c_{56} \neq 0$.

If $\alpha \neq 0$, (9.2.5),(9.2.6) may be further reduced, by scaling, to

$$\dot{x} = \lambda_1 + \lambda_3 + \lambda_2 y + \lambda_1 xy - (sgnc_{56}) y^2, \qquad (9.2.7)$$
$$\dot{y} = 1 + xy, \qquad (9.2.8)$$

where $\lambda_1 \equiv |c_{56}|^{-\frac{1}{2}} (a_{11} + b_{02}) sgn\alpha$, $\lambda_2 \equiv |c_{56}|^{-\frac{3}{4}} |\alpha|^{-\frac{1}{2}} (b_{11} c_{35} + 2 b_{10} c_{56})$, $\lambda_3 \equiv -b_{11}^4 |c_{56}|^{-\frac{1}{2}} |\alpha|^{-1} \gamma_2$, $-\infty < \lambda_1 < \infty, -\infty < \lambda_2 < \infty, \lambda_3 \neq 0$.

Note that from (9.2.7),(9.2.8) it follows that we may write $\frac{dx}{dy} = \lambda_1 + f(x, y; \lambda_2, \lambda_3)$, which shows that λ_1 is a parameter rotating the vector field $(P(x, y), Q(x, y))$; increasing λ_1 yields a clockwise rotation.

If $\alpha = 0$, (9.2.5),(9.2.6) may be reduced, by scaling, to

$$\dot{x} = \lambda_3 + \lambda_2 y + \lambda_1 xy - (sgnc_{56}) y^2, \qquad (9.2.9)$$
$$\dot{y} = xy, \qquad (9.2.10)$$

where now $\lambda_1 \equiv |c_{56}|^{-\frac{1}{2}} (a_{11} + b_{02}), \lambda_2 \equiv |c_{56}|^{-\frac{1}{2}} b_{11} (b_{11} c_{35} + 2 b_{10} c_{56}), \lambda_3 \equiv -\gamma_2 b_{11}^6, -\infty < \lambda_1 < \infty, -\infty < \lambda_2 < \infty, \lambda_3 \neq 0$.

Note that for all values of α, $\lambda_3 < 0 (\lambda_3 > 0)$ corresponds to a third order infinite critical point $M_{2,1}^1 (M_{2,1}^{-1})$.

Moreover, if the finite critical points are both saddles(antisaddles) so $i_f = -2 (i_f = 2)$, Theorem 3.15 yields $c_{56} < 0, (c_{56} > 0)$, whereas in the proof of that theorem it is shown that $c_{26}^2 + c_{20} c_{35} - c_{15} c_{30} - \gamma_2 c_{56} b_{11}^2 < 0$ so $c_{56} \lambda_3 > 0$, and $\lambda_3 < 0 (\lambda_3 > 0)$. Index theory then yields for $i_f = -2$ the combination $e^{-1} e^{-1} E^1 E^1 M_{2,1}^1$ in the class 1 of Table 3 of Chapter 3 and for $i_f = 2$ the combinations $e^1 e^1 E^{-1} E^1 M_{2,1}^{-1}, e^1 e^1 M_{0,2}^0 M_{2,1}^{-1}$ and $e^1 e^1 C_{0,1}^0 C_{0,1}^0 M_{2,1}^{-1}$ of class 3 in that table. In the case that $i_f = 0$ it follows from the proof of Theorem 3.15 that $c_{26}^2 + c_{26} c_{35} - c_{15} c_{36} = \gamma_2 c_{56} b_{11}^2 > 0$ so $c_{56} \lambda_3 < 0$ and either

$c_{56} > 0, \lambda_3 < 0$ or $c_{56} < 0, \lambda_3 > 0$. For $c_{56} > 0, \lambda_3 < 0$ the combinations of finite critical points $e^{-1}e^1, c_1^0 c_1^0$ and the point m_2^0 can be combined with the combinations of infinite critical points $E^{-1}E^1 M_{2,1}^1$, $M_{0,2}^0 M_{2,1}^1$ and $C_{0,1}^0 C_{0,1}^0 M_{2,1}^1$ and for $c_{56} < 0, \lambda_3 > 0$ these finite critical points can be combined with combination $E^1 E^1 M_{2,1}^{-1}$. All these combinations belong to class 2 in Table 3 in Chapter 3.

Systems with finite index $i_f = -2$

As stated above there exists only one class: $e^{-1}e^{-1}E^1 E^1 M_{2,1}^1$; it occurs for $c_{56} < 0, \lambda_3 < 0$.

First consider $\alpha \neq 0$, using (9.2.7),(9.2.8). It can be assumed that $\lambda_2 \geq 0$; if not apply the transformation $(x, y, t) \rightarrow (-x, -y, -t)$. The relative positions of the separatrices of the two saddles can be derived by the flow across the line connecting the two saddles $l \equiv y - \lambda_3 x + \lambda_2 {=} 0$, and is given by $\frac{dl}{dt}|_{l=0} = (\dot{y} - \lambda_3 \dot{x})|_{l=0} = (1{-}\lambda_1 \lambda_3 - \lambda_2^2)(1 - \lambda_2 x + \lambda_3 x^2)$. We find three possibilities, depending on the sign of $\sigma \equiv 1 - \lambda_1 \lambda_3 - \lambda_3^2$; see Figure 9.1 a, b and c.

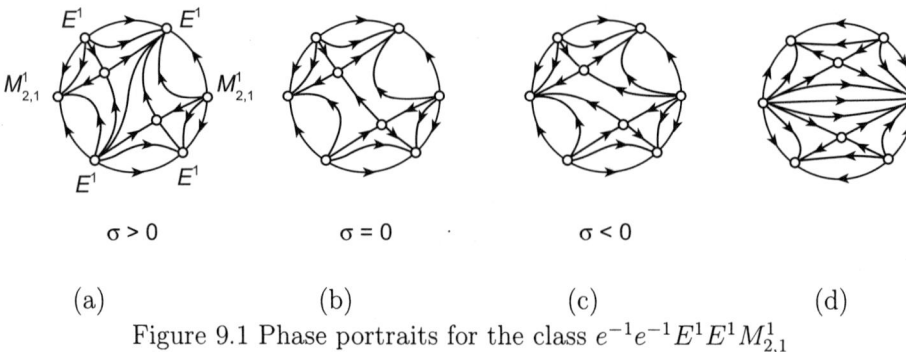

$\sigma > 0$ $\sigma = 0$ $\sigma < 0$

(a)   (b)   (c)   (d)

Figure 9.1 Phase portraits for the class $e^{-1}e^{-1}E^1 E^1 M_{2,1}^1$

Similarly, for $\alpha{=}0$ we may use (9.2.9),(9.2.10). The phase portrait is given in Figure 9.1 d. A saddle connection is not possible since $y \equiv 0$ is an orbit of the system and from $\lambda_3 + y_i + y_i^2 = 0$ (i=1,2) and $\lambda_3 < 0$ it follows that the saddle points in $(0, y_1)$ and $(0, y_2)$ are located on either side of this line. Note that the phase portraits a, c and d are topologically equivalent if the bifurcational properties of E^1 and $M_{2,1}^1$ are ignored.

Systems with finite index $i_f = 0$

There exist the possibilities $c_{56} > 0$, $\lambda_3 < 0$ and $c_{56} < 0$, $\lambda_3 > 0$.

Case 1. $c_{56} > 0, \lambda_3 < 0$

For this case there exist the classes $e^{-1}e^1 E^{-1} E^1 M_{2,1}^1$, $m_2^0 E^{-1} E^1 M_{2,1}^1$, $c_1^0 c_1^0 E^{-1} E^1 M_{2,1}^1$, $e^{-1}e^1 M_{0,2}^0 M_{2,1}^1$, $m_2^0 M_{0,2}^0 M_{2,1}^1$, $c_1^0 c_1^0 M_{0,2}^0 M_{2,1}^1$, $e^{-1}e^1 C_{0,1}^0 C_{0,1}^0 M_{2,1}^1$, $m_2^0 C_{0,1}^0 C_{0,1}^0 M_{2,1}^1$ and $c_1^0 c_1^0 C_{0,1}^0 C_{0,1}^0 M_{2,1}^1$.

System (9.2.7),(9.2.8) now reads

$$\dot{x} = \lambda_1 + \lambda_3 + \lambda_2 y + \lambda_1 xy + -y^2, \qquad (9.2.11)$$
$$\dot{y} = 1 + xy, \qquad (9.2.12)$$

with $-\infty < \lambda_1 < \infty, -\infty < \lambda_2 < \infty, \lambda_3 < 0$, where we again take $\lambda_2 \geq 0$ without losing generality.

For the study of the infinite critical points of (9.2.11),(9.2.12), we may again use (9.2.3),(9.2.4). Apart from the $M_{2,1}^1$ point in $u = 0$ these points are located in $u_\pm = \frac{1}{2}(\lambda_1 \pm \sqrt{\lambda_1^2 - 4})$ and there follows from $\Omega(0, u) \equiv \bar{P}_z(0, u)\bar{Q}_u(0, u) - \bar{P}_u(0, u)\bar{Q}_z(0, u) = u(u - \lambda_1)(1 - 2\lambda_1 u + 3u^2)$, that $\Omega(0, u_\pm) = \mp u_\pm \sqrt{\lambda_1^2 - 4}$. So for $\lambda_1 > 2$ there is a node E^1 in u_- and a saddle E^{-1} in u_+, for $\lambda_1 - 2$ a saddle node $M_{0,2}^0$ in $u = 1$, for $-2 < \lambda_1 < 2$ there are two complex critical points $C_{0,1}^0$, for $\lambda_1 = -2$ a saddle node $M_{0,2}^0$ in $u = -1$, whereas for $\lambda_1 < -2$ there is a saddle in u_- and a node in u_+.

The finite critical points are located in (x_1, y_1) and (x_2, y_2) where $x_i = -\frac{1}{y_i}$, (i=1,2), $y_1 = y_+, y_2 = y_-$ in $y_\pm = \frac{1}{2}(\lambda_2 \pm \sqrt{\lambda_2^2 + 4\lambda_3})$. So for $\lambda_2^2 + 4\lambda_3 > 0$ we have the combination $e^{-1}e^1$, for $\lambda_2^2 + 4\lambda_3 = 0$ a critical point m_2^0 and for $\lambda_2^2 + 4\lambda_3 < 0$ two complex points $c_1^0 c_1^0$. In the combination $e^{-1}e^1$, the antisaddle is located in (x_+, y_+) and the saddle in x_-, y_-. They are weak critical points under the condition

$$\lambda_1(\lambda_3 + \frac{1}{2}\lambda_2^2 \pm \frac{1}{2}\lambda_2\sqrt{(\lambda_2^2 + 4\lambda_3)}) = 1, \qquad (9.2.13)$$

where the $-$ sign corresponds to a weak saddle in (x_-, y_-) and the +sign to a weak focus in (x_+, y_+). Also it results that the m_2^0 point for $\lambda_2^2 + 4\lambda_3 = 0$ is a cusp (saddle node) for $-\lambda_1\lambda_3 = \frac{1}{4}\lambda_1\lambda_2^2 = 1(\neq 1)$.

The relation separating the nodes from the foci is

$$\{1 + \lambda_3[\lambda_1\{-2 + \lambda_1(\lambda_3 + \lambda_2^2)\} - 4(\lambda_2^2 + 2\lambda_3)]\} +$$
$$\{\lambda_1\lambda_2[-2 + \lambda_1(2\lambda_3 + \lambda_2^2)] - 4\lambda_2(\lambda_2^2 + 3\lambda_3)\}(\frac{1}{2} + \frac{1}{2}\sqrt{(\lambda_3^2 + 4\lambda_1)}) = 0. (9.2.14)$$

With regard to the limit cycles it may be remarked that $(9.2.11)$, $(9.2.12)$ can be transformed to

$$\dot{x} = dx - y + lx^2 + mxy, \qquad\qquad (9.2.15)$$
$$\dot{y} = x + x^2, \qquad\qquad (9.2.16)$$

where now

$$d = \frac{-1+\lambda_1[\frac{1}{2}\lambda_2+\frac{1}{2}(\lambda_2^2+4\lambda_3)^{\frac{1}{2}}]^2}{(\lambda_2^2+4\lambda_3)^{\frac{1}{4}}[\frac{1}{2}\lambda_2+\frac{1}{2}(\lambda_2^2+4\lambda_3)^{\frac{1}{2}}]^{\frac{3}{2}}},$$

$$l = \frac{\lambda_1(\lambda_2^2+4\lambda_3)^{\frac{1}{4}}}{(\frac{1}{2}\lambda_2+\frac{1}{2}(\lambda_2^2+4\lambda_3)^{\frac{1}{2}})^{\frac{1}{2}}},$$

$$m = \frac{-(\lambda_2^2+4\lambda_3)^{\frac{1}{2}}}{\frac{1}{2}\lambda_2+\frac{1}{2}(\lambda_2^2+4\lambda_3)^{\frac{1}{2}}},$$

thus $-\infty < d < \infty, -\infty < l < \infty, -1 < m < 0$. It is known that this system has at most one limit cycle, and, if it exists is hyperbolic [94,Z]. At the parameter values corresponding to a weak focus there is a Hopf bifurcation, indicated by Hopf in the bifurcation diagrams to be given later. The expansion of the generated limit cycle can be traced by changing λ_1 in the proper direction, keeping λ_2 and λ_3 constant, until it is adsorbed in a unique saddle loop (or homoclinic cycle), indicated by Hom. Another bifurcation occurs at a heteroclinic saddle connection between the finite saddle and an infinite saddle or saddle node; it will be indicated by Het.

Also the possibility of a connection between diametrically opposite infinite saddles should be considered. Infinite saddles occur for $|\lambda_1| > 2$.

(i)For $\lambda_1 > 2$ consider the line $l_+ \equiv y-\frac{1}{2}(\lambda_1+\sqrt{\lambda_1^2-4})x-\lambda_2=0$ connecting two opposite saddles. An elementary calculation shows that we may define

$$\frac{dl_+}{dt}\big|_{l_+=0} = (\dot{y} - \tfrac{1}{2}(\lambda_1 + \sqrt{\lambda_1^2 - 4})\dot{x})\big|_{l_+=0}=$$
$$1-\tfrac{1}{2}\lambda_1^2 - \tfrac{1}{2}\lambda_1\lambda_3 - \tfrac{1}{2}(\lambda_1 + \lambda_3)\sqrt{\lambda_1^2 - 4} \equiv \sigma_+.$$

Then if $\lambda_3 > -1$, then $\sigma_+ < 0$; if $\lambda_3 = -1$, then $\sigma_+ = 0$ if $\lambda_1 = 2$ and $\sigma_+ < 0$ for $\lambda_1 > 2$; whereas for $\lambda_3 < -1$ there exists a unique value of $\lambda_1 = \bar{\lambda}_1 > 2$ such that $\sigma_+ > 0$ for $\lambda_1 < \bar{\lambda}_1, \sigma_+ = 0$ for $\lambda_1 = \bar{\lambda}_1$ and $\sigma_+ < 0$ for $\lambda_1 > \bar{\lambda}_1$. In Figure 9.2a the function $\lambda_3 = g(\lambda_1)$ is sketched, obtained by solving $\sigma_+ = 0$ with respect to λ_3.

(ii)For $\lambda_1 < -2$ consider the line $l_- \equiv y - \frac{1}{2}(\lambda_1 - \sqrt{\lambda_1^2 - 4})x - \lambda_2=0$, connecting two opposite saddles. Then define

$$\frac{dl_-}{dt}\Big|_{l_-=0} = (\dot{y} - \tfrac{1}{2}(\lambda_1 - \sqrt{\lambda_1^2 - 4})\dot{x})\Big|_{l_-=0} =$$
$$1 - \tfrac{1}{2}\lambda_1^2 - \tfrac{1}{2}\lambda_1\lambda_3 + \tfrac{1}{2}(\lambda_1 + \lambda_3)\sqrt{\lambda_1^2 - 4} \equiv \sigma_-.$$

Then if $\lambda_3 < 0$, $\lambda_1 < -2$ it follows that $\sigma_- < 0$ and no saddle connection can occur. In Figure 9.2b the function $\lambda_3 = h(\lambda_1)$ is sketched; it is obtained by solving $\sigma_- = 0$ with respect to λ_3.

We will now consider the phase portraits and bifurcation diagrams taking λ_1 and λ_2 as parameters and λ_3 constant in each of the intervals (i)$-\tfrac{1}{2} < \lambda_3 < 0$, (ii)$\lambda_3 = -\tfrac{1}{2}$, (iii)$-1 < \lambda_3 < -\tfrac{1}{2}$, (iv)$\lambda_3 = -1$, and (v)$\lambda_3 < -1$.

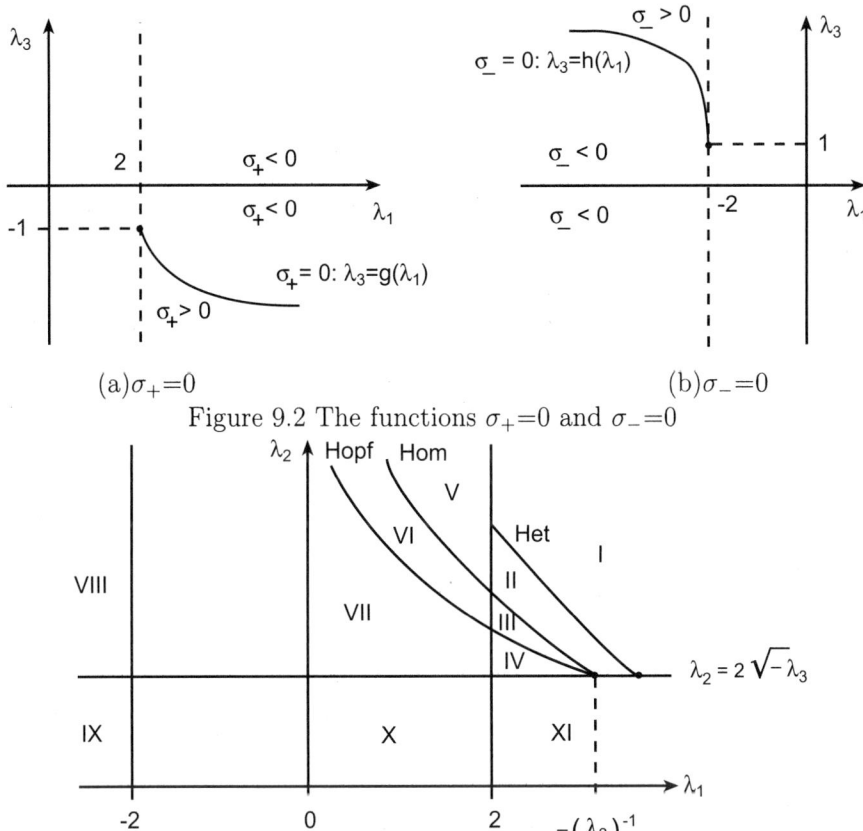

(a)$\sigma_+ = 0$ (b)$\sigma_- = 0$

Figure 9.2 The functions $\sigma_+ = 0$ and $\sigma_- = 0$

Figure 9.3 Bifurcation diagram for $-\tfrac{1}{2} < \lambda_3 < 0$

(i)$-\tfrac{1}{2} < \lambda_3 < 0$. The bifurcation diagram is given in Figure 9.3. The location of the curve Het was determined numerically, as was the curve Hom yielding a very narrow intermediate region II. Notice that the curves Hom and Hopf intersect at the point $\lambda_1 = -\frac{1}{\lambda_3}$, $\lambda_2 = 2\sqrt{-\lambda_3}$, which represents a system with a cusp point. The phase portraits are given in Figure 9.4. The unique (hyperbolic) limit cycle only appears in the regions III and VI and on the line segment between III and VI.

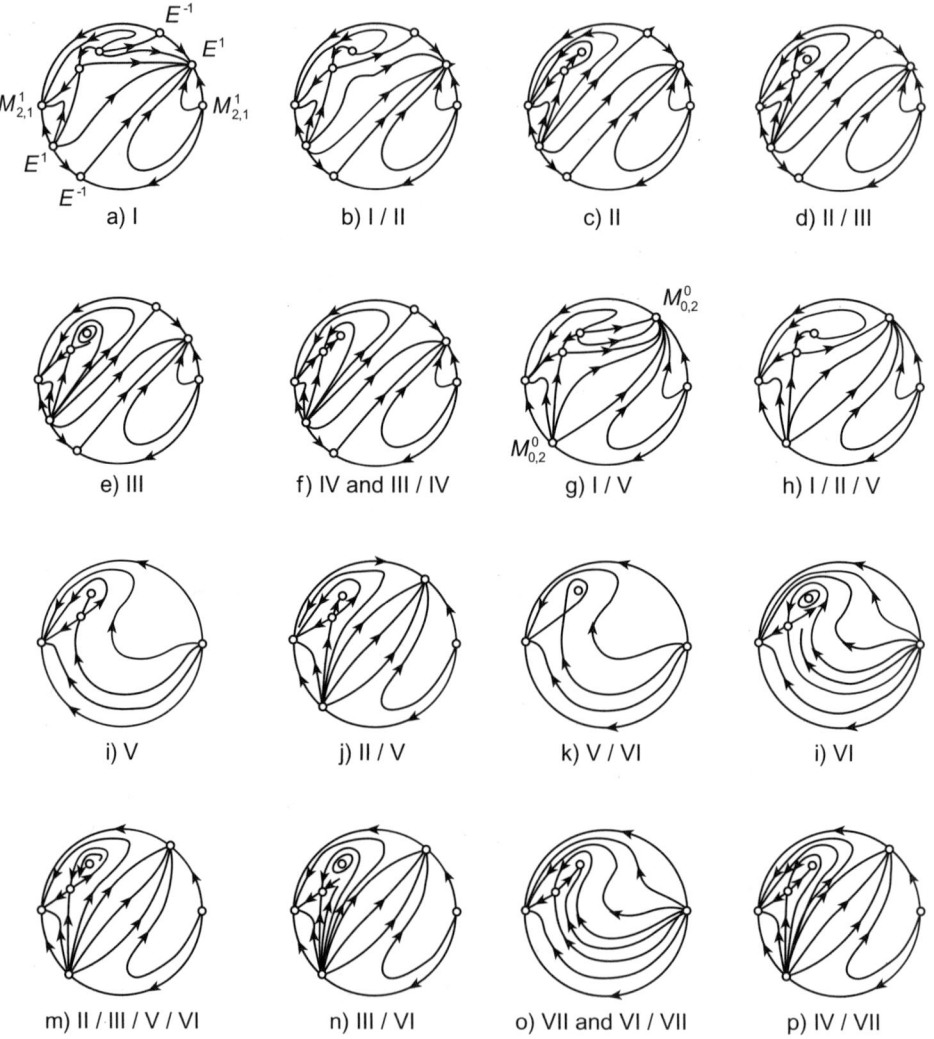

Figure 9.4(i) Phase portraits for $-\frac{1}{2} < \lambda_3 < 0$

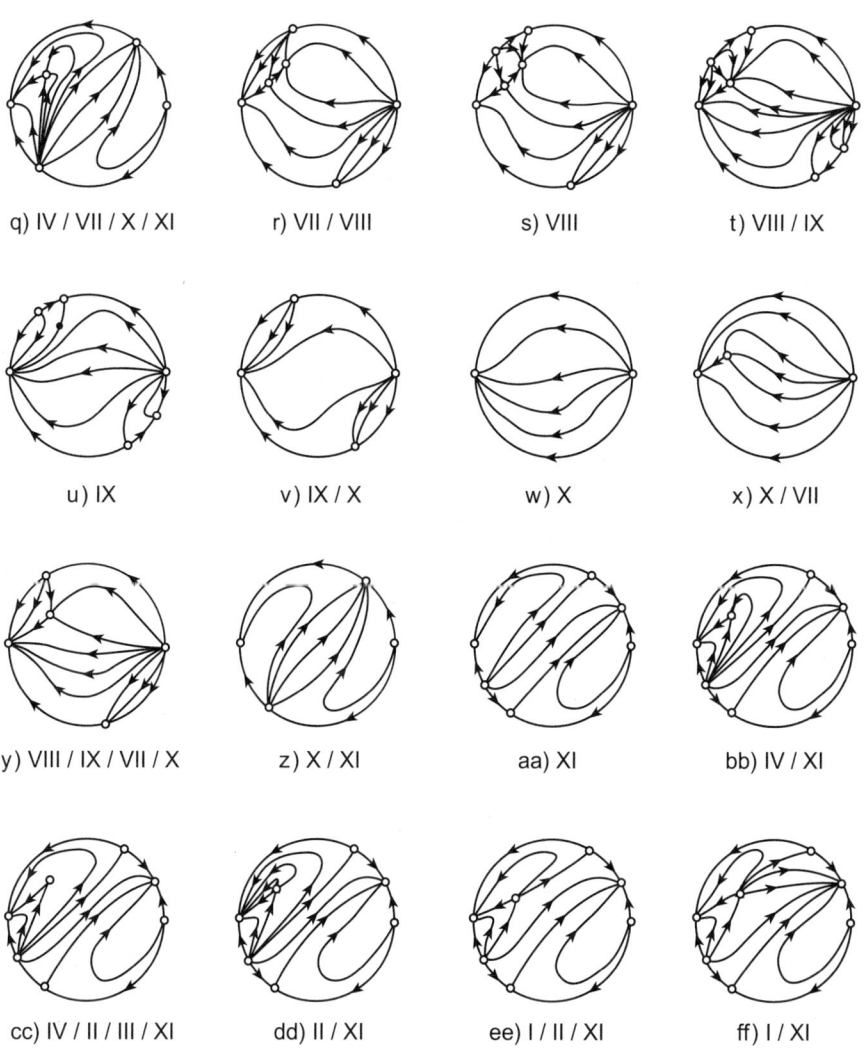

q) IV / VII / X / XI   r) VII / VIII   s) VIII   t) VIII / IX

u) IX   v) IX / X   w) X   x) X / VII

y) VIII / IX / VII / X   z) X / XI   aa) XI   bb) IV / XI

cc) IV / II / III / XI   dd) II / XI   ee) I / II / XI   ff) I / XI

Figure 9.4(ii) Phase portraits for $-\frac{1}{2} < \lambda_3 < 0$

(ii)$\lambda_3 = -\frac{1}{2}$. The curves Hom and Hopf now intersect at $\lambda_1{=}2$, $\lambda_2 = \sqrt{2}$, wherein the system shows a simultaneous occurrence of a finite cusp point and an infinite saddle node $M_{0,2}^0$; the phase portrait is given in Figure 9.6, which is the only additional phase portrait to those already given by the previous case. The bifurcation diagram is given in Figure 9.5; Figures 9.4 and 9.6 show the phase portraits. As a result of numerical calculations it appears that the curve Het and a still smaller region II still exists.

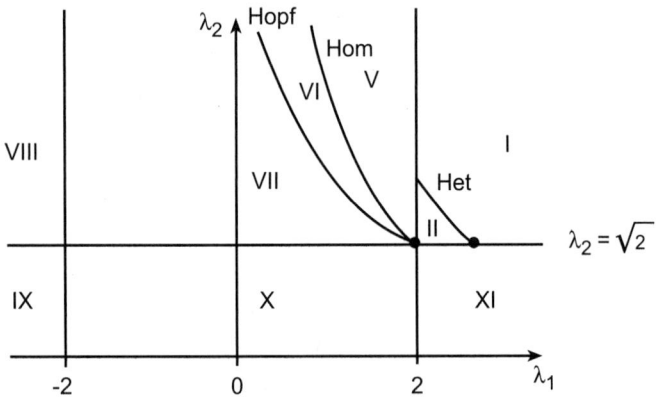

Figure 9.5 Bifurcation diagram for $\lambda_3 = -1$

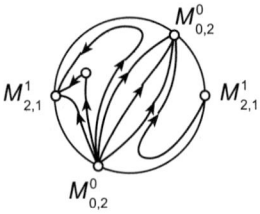

Figure 9.6 Phase portrait for $\lambda_1 = 2, \lambda_2 = \sqrt{2}, \lambda_3 = -1$

(iii)$-1{<} \lambda_3 < -\frac{1}{2}$. Compared with the previous case, due to the decrease of λ_3, the intersection point of the curves Hom and Hopf, indicating the system with a finite cusp point, moves further to the left in the bifurcation diagram to be located now in the interval $1< \lambda_1 <2$. Also a further decrease of the area of region II resulting in a contraction of the curve Het onto the point $\lambda_1=2, \lambda_2 = 2\sqrt{} - \lambda_3$ occurs. In the system corresponding to this point, indicated by $\lambda_3 = \bar{\lambda}_3$, one of the separatrices of the infinite saddle node $M^0_{0,2}$ is also a separatrix of the finite saddle node m^0_2. The phase portrait of this system is given in Figure 9.7.

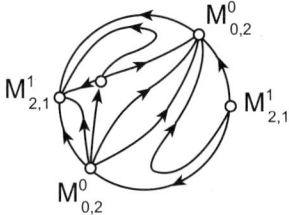

Figure 9.7 Phase portrait for $\lambda_1=2$, $\lambda_2 = 2\sqrt{2}, \lambda_3 = \bar{\lambda}_3$

It may be seen that $-1< \bar{\lambda}_3 < -\frac{1}{2}$. This follows from considering the phase portraits corresponding to $\lambda_1=2$, $\lambda_2=2\sqrt{} - \lambda_3$, $\lambda_3=-\frac{1}{2} - \varepsilon$, $0< \varepsilon \ll 1$ and $\lambda_3 = -1$, respectively, given in Figures 9.8 and 9.9, respectively.

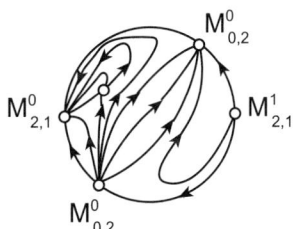

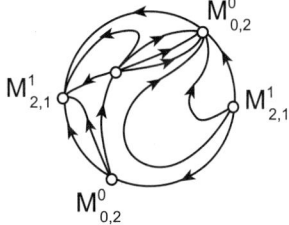

Figure 9.8 Phase portrait for $\lambda_1=2$, $\lambda_2=2\sqrt{} - \lambda_3, \lambda_3 = -\frac{1}{2} - \varepsilon$

Figure 9.9 Phase portrait for $\lambda_1=2$, $\lambda_2=2, \lambda_3 = -1$

The construction of the phase portrait for $\lambda_1 = \lambda_2=2$, $\lambda_3 = -1$ is facilitated by the appearance of two straight orbits on $y = x + 2$, combined to form the separatrix of the $M^0_{0,2}$ point and one of them also being the separatrix separating the hyperbolic sectors in the finite saddle node. Comparing the phase portraits for $\lambda_3 = -\frac{1}{2}$ and $\lambda_3 = -1$ shows, by continuity, the existence of at least one value $\lambda_3 = \bar{\lambda}_3$; however, if more than one value exists this does not affect the number of possible phase portraits. Assume now that $\bar{\lambda}_3$

is unique. Then the bifurcation diagram for $\bar{\lambda}_3 < \lambda_3 < -\frac{1}{2}$ is given in Figure 9.10.

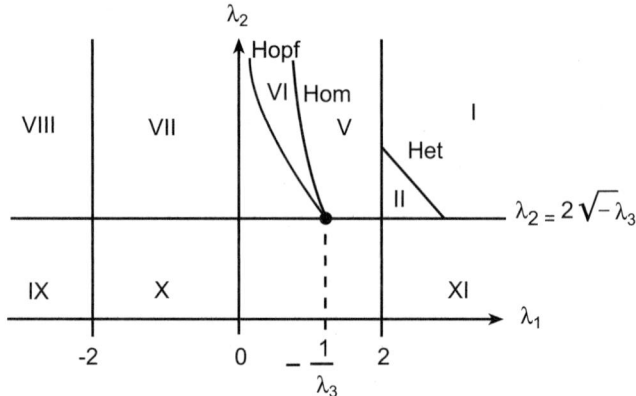

Figure 9.10 Bifurcation diagram for $\bar{\lambda}_3 < \lambda_3 < -\frac{1}{2}$

The additional phase portraits can be found in Figure 9.11. Note that Figure 9.11c corresponds to Figure 9.8.

a) V / X   b) V / VI / VII / X   c) II / V / X / XI

Figure 9.11 Additional phase portraits for $\bar{\lambda}_3 < \lambda_3 < -\frac{1}{2}$

The bifurcation diagram for $\lambda_3 = \bar{\lambda}_3$ is given in Figure 9.12. The corresponding phase portraits can be found in Figures 9.4 and 9.11 with the exception for $\lambda_2 = 2$, $\lambda_2 = 2\sqrt{-\lambda_3}$ which is drawn in Figure 9.7. The bifurcation diagram for $-1 < \lambda_3 < \bar{\lambda}_3 < -\frac{1}{2}$ is the same as for $\lambda_3 = \bar{\lambda}_3$. All corresponding phase portraits can be found in Figures 9.4 and 9.11 except that for $\lambda_1 = 2$, $\lambda_2 = 2\sqrt{-\lambda_3}$ Figure 9.13 applies.

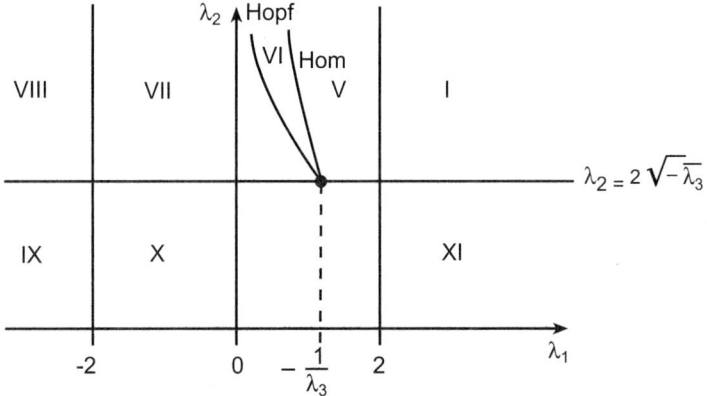

Figure 9.12 Bifurcation diagram for $\lambda_3 = \bar{\lambda}_3$

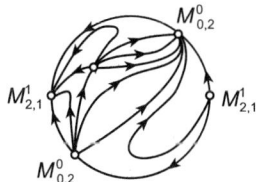

Figure 9.13 Phase portraits for $\lambda_1 = 2, \lambda_2 = 2\sqrt{-\lambda_3}, -1 < \lambda_3 < \bar{\lambda}_3$

(iv) The bifurcation diagram for $\lambda_3 = -1$ is the same as for $-1 < \lambda_3 < \bar{\lambda}_3 < -\frac{1}{2}$ but now for $\lambda_1 = 2$ the system has a straight orbit, because for $\lambda_1 = 2$, $\lambda_3 = -1$ there is $\sigma_+ = 0$. This results in the phase portraits shown in Figure 9.14. Note that Figure 9.14 b corresponds to Figure 9.9.

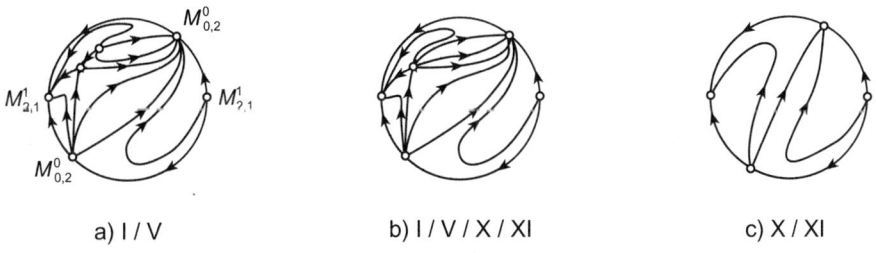

a) I / V   b) I / V / X / XI   c) X / XI

Figure 9.14 Additional phase portraits for $\lambda_1 = 2, \lambda_3 = -1$

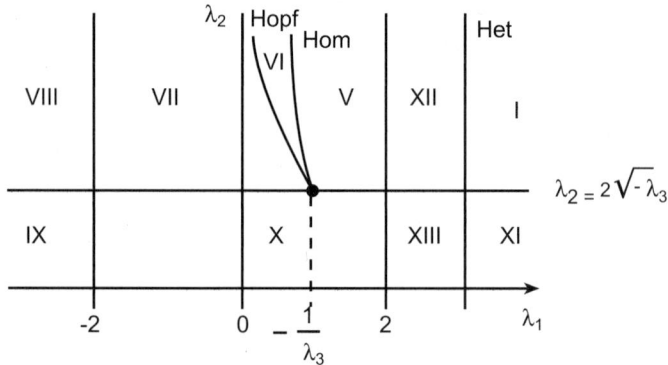

Figure 9.15 Bifurcation diagram for $\lambda_3 < -1$

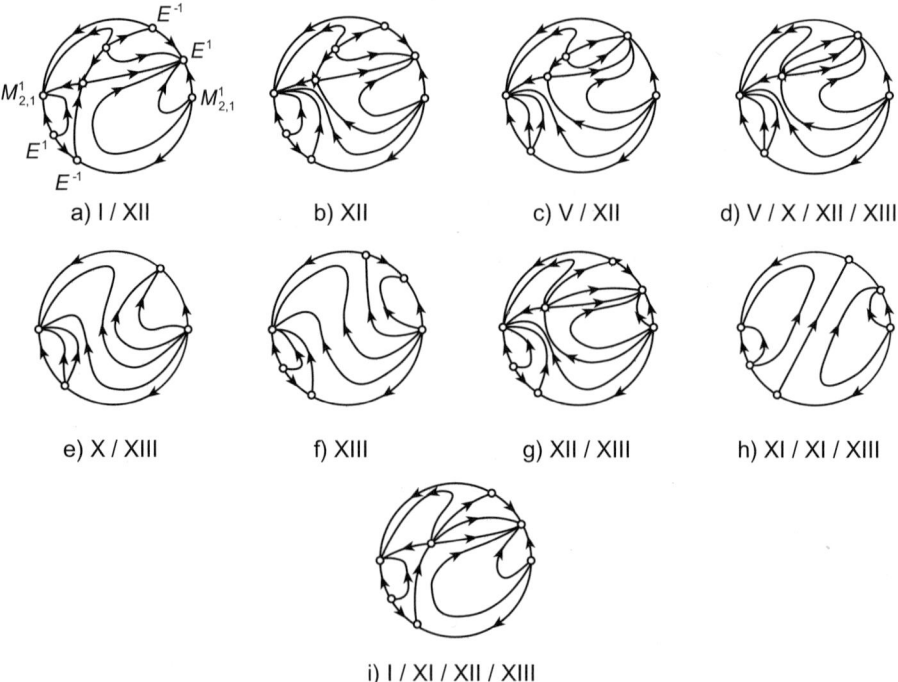

Figure 9.16 Additional phase portraits for $\lambda_3 < -1$

(v) Finally we draw the bifurcation diagram when $\lambda_3 < -1$, which is given in Figure 9.15. For a given value of $\lambda_3 < -1$, there is $\sigma_+ \equiv 1 - \frac{1}{2}\lambda_1^2 - \frac{1}{2}\lambda_1\lambda_3 - \frac{1}{2}(\lambda_1 + \lambda_3)\sqrt{\lambda_1^2 - 4} = 0$ for a unique value $\lambda_1 = \bar{\lambda}_1 > 2$; so there exists a straight orbital connection between the infinite saddle points at $\lambda_1 = \bar{\lambda}_1$. The occurrence of this phenomenon is indicated by Het in Figure 9.15. The phase portraits for $\lambda_3 < -1$ are given in Figures 9.4, 9.11 and 9.16.

We now turn to system (9.2.9),(9.2.10) which now can be put in the form with $-\infty < \lambda_1 < \infty, -\infty < \lambda_2 < \infty, \lambda_3 < 0$, where we may take $\lambda_1 \geq 0$, if not apply the transformation $(x,t) \rightarrow (-x,-t)$ and $\lambda_2 \geq 0$, if not let $y \rightarrow -y$.

As for (9.2.11),(9.2.12) the infinite critical points may be shown to be located in $u_\pm = \frac{1}{2}(\lambda_1 \pm \sqrt{\lambda_1^2 - 4})$, whereas for $\lambda_1 > 2$ there is again a node E^1 in u_- and a saddle E^{-1} in u_+; for $\lambda_1 = 2$ a saddle node $M_{0,2}^0$ in u=1; and for $0 \leq \lambda_1 < 2$ there exist two complex critical points $C_{0,1}^0$.

The finite critical points are located in $(0, y_\pm)$, where $y_\pm = \frac{1}{2}(\lambda_2 \pm \sqrt{\lambda_2^2 + 4\lambda_3})$. Then for $\lambda_2^2 + 4\lambda_3 > 0$ there exists a saddle in y_- and an antisaddle in y_+ ; for $\lambda_2^2 + 4\lambda_3 = 0$ a critical point m_2^0 and for $\lambda_2^2 + 4\lambda_3 < 0$ two complex critical points $c_1^0 c_1^0$. The real critical points are weak for $\lambda_1 = 0$; then the phase portrait is symmetric around the y axis and there exists either a weak saddle and a center point for $\lambda_2^2 + 4\lambda_3 > 0$, or a cusp for $\lambda_2^2 + 4\lambda_3 = 0$. For $\lambda_1 > 0$ the m_2^0 point is a second order saddle node.

There is no limit cycle, as on the line div $(P(x,y), Q(x,y)) = x + \lambda_1 y = 0$ there is $\dot{y} = -\frac{1}{\lambda_1}x^2$ and there cannot be a crossing back and forth across this line by a limit cycle.

The bifurcation diagram is given in Figure 9.17 and the corresponding phase portraits in Figure 9.18.

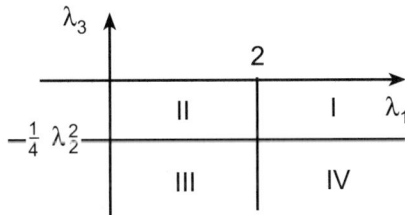

Figure 9.17 Bifurcation diagram for (9.2.17),(9.2.18)

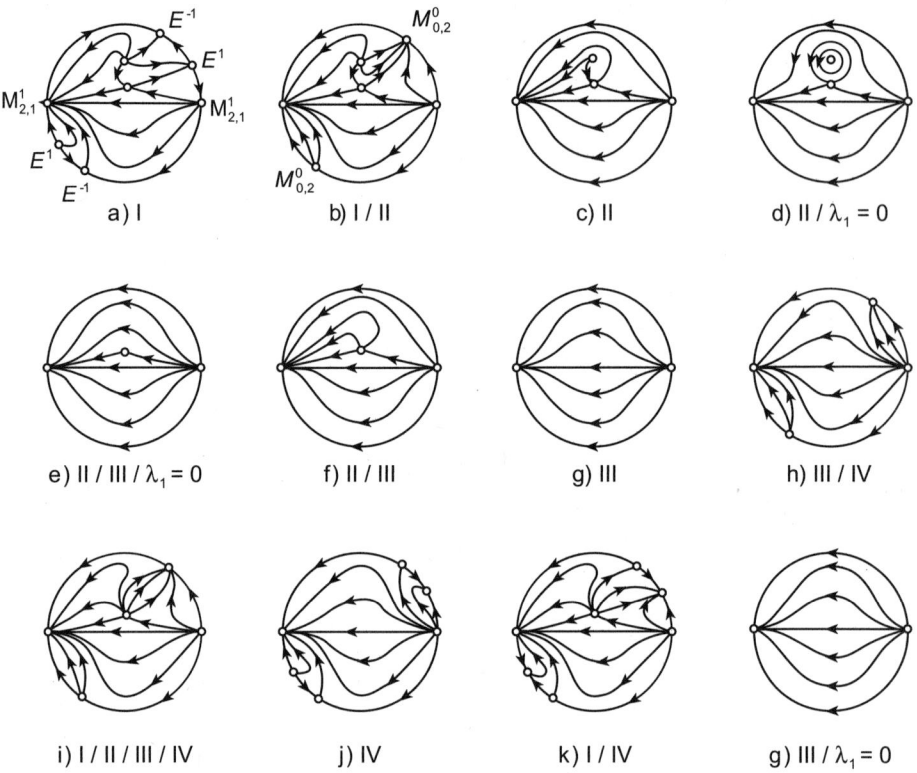

Figure 9.18 Phase portraits for (9.2.17),(9.2.18)

Case2: $c_{56} < 0, \lambda_3 > 0$

For this case there exist the classes $e^{-1}e^1 E^1 E^1 M_{2,1}^{-1}, m_2^0 E^1 E^1 M_{2,1}^{-1}$ and $c_1^0 c_1^0 E^1 E^1 M_{2,1}^{-1}$.

System (9.2.7),(9.2.8) now reads

$$\dot{x} = \lambda_1 + \lambda_3 + \lambda_2 y + \lambda_1 xy + y^2, \tag{9.2.17}$$
$$\dot{y} = 1 + xy, \tag{9.2.18}$$

with $-\infty < \lambda_1 < \infty, -\infty < \lambda_2 < \infty, \lambda_3 > 0$. We take $\lambda_2 \geq 0$; if not satisfied apply $(x, y, t) \rightarrow (-x, -y, -t)$.

We use (9.2.3),(9.2.4) again to see that the infinite critical points are located at $u_\pm = \frac{1}{2}(-\lambda_1 \pm \sqrt{\lambda_1^2 + 4})$ whereas index theory tells these points are elementary nodes E^1, since $i_f = 0$ and for $\lambda_3 > 0$ the point $M_{2,1}^i$ is a saddle $M_{2,1}^{-1}$.

The finite critical points are located in (x_1, y_1) and (x_2, y_2), where $x_i = -\frac{1}{y_i}$ (i=1,2), $y_1 = y_+, y_2 = y_-$ in $y_\pm = \frac{1}{2}(-\lambda_2 \pm \sqrt{\lambda_2^2 - 4\lambda_3})$. So for $\lambda_2^2 - 4\lambda_3 > 0$ there exists the combination $e^{-1}e^1$, for $\lambda^2 - 4\lambda_3 = 0$ a critical point m_2^0 and for $\lambda_2^2 - 4\lambda_3 < 0$ there exist two complex critical points.

In the combination $e^{-1}e^1$ the antisaddle is in (x_+, y_+) and the saddle in (x_-, y_-). They are weak critical points under the condition

$$-\lambda_1(\lambda_3 - \frac{1}{2}\lambda_2^2 \pm \frac{1}{2}\lambda_2\sqrt{\lambda_2^2 - 4\lambda_3}) = 1, \qquad (9.2.19)$$

where the +sign corresponds to a weak focus in (x_+, y_+) and the −sign to a weak saddle in (x_-, y_-). The weak focus is of first order, except for $\lambda_1 = \frac{3}{\lambda_3}, \lambda_2 = \frac{4}{3}\sqrt{3\lambda_3}$; then it is of second order. In the bifurcation diagram, given in Figure 9.19 the weak focus is indicated by the curve Hopf. Also results that the m_2^0 point for $\lambda_2^2 - 4\lambda_3 = 0$ is a cusp (saddle node) for $\lambda_1\lambda_3 = \frac{1}{4}\lambda_1\lambda_2^2 = 1(\neq 1)$.

The relation separating the nodes from the foci is

$$\Delta \equiv (1 + \lambda_3[\lambda_1[2 + \lambda_1(\lambda_3 - \lambda_2^2)] - 4(\lambda_2^2 - 2\lambda_3)])$$
$$+[\lambda_1\lambda_2[2 + \lambda_1(2\lambda_3 - \lambda_2^2)] - 4\lambda_2(\lambda_2^2 - 3\lambda_3)](-\frac{1}{2}\lambda_2 + \frac{1}{2}\sqrt{\frac{1}{2}} - 4\lambda_3) = 0.$$

With relation to the limit cycle problem it may be remarked that (9.2.19), (9.2.20) can be transformed to

$$\dot{x} = dx - y + lx^2 + mxy, \qquad (9.2.20)$$
$$\dot{y} = x + x^2, \qquad (9.2.21)$$

where now

$$d = \frac{1 - \frac{1}{4}\lambda_1(\lambda_2 - \sqrt{\lambda_2^2 - 4\lambda_3})^2}{\lambda_2^2 - 2^{\frac{1}{4}}\lambda_3(\lambda_2 \sqrt{\lambda_2^2 4\lambda_3})^{\frac{3}{2}})^{\frac{1}{4}}},$$

$$l = \frac{2^{\frac{1}{2}}\lambda_1(\lambda_2^2 - 4\lambda_3)^{\frac{1}{4}}}{(\lambda_2 - \sqrt{\lambda_2^2 - 4\lambda_3})^{\frac{1}{2}}},$$

$$m = \frac{2\sqrt{\lambda_2^2 - 4\lambda_3}}{\lambda_2 - \sqrt{\lambda_2^2 - 4\lambda_3}},$$

thus $-\infty < d < \infty, -\infty < l < \infty, 0 < m < \infty$.

We will discuss the limit cycle problem making use of the known results for (9.2.22),(9.2.23), indicated in the classification of Ye Yanqian [65,Y] by $(II)_{n=0,m=>0}$, and discussing simultaneously the bifurcation diagram for (9.2.19),(9.2.20), given in Figure 9.19 and the phase portraits as given in Figure 9.20.

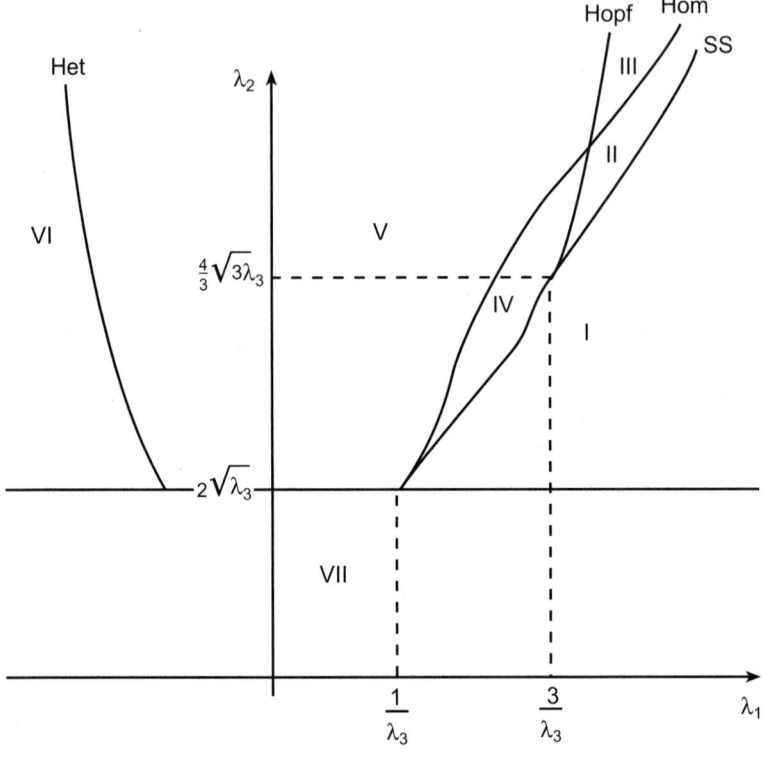

Figure 9.19 Bifurcation diagram for $\lambda_3 > 0$: eqs.(9.2.19),(9.2.20)

In relation to the limit cycle problem, (9.2,22),(9.2.23) was studied by various authors when $|d| < 2$, since then the origin is a focus or a center point, and for all values of l and m. In the present context the anti-saddle of (9.2.19),(9.2.20) corresponds to the origin in (9.2.22),(9.2.23) as a result of which we are only interested in the limit cycles around (0,0) in (9.2.22),(9.2.23). A comprehensive review of the knowledge of the limit cycles of $II_{n=0}$ around (0,0) was given by Zegeling. See also [02,R].

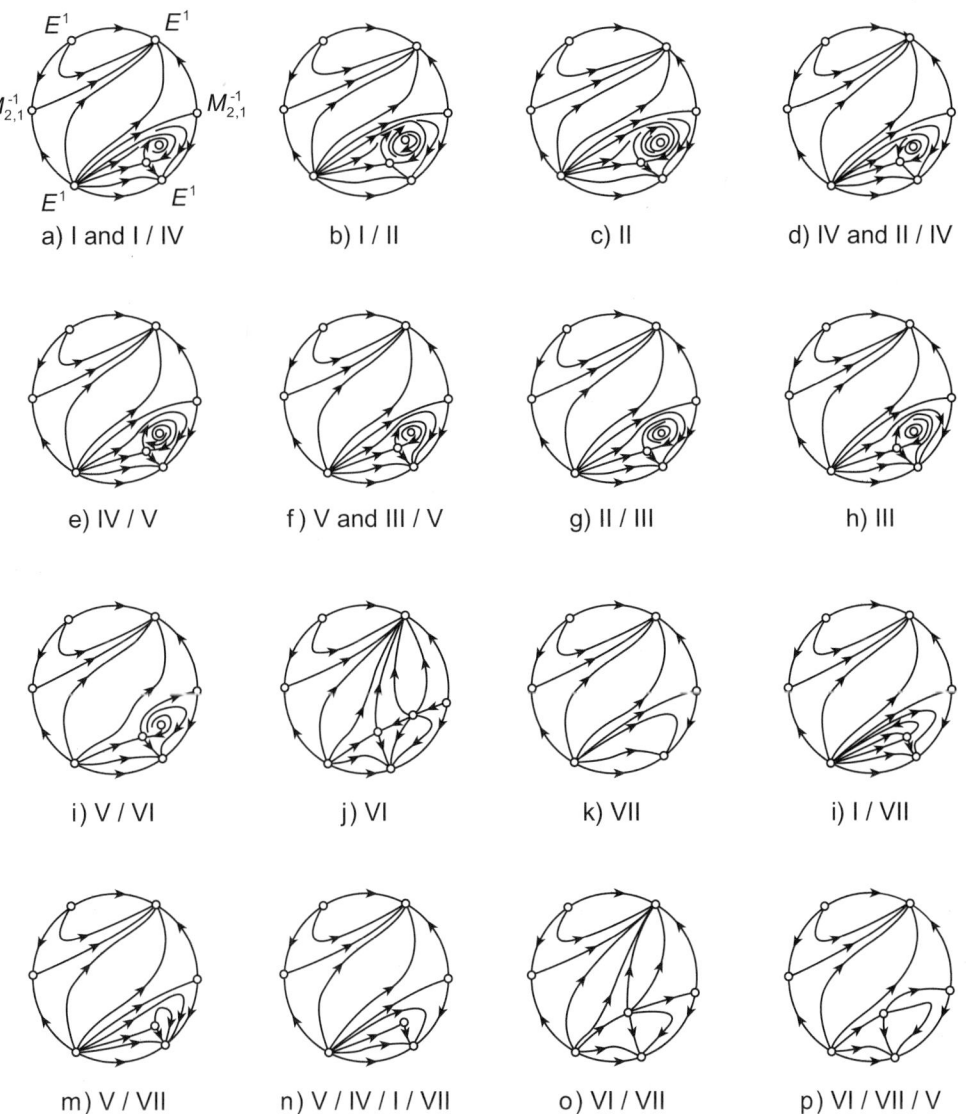

Figure 9.20 Phase portraits for $\lambda_3 > 0$: eqs.(9.2.19),(9.2.20)

We will subsequently consider the intervals $0 < \lambda_2 \leq 2\sqrt{\lambda_3}$, $2\sqrt{\lambda_3} < \lambda_2 < \frac{4}{3}\sqrt{3\lambda_3}$ and $\frac{4}{3}\sqrt{3\lambda_3} < \lambda_2 < \infty$.

(i)If $0 < \lambda_2 \leq 2\sqrt{\lambda_3}$, indicated by region VII in Figure 9.19, (9.2.19),(9.2.20) belongs to class $c_1^0 c_1^0 E^1 E^1 M_{2,1}^{-1}$. The phase portrait is given in Figure 9.20 (k). If $\lambda_2 = 2\sqrt{\lambda_3}, \lambda_1 \lambda_3 = 1, \frac{1}{4}\lambda_1 \lambda_2^2 = 1$, the system belongs to class $m_2^0 E^1 E^1 M_{2,1}^{-1}$ where m_2^0 is a cusp; the phase portrait is given in Figure 9.20 (n). If $\lambda_2 = 2\sqrt{\lambda_3}, \lambda_1 \lambda_3 \neq 1$ so $\frac{1}{4}\lambda_1 \lambda_2^2 \neq 1$, the system also belongs to class $m_2^0 E^1 E^1 M_{2,1}^{-1}$, where m_2^0 now is a saddle node; the phase portraits being given in figures 9.20 (l,m,o) and p and their separatrix structure may be understood from the transition across $\lambda_2 = 2\sqrt{\lambda_3}$.

(ii)If $2\sqrt{\lambda_3} < \lambda_2 < \frac{4}{3}\sqrt{3\lambda_3}$ and $\lambda_1 > 0$ large enough, so region I prevails, the antisaddle is a stable node and Figure 9.20 (a) yields the phase portrait. Lowering the value of λ_1 yields that on $\Delta = 0$ the antisaddle becomes a focus and subsequently a weak focus on the Hopf bifurcation line and a unique hyperbolic limit cycle is bifurcated into region IV leading to Figure 9.19 (d). The uniqueness of the limit cycle follows from the fact that region IV between the curves Hopf and Hom corresponds to $d > 0$ in (9.2.22),(9.2.23) for which case a proof exists, see [94,Z]. This limit cycle is adsorbed by a separatrix cycle (homoclinic cycle), indicated by Hom, and the corresponding phase portrait is given in Figure 9.20 (e). The curve Hom is unique since λ_1 is a parameter rotating the vector field . If λ_1 is further decreased the homoclinic cycle is broken; see Figure 9.20 (f) and the region V is entered. Further decrease of λ_1 leads to a heteroclinic saddle connection between the finite saddle and the infinite saddle $M_{2,1}^{-1}$, as illustrated in Figure 9.20 (i) and indicated by Het in Figure 9.19. The existence of this curve follows from the relative positions of the separatrices for $-\lambda_1$ sufficiently large, as given in region VI and the phase portrait in Figure 9.20 (j). Its uniqueness follows again from the fact that λ_1 is a parameter rotating the field. The precise location of Het has yet to be determined numerically.

(iii) The interval $\frac{4}{3}\sqrt{3\lambda_3} \leq \lambda_2 < \infty$ corresponds to $m \geq 2$ in (9.2.22),(9.2.23). A characteristic feature is now that the weak focus in $\lambda_1 = \frac{3}{\lambda_3}, \lambda_2 = \frac{4}{3}\sqrt{3\lambda_3}$ is of second order and two limit cycles bifurcate into region II, yielding the phase portrait in Figure 9.20 (c), or a second order semi-stable limit cycle onto the curve I- II indicated by SS, and located to the right of the curve Hom. The phase portrait on I- II is given in Figure 9.20 (b). Using the argument that λ_1 is a parameter rotating the field, the existence of SS can be shown, whereas numerical evidence suggests that this curve is unique [86,Ya]. It was pointed out by Liang Zhaojun and Ye Yanqian [85,LY], who studied $II_{d=n=0}$, that the curves Hom and SS approach each other for $\lambda_2 \to \infty$, that is, beyond the intersection point of the curves Hom and Hopf. This was confirmed by numerical results of Yan Zhong [86,Ya] and by Zegeling [94,Z], who studied the asymptotic behavior of the curves Hopf, Hom and SS for larger values of l and m. It can be shown that the phase portrait , given in Figure 9.20 h corresponding to region III between the curves Hopf and Hom contains precisely one (hyperbolic) limit cycle [94,Z]. Although all evidence so far points in the direction that in region II there exist at most two limit cycles, at present there is no proof yet and in fact the only open problem in relation to the conjecture that a quadratic system of finite multiplicity $m_f=2$ contains at most two limit cycles.

We now turn to system (9.2.9),(9.2.10) which now can be written in the form

$$\dot{x} = \lambda_3 + \lambda_2 y + \lambda_1 xy + y^2, \tag{9.2.22}$$

$$\dot{y} = xy, \tag{9.2.23}$$

with $-\infty < \lambda_1 < \infty, -\infty < \lambda_2 < \infty, \lambda_3 > 0$, where we may take $\lambda_1 \geq 0$ and $\lambda_2 \geq 0$ without losing generality.

As for (9.2.19),(9.2.20), the infinite critical points may be shown to be located in $u_\pm = \frac{1}{2}(-\lambda_1 \pm \sqrt{\lambda_1^2 + 4})$, being both elementary nodes E^1.

The finite critical points are located in $(0, y_\pm)$ where $y_\pm = \frac{1}{2}(-\lambda_2 \pm \sqrt{\lambda_2^2 - 4\lambda_3})$. Then for $\lambda_2^2 - 4\lambda_3 < 0$ there exists a saddle in y_- and an antisaddle in y_+, for $\lambda_2^2 - 4\lambda_3 = 0$ a critical point m_2^0 and for $\lambda_2^2 - 4\lambda_3 < 0$ two complex critical points $c_1^0 c_1^0$. The critical points are weak for $\lambda_1 = 0$, in which case the phase portraits are symmetric around the y axis. Then there exists a weak saddle and a center point for $\lambda_2^2 - 4\lambda_3 > 0$, or a cusp for $\lambda_2^2 - 4\lambda_3 = 0$. For $\lambda_1 > 0$ the critical points are not weak and the m_2^0 is a saddle node. No limit cycles are possible. This can be seen by putting $d\tau = ydt$, then (9.2.24),(9.2.25) becomes

$$x' = \frac{\lambda_3}{y} + \lambda_2 + \lambda_1 x + y \equiv P(x,y), \tag{9.2.24}$$

$$y' = x \equiv Q(x,y), \tag{9.2.25}$$

and $\text{div} P(x,y), Q(x,y) \equiv \lambda_1 > 0$, not allowing the divergence integral over the interior of a possible limit cycle to vanish as it should.

The phase portraits are given in Figure 9.21, where there is taken $\lambda_2 = 1$ since for $\lambda_2 > 0$ this does not lose generality and for $\lambda_2 = 0$ no additional phase portraits exist.

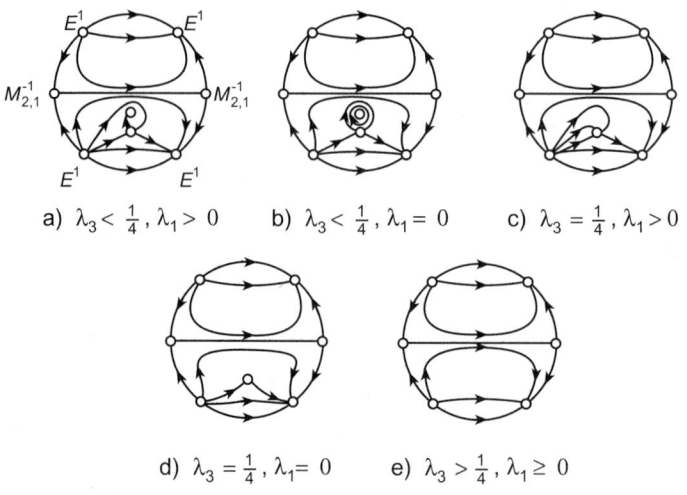

a) $\lambda_3 < \frac{1}{4}, \lambda_1 > 0$ b) $\lambda_3 < \frac{1}{4}, \lambda_1 = 0$ c) $\lambda_3 = \frac{1}{4}, \lambda_1 > 0$

d) $\lambda_3 = \frac{1}{4}, \lambda_1 = 0$ e) $\lambda_3 > \frac{1}{4}, \lambda_1 \geq 0$

Figure 9.21 Phase portraits
for $\lambda_3 > 0$: eqs.(9.2.24),(9.2.25)

Systems with finite index $i_f = 2$

As stated in the beginning of section 9.2.1 there exist the combinations $e^1 e^1 E^{-1} E^1 M_{2,1}^{-1}$, $e^1 e^1 M_{0,2}^0 M_{2,1}^{-1}$ and $e^1 e^1 C_{0,1}^0 C_{0,1}^0 M_{2,1}^{-1}$; they occur for $c_{56} > 0$ and $\lambda_3 > 0$.

We work again with (9.2.5),(9.2.6), taking $c_{56} > 0$,

$$\dot{x} = \lambda_1 + \lambda_3 + \lambda_2 y + \lambda_1 xy - y^2, \qquad (9.2.26)$$

$$\dot{y} = 1 + xy, \qquad (9.2.27)$$

with $-\infty < \lambda_1 < \infty, -\infty < \lambda_2 < \infty, \lambda_3 > 0$; where again we take $\lambda_2 \geq 0$ without losing generality.

As for (9.2.11),(9.2.12) it may be seen that the infinite critical points are located in $u_\pm = \frac{1}{2}(\lambda_1 \pm \sqrt{\lambda_1^2 - 4})$. For $\lambda_1 > 2$ there is a node E^1 in u_- and a saddle in u_+, for $\lambda_1 = 2$ a saddle node in $u=1$, for $-2 < \lambda_1 < 2$ there are two complex critical points $C_{0,1}^0$, for $\lambda_1 = -2$ a saddle node in $M_{0,2}^0$ in $u=-1$, whereas for $\lambda_1 < -2$ there is a saddle in u_- and a node in u_+.

Also it may be seen that the finite critical points are located in (x_+, y_+) and (x_-, y_-), where $x_\pm = -\frac{1}{y_\pm}$ and $y_\pm = \frac{1}{2}(\lambda_2 \pm \sqrt{\lambda_2^2 + 4\lambda_3})$. These points are antisaddles, being weak foci if

$$\lambda_1(\lambda_3 + \tfrac{1}{2}\lambda_2^2 \pm \tfrac{1}{2}\sqrt{\lambda_2^2 + 4\lambda_3}) = 1,$$

where the $-$ sign corresponds to an unstable first order focus in (x_-, y_-) and the $+$ sign to a stable first order weak focus in (x_+, y_+). They correspond in the bifurcation diagrams to the curves $Hopf_-$ and $Hopf_+$, respectively, intersecting in $\lambda_1 = \frac{1}{\lambda_3}, \lambda_2 = 0$, which point represents a centro-symmetric system, which are also given by $\lambda_2 = 0$. The location of the point $\lambda_1 = \frac{1}{\lambda_3}, \lambda_2 = 0$ with respect to the line $\lambda_1 = 1$ is important for the construction of the bifurcation diagrams, We thus distinguish three cases (i)$0 < \lambda_3 < \frac{1}{2}$,(ii)$\lambda_3 = \frac{1}{2}$,(iii)$\lambda_3 > \frac{1}{2}$. Bifurcation diagrams will be constructed for constant values of λ_3.

The limit cycle behavior may again be studied using $II_{n=0}$, as given in (9.2.22),(9.2.23), now with $-\infty < d < \infty, -\infty < l < \infty, -2 < m < -1$. It is known that this system has at most one limit cycle around any of its foci [94,Z]. Moreover, for some values of the parameters a $(1,1)$ limit cycle distribution exists.

Starting from the curves $Hopf_\pm$ and increasing λ_1, unique limit cycles will appear around $(x_\pm, y_\pm)$. Each limit cycle will expand when λ_1 increases, since λ_1 rotates the vector field, and subsequently disappear in an unbounded heteroclinic separatrix cycle connecting E^1 and $M_{2,1}^{-1}$. These cycles are indicated in the bifurcation diagram by $Het_\pm$. Their precise location can only be found numerically. However, since for $|\lambda_1| < 2$ there are no infinite saddles, they must be located in the region $\lambda_1 > 2$. The curves $Het_\pm$ intersect in a point on the λ_1 axis, representing a centro-symmetric system.

It may be seen that there exist no saddle connections between the infinite saddles, as $l_+ \equiv y - \frac{1}{2}(\lambda_1 + \sqrt{\lambda_1^2 - 4})x - \lambda_3 = 0$ is a line without contact. Here $l_+ \equiv 0$ connects the diametrically opposite points that form the saddle points E^{-1} if $\lambda_1 > 2$. In fact, there follows, with

$$\tfrac{dl_+}{dt}\big|_{l=0} = 1 - \tfrac{1}{2}\lambda_1^2 - \tfrac{1}{2}\lambda_1\lambda_3 - \tfrac{1}{2}(\lambda_1 + \lambda_3)\sqrt{\lambda_1 - 4} \equiv \sigma_+,$$

for $\lambda_1 \geq 2, \lambda_3 > 0$ that $\sigma_+ < 0$ (see Figure 9.2a).

The phase portraits for (9.2.28),(9.2.29), $0 < \lambda_3 < \frac{1}{2}$, corresponding to the bifurcation diagram in Figure 9.22, are given in Figure 9.23. The bifurcation diagram for $\lambda_3 = \frac{1}{2}$ is given in Figure 24; compared to Figure 22 region VI has disappeared and all phase portraits can still be found in Figure 9.23. The bifurcation diagram for $\lambda_3 > \frac{1}{2}$ is given in Figure 9.25; compared to Figure 9.24 a new region (region X) appears which leads to two new phase portraits, as given in Figure 9.26.

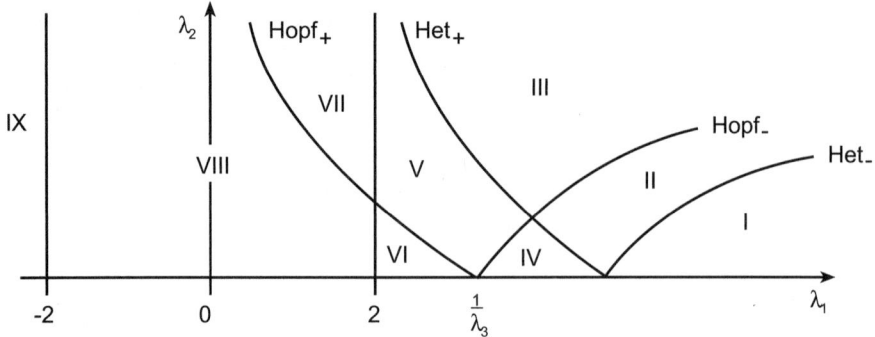

Figure 9.22 Bifurcation diagram for $0 < \lambda_3 < \frac{1}{2}$

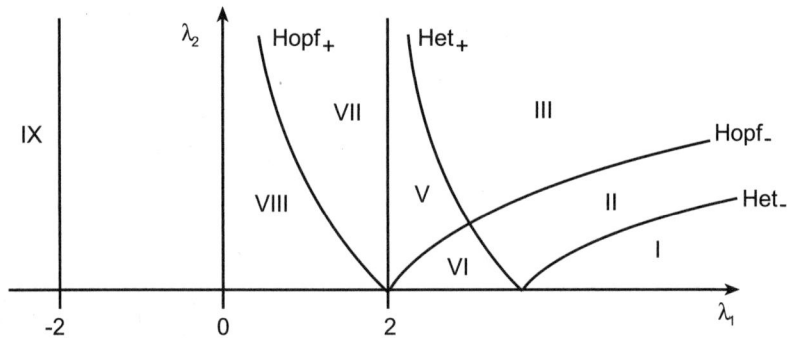

Figure 9.24 Bifurcation diagram for $\lambda_3 = \frac{1}{2}$

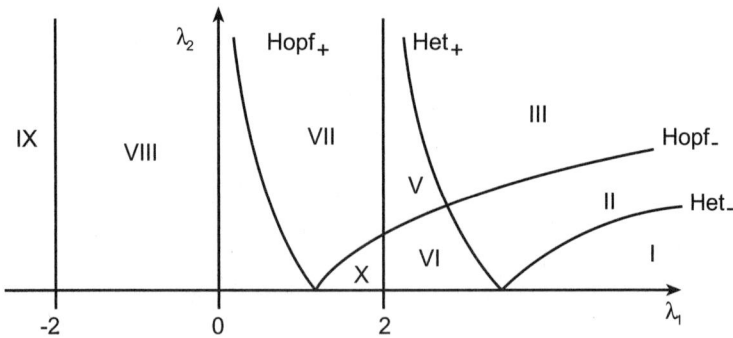

Figure 9.25 Bifurcation diagram for $\lambda_3 > \frac{1}{2}$

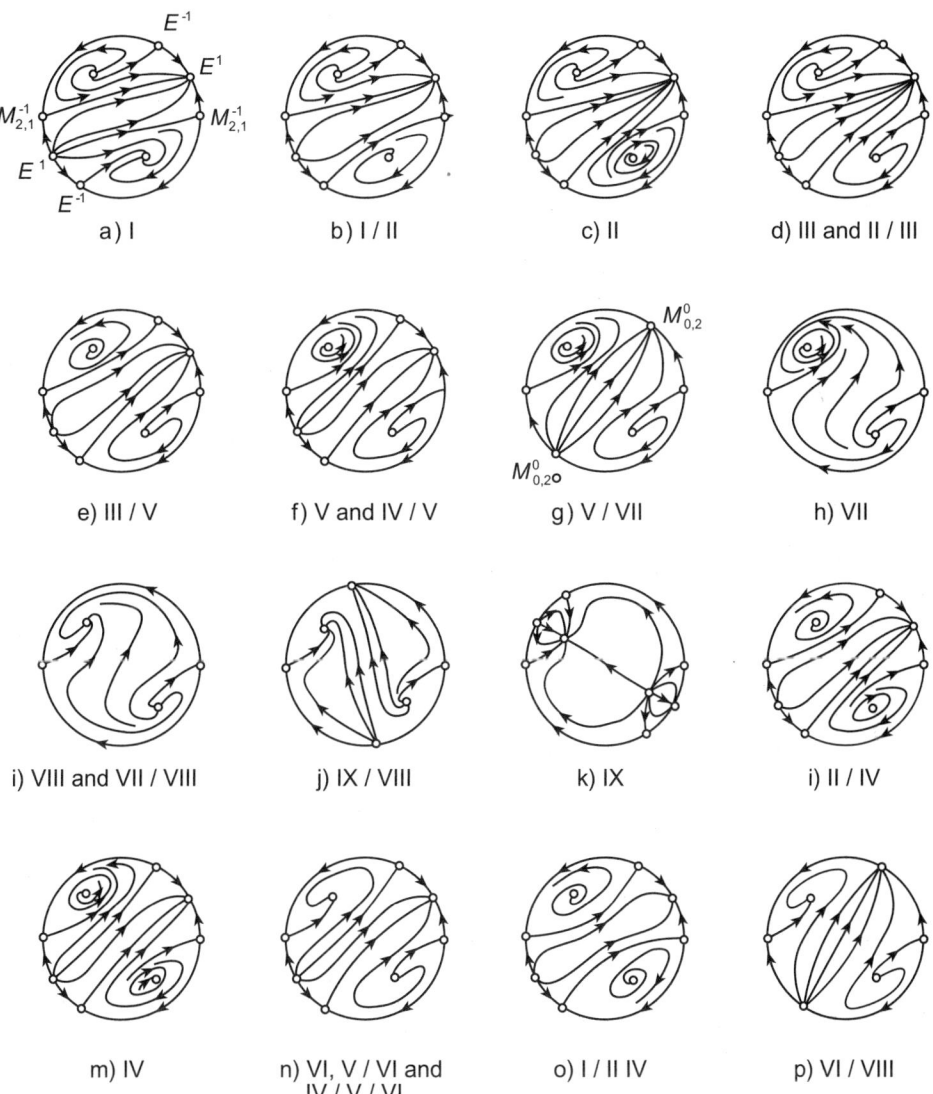

Figure 9.23 Phase portraits for $0 < \lambda_3 < \frac{1}{2}$

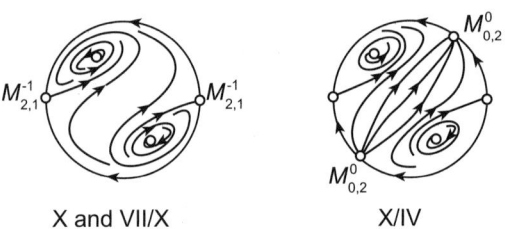

X and VII/X   X/IV

Figure 9.26 Phase portraits for $\lambda_3 > \frac{1}{2}$

We also again consider (9.2.9),(9.2.10)

$$\dot{x} = \lambda_3 + \lambda_2 y + \lambda_1 xy - y^2, \qquad (9.2.28)$$
$$\dot{y} = xy, \qquad (9.2.29)$$

where now $-\infty < \lambda_1 < \infty, -\infty < \lambda_2 < \infty, \lambda_3 > 0$, and we may take $\lambda_1 \geq 0$, $\lambda_2 \geq 0$ without losing generality.

The infinite critical points are located in $u_\pm = \frac{1}{2}(\lambda_1 \pm \sqrt{\lambda_1^2 - 4})$, and for $\lambda_1 > 2$ are a node E^1 in u_- and a saddle E^{-1} in u_+; for $\lambda_1{=}2$ a saddle node $M_{0,2}^0$ in $u{=}1$, and for $0 \leq \lambda_1 < 2$ two complex critical points.

The two finite critical antisaddles are located in $(0, y_\pm)$ where $y_\pm = \frac{1}{2}(\lambda_2 \pm \sqrt{\lambda_2^2 + 4\lambda_3})$; weak if $\lambda_1{=}0$ and then center points because of the symmetry around the y axis.

No limit cycles can occur, as may be seen using the divergence argument given in relation to (9.2.26),(9.2.27).

The phase portraits are given in Figure 9.27.

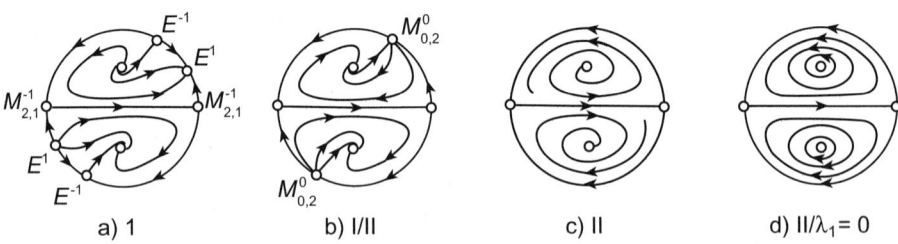

a) 1   b) I/II   c) II   d) II/$\lambda_1{=}0$

Figure 9.27 Phase portraits for (9.2.30),(9.2.31)

9.2.2 Systems with a fourth order infinite critical point

For these systems we start again with (9.2.1),(9.2.2)

$$\dot{x} = a_{00} + a_{10}x + a_{01}y + a_{11}xy + a_{02}y^2, \qquad (9.2.30)$$

$$\dot{y} = b_{00} + b_{10}x + b_{01}y + b_{11}xy + b_{02}y^2, \qquad (9.2.31)$$

with $a_{11}^2 + b_{11}^2 \neq 0$, $a_{02}^2 + b_{02}^2 \neq 0$, whereas $c_{56} \neq 0$ yields that the transversally non-hyperbolic infinite critical point is "at the ends of the x axis". It belongs to $m_f = 2$ iff $c_{25} = 0$, $c_{26}^2 + c_{26}c_{35} - c_{15}c_{56} \neq 0$. Moreover for an $M_{1,2}^i$ point we require $b_{11} = 0$, $a_{11} - b_{02} \neq 0$.

Use will again be made of the classification of infinite critical points by Reyn and Kooij [95,RK] to determine the possible critical points $M_{2,2}^i$ in the present case. From $c_{25} \equiv a_{10}b_{11} - a_{11}b_{10} = -a_{11}b_{10} = 0$ and $c_{56} \equiv a_{11}b_{02} - a_{02}b_{11} = a_{11}b_{02} \neq 0$ it follows that $b_{10} = 0$ so that $M_{2,2}^i$ is a highly degenerate critical point. Moreover there can be written that $c_{26}^2 + c_{26}c_{35} - c_{15}c_{56} = b_{02}(a_{10}^2b_{02} - a_{10}a_{11}b_{01} + a_{11}^2b_{00}) = b_{02}\gamma = \frac{1}{4}[(2a_{10}b_{02} - a_{11}b_{01})^2 - a_{11}^2\Delta] \neq 0$, which defines γ and where $\Delta \equiv b_{01}^2 - b_{00}b_{02}$. Figures 7,8 and 9 of [95,RK] should now be consulted. For $\Delta < 0$ Figure 7 yields for $b_{02}(a_{11} - b_{02}) < 0$ a point $M_{2,2}^0$ and for $b_{02}(a_{11} - b_{02}) > 0$ a point $M_{2,2}^2$. For $\Delta > 0$ Figure 9 yields for $b_{02}(a_{11} - b_{02}) > 0$ a point $M_{2,2}^2$ if $b_{02}\gamma > 0$ and a point $M_{2,2}^0$ if $b_{02}\gamma < 0$, whereas for $b_{02}(a_{11} - b_{02}) < 0$ a point $M_{2,2}^0$ if $b_{02}\gamma > 0$ and, if $b_{02}\gamma < 0$ a point $M_{2,2}^{-2}$ if $a_{11}b_{02} > 0$ and a point $M_{2,2}^2$ if $a_{11}b_{02} < 0$.

We also use (9.2.3),(9.2.4) again, with $b_{10} = b_{11} = 0$ taking the form

$$z' = -z(a_{10}z + a_{11}u + a_{00}z^2 + a_{01}zu + a_{02}u^2) \equiv \bar{P}(z,u), \qquad (9.2.32)$$

$$u' = b_{00}z^2 + (-a_{10} + b_{01})zu + (-a_{11} + b_{02})u^2 - a_{01}zu^2$$
$$-a_{00}z^2u - a_{02}u^3 \equiv \bar{Q}(z,u), \qquad (9.2.33)$$

Apart from $M_{2,2}^i$ there exists an elementary infinite critical point in $u_e = -\frac{a_{11}-b_{02}}{a_{02}} \neq 0$; its character is given through $\Omega(0,u_e) \equiv \bar{P}_z(0,u_e)\bar{Q}_u(0,u_e) - \bar{P}_u(0,u_e)\bar{Q}_z(0,u_e) = \frac{b_{02}^4}{a_{02}^2}(1 - \frac{a_{11}}{b_{02}})^3$, to be a node (saddle) for $\frac{a_{11}}{b_{02}} < 1$ (<1).

The final critical points are located on the line $y_{\pm} = \frac{1}{2b_{02}}(-b_{01} \pm \sqrt{\Delta})$ and there exist two complex critical points for $\Delta < 0$, a saddle node for $\Delta = 0$ and two real elementary points for $\Delta > 0$. As follows from the proof of Theorem 3.15 there exist for $b_{02}\gamma < 0$ a saddle and a node, and for $b_{02}\gamma < 0$ either two saddles if $a_{11}b_{02} < 0$ or two nodes if $a_{11}b_{02} > 0$.

In order to study saddle connections we consider the line $l \equiv y + \alpha x + \beta = 0$, where $\alpha = -\frac{b_{02}}{a_{02}}(1 - \frac{a_{11}}{b_{02}})$ and $\beta = -\frac{1}{a_{11}}(a_{01}a_{11} - a_{10}a_{02} + a_{02}b_{01} - a_{01}b_{02})$ connecting the elementary infinite critical points. Then $\frac{dl}{dt}|_{l=0} = b_{00} + \alpha a_{00} - \beta b_{01} - \alpha \beta a_{01} + \beta^2 b_{02} + \alpha \beta^2 a_{02} \equiv \sigma$. For $\sigma \neq 0$, l is a line without contact. For $\sigma=0$ the line l consists of (a) straight line orbit(s) either connecting the diametrically opposite parts of the infinite saddle or a finite saddle connection or a connection between a finite saddle or saddle node with an infinite saddle. The finite saddle connection occurs when the elementary infinite critical point is a node; since $l(x,y)=0$ for $\sigma=0$ is a (degenerate) isocline it must, however, contain the two finite saddle points.

The phase portraits for $\Delta < 0$ are given in Figure 9.28; they cover the classes $c_1^0 c_1^0 E^1 M_{0,2}^0$ and $c_1^0 c_1^0 E^{-1} M_{2,2}^2$ in class 2 of Table 3 of Chapter 3. The phase portraits for $\Delta=0$ are given in Figure 9.29; they cover the classes $m_2^0 E^1 M_{2,2}^0$ and $m_2^0 E^{-1} M_{2,2}^2$ in class 2 of Table 3 in Chapter 3. The phase portraits for $\Delta > 0$ are given in Figure 9.30; for (a) and (b) they cover the classes $e^{-1}e^{-1}E^1 M_{2,2}^2$ in class 1 of Table 3 of Chapter 3; for (c) and (d) they cover the classes $e^1 e^1 E^{-1} M_{2,2}^0$ and $e^1 e^1 E^1 M_{2,2}^{-2}$ in class 3 of Table 3 of Chapter 3; for (e),(f) and (g) they cover the classes $e^{-1}e^1 E^{-1} M_{2,2}^2$ and $e^{-1}e^1 E^1 M_{2,2}^0$ in class 2 of Table 3 in Chapter 3.

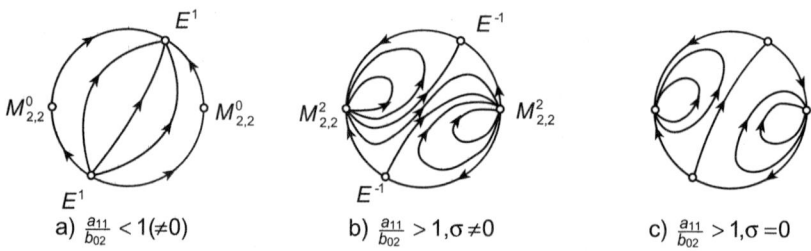

a) $\frac{a_{11}}{b_{02}} < 1 (\neq 0)$ b) $\frac{a_{11}}{b_{02}} > 1, \sigma \neq 0$ c) $\frac{a_{11}}{b_{02}} > 1, \sigma = 0$

Figure 9.28 Phase portraits for $\Delta < 0$; $M_{p,q}^i = M_{2,2}^i$

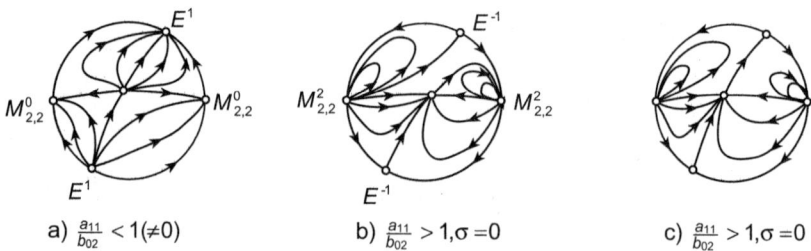

a) $\frac{a_{11}}{b_{02}} < 1 (\neq 0)$ b) $\frac{a_{11}}{b_{02}} > 1, \sigma = 0$ c) $\frac{a_{11}}{b_{02}} > 1, \sigma = 0$

Figure 9.29 Phase portraits for $\Delta = 0$; $M_{p,q}^i = M_{2,2}^i$

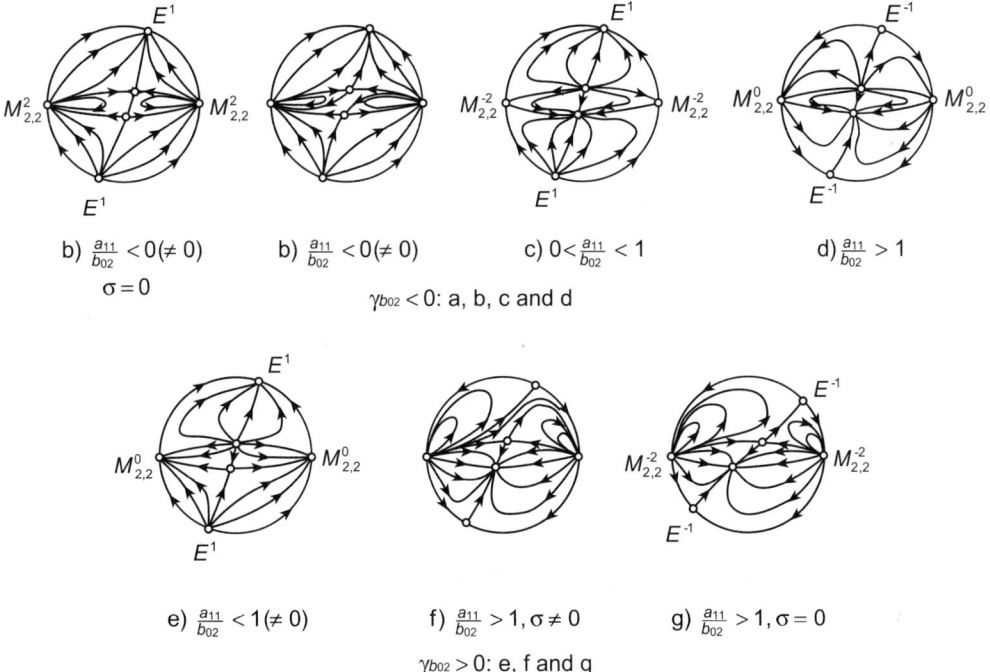

b) $\frac{a_{11}}{b_{02}} < 0 (\neq 0)$ b) $\frac{a_{11}}{b_{02}} < 0 (\neq 0)$ c) $0 < \frac{a_{11}}{b_{02}} < 1$ d) $\frac{a_{11}}{b_{02}} > 1$

$\sigma = 0$

$\gamma b_{02} < 0$: a, b, c and d

e) $\frac{a_{11}}{b_{02}} < 1 (\neq 0)$ f) $\frac{a_{11}}{b_{02}} > 1, \sigma \neq 0$ g) $\frac{a_{11}}{b_{02}} > 1, \sigma = 0$

$\gamma b_{02} > 0$: e, f and g

Figure 9.30 Phase portraits for $\Delta > 0$; $M_{p,q}^i = M_{2,2}^i$

9.2.3 Systems with a fifth order infinite critical point

The analysis is similar to that given in the previous section. Equations (9.2.32),(9.2.33) can again be used by slightly changing the stated conditions; now, in order for an $M_{2,3}^i$ point to exist, it is required that $b_{11} = a_{11} - b_{02} = 0, a_{02} \neq 0$. From [95,RK] it follows that Figures 12 and 13 should be consulted. From Figure 12 then follows that if $\Delta < 0$, the $M_{2,3}^i$ point is an $M_{2,3}^i$ node and if $\Delta = 0$ (so $b_{02}\gamma = (-a_{10} + \frac{1}{2}b_{01})^2 a_{11}^2 \neq 0$ a point $M_{2,3}^i$. From Figure 13 it follows that for $\Delta > 0$ the $M_{2,3}^i$ point is an $M_{2,3}^1$ point if $b_{02}\gamma > 0$ and an $M_{2,3}^{-1}$ point if $b_{02}\gamma < 0$. The phase portraits can now easily be constructed and are given in Figure 9.31.

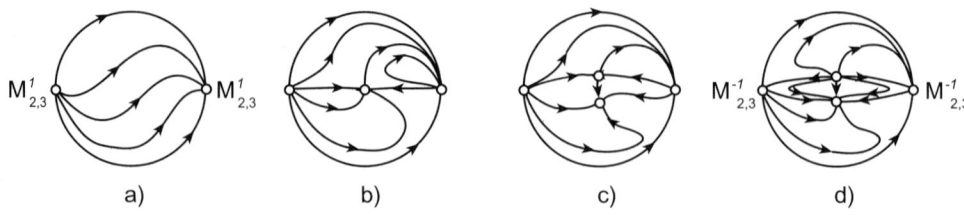

a)$\Delta < 0$ b)$\Delta = 0$ c)$\Delta > 0, \gamma b_{02} > 0$ d)$\Delta > 0, \gamma b_{02} < 0$

Figure 9.31 Phase portraits for the class with an $M_{2,3}^i$ point

9.2.4 Systems with infinitely many infinite critical points

In this case $b_{20} = b_{11} - a_{20} = b_{02} - a_{11} = a_{02}=0$ and (9.2.1),(9.2.2) read

$$\dot{x} = a_{00} + a_{10}x + a_{01}y + xy, \qquad (9.2.34)$$
$$\dot{y} = b_{00} + b_{10}x + b_{01}y + xy, \qquad (9.2.35)$$

where $b_{10}=0$ and $C_1 \equiv a_{10}^2 - a_{10}b_{01} + b_{00} \neq 0$ in order to have $m_f=2$.

For $\Delta \equiv b_{01}^2 - 4b_{00} > 0$ there are two real distinct elementary critical points, for $\Delta=0$ an m_2^0 point, and for $\Delta < 0$ two complex critical points. From the analysis in section 6.1 of Chapter 3 it follows that for $\Delta > 0$ and $C_1 < 0$ there exist two antisaddles so $i_f=2$ and for $C_1 > 0$ there is a saddle and an antisaddle so $i_f=0$; the systems belong to class 3 and 2 of Table 3 in Chapter 3, respectively. If $\Delta \leq 0$, then $i_f=0$ and the system also belongs to class 2 of this table.

The behaviour at infinity can be studied using the system

$$z' = -a_{10}z - u - a_{00}z^2 - a_{01}zu, \qquad (9.2.36)$$
$$u' = b_{00}z + (-a_{10} + b_{01})u - a_{01}u^2 - a_{00}zu. \qquad (9.2.37)$$

For this system u=z=0 is a saddle if $C_1 < 0$ and an antisaddle if $C_1 > 0$. There exist four phase portraits: for $\Delta > 0$, $C_1 < 0$ and $C_1 > 0$, for $\Delta=0$, $C_1 > 0$, and for $\Delta < 0$, $C_1 > 0$ (note that for $\Delta < 0$, $C_1 < 0$ is not possible); they are given in figure 9.32, They confirm those given in the paper by Gasull and Prohens [96,GP1].

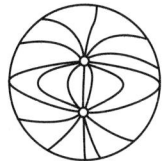   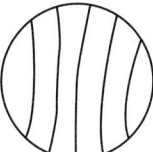

$\Delta > 0, C_1 < 0$ $\Delta > 0, C_1 > 0$ $\Delta = 0, C_1 > 0$ $\Delta < 0, C_1 > 0$

Figure 9.32 Phase portraits for systems with infinitely many infinite critical points

9.2.5   Conclusion

Theorem 9.1 *There exist 131 possible phase portraits for quadratic systems with finite multiplicity $m_f = 2$ and one transversally non-hyperbolic infinite critical point. They are given in Figures 9.1-9.32. It is conjectured that this class has at most two limit cycles.*

9.3   Two transversally non-hyperbolic critical points

If there are two transversally non-hyperbolic critical points at infinity they can be either real different, coinciding or complex. Corresponding, the quadratic terms in the system have two common linear factors, real different, coinciding or complex, respectively. It follows from Chapter 3 that $i_f = 0$ for these systems; they belong to class 4 in Table 3.

9.3.1   Two real, different transversally non-hyperbolic infinite critical points

At infinity there exist two cases: either in two directions there exists an $M_{1,1}^0$ saddle node and in a third direction an elementary critical point E^i or at one direction there is an $M_{1,2}^i$ point and in another direction an $M_{1,1}^0$ saddle node.

Two infinite saddle nodes $M_{1,1}^0$

The non-hyperbolic infinite critical points may be thought to be located "at the ends of the x and y axis". Then $a_{20} = a_{02} = b_{20} = b_{02} = 0, a_{11} \neq 0, b_{11} \neq 0$, and in order to have $m_f=2$ there should be $c_{23}c_{35} \neq 0$, as follows from Theorem 2.3. Figure 2 in [95,RK] then tells that the $M_{1,1}^i$ points are second order semilinear saddle nodes $M_{1,1}^0$.

Using these conditions and a shift of the origin leads to the system

$$\dot{x} = \frac{\beta}{b_{11}^2} + \frac{c_{25}}{b_{11}}x + \frac{c_{35}}{b_{11}}y + a_{11}xy, \qquad (9.3.1)$$

$$\dot{y} = \frac{\alpha}{b_{11}} + b_{11}xy, \qquad (9.3.2)$$

where $\alpha = b_{00}b_{11} - b_{10}b_{01}, \beta = a_{00}b_{11}^2 - a_{10}b_{01}b_{11} - a_{01}b_{10}b_{11} + a_{11}b_{00}b_{01}$.
   If $\alpha \neq 0$, a scaling of (9.3.1),(9.3.2), gives

$$\dot{x} = \lambda_1 + \lambda_2 + x + \lambda_3 y + \lambda_1 xy \equiv P(x,y), \qquad (9.3.3)$$

$$\dot{y} = 1 + xy \equiv Q(x,y), \qquad (9.3.4)$$

where $\lambda_1 = \frac{\alpha a_{11}b_{11}}{c_{25}^2} \neq 0, \lambda_2 = \frac{(\beta - \alpha a_{11})b_{11}}{c_{25}^2}$ and $\lambda_3 = \frac{\alpha b_{11}^2 c_{35}}{c_{25}^3} \neq 0$.
   If $\alpha=0$, (9.3.1),(9.3.2) may further be reduced to

$$\dot{x} = \lambda_2 + x + y + \lambda_1 xy \equiv P(x,y), \qquad (9.3.5)$$

$$\dot{y} = xy \equiv Q(x,y), \qquad (9.3.6)$$

where $\lambda_1 = \frac{a_{11}c_{25}}{b_{11}c_{35}} \neq 0$ and $\lambda_2 = \frac{\beta b_{11}}{c_{25}^2}$.
   (i)$\alpha \equiv b_{00}b_{11} - b_{10}b_{01} \neq 0$.
   The finite critical points are located in $(x_\pm, y_\pm)$, where $x_\pm = \frac{1}{2}(-\lambda_2 \pm \sqrt{\lambda_2^2 + 4\lambda_3}), y_\pm = -\frac{1}{x_\pm}$. So there are two elementary critical points if $\lambda_2^2 + 4\lambda_3 >0$, a second order point m_2^0 if $\lambda_2^2+4\lambda_3=0$ and two complex critical points if $\lambda_2^2 + 4\lambda_3 <0$. Furthermore $\Omega(x,y) \equiv P_x(x,y)Q_y(x,y) - P_y(x,y)Q_x(x,y) = \frac{\lambda_3+x^2}{x}$, and $\Omega(x_+,y_+)\Omega(x_-,y_-) = -(\lambda_2^2 + 4\lambda_3)$. Then , if there are two elementary critical points there is a saddle in (x_-,y_-) and an antisaddle in (x_+,y_+), and $i_f=0$ as in the remaining two cases. Index theory then tells us that the third infinite critical point is an elementary node E^1 in $u = \frac{1}{\lambda_1}$.
   The system for $\alpha \neq 0$ thus belongs to the classes $e^{-1}e^1 E^1 M_{1,1}^0 M_{1,1}^0$, $m_2^0 E^1 M_{1,1}^0 M_{1,1}^0$ and $c_1^0 c_1^0 E^1 M_{1,1}^0 M_{1,1}^0$.
   There are weak critical points under the condition

$$-\lambda_1 - \tfrac{1}{2}\lambda_2 + \lambda_3 + \tfrac{1}{2}\lambda_2^2 \pm \tfrac{1}{2}(1 - \lambda_2)\sqrt{\lambda_2^2 + 4\lambda_3} = 0.$$

So for $\lambda_2^2 + 4\lambda_3 > 0$ the $+$ sign corresponds to a weak focus and the $-$ sign to a weak saddle. For $\lambda_2^2 + 4\lambda_3 = 0$ the m_2^0 point is a cusp; thus for $\lambda_1 = -\lambda_3 \mp \sqrt{-\lambda_3}, \lambda_2 = \pm 2\sqrt{-\lambda_3}$.

The relation separating the nodes from the foci is given by

$$\lambda_3 - 3\lambda_2\lambda_3 + (\lambda_1 - \lambda_3)^2 + \tfrac{1}{2}[\lambda_2(1 - 3\lambda_2) + 2(\lambda_1 - \lambda_3)(1 - \lambda_2)][\lambda_2 - \sqrt{\lambda_2^2 + 4\lambda_3}] = 0.$$

With regard to the limit cycles, it may be remarked that for $\lambda_1 + \lambda_3 \neq 0$, (9.3.3),(9.3.4) can be transformed to

$$\dot{x} = dx - y + lx^2 + mxy + ny^2, \tag{9.3.7}$$
$$\dot{y} = x, \tag{9.3.8}$$

where now

$$d = \frac{2\lambda_3 - \lambda_1\lambda_2 - \lambda_2\lambda_3 - (\lambda_1 - \lambda_3)\sqrt{\lambda_2^2 + 4\lambda_3}}{\sqrt[4]{\lambda_2^2 + 4\lambda_3}},$$
$$l = -\frac{\lambda_1}{\lambda_1 + \lambda_3},$$
$$m = \frac{\lambda_1 - \lambda_3}{(\lambda_1 + \lambda_3)\sqrt[4]{\lambda_2^2 + 4\lambda_3}},$$
$$n = \frac{\lambda_3}{(\lambda_1 + \lambda_3)\sqrt{\lambda_2^2 + 4\lambda_3}}.$$

For $|d| < 2$, the origin is a focus or a center point and the system belongs to class I of the classification of Ye Yanqian [65,Y]. The limit cycle problem was studied by various authors, see for instance Ye Yanqian and Chen Lansun [75,YC], Coll [87,Col] and Coppel[89,Co]. It is known that (9.3.7),(9.3.8) has at most one limit cycle and if it exists it is hyperbolic; see Coppel[89,Co].

If $\lambda_1 + \lambda_3 = 0$ there exists no limit cycle as the antisaddle is a node, located on the line $-x + \lambda_1 y - \lambda_2 = 0$, which consists of straight line orbits and the two critical points.

We will now discuss the bifurcation diagrams and phase portraits for constant values of λ_3. An important feature thereby appears to be the location of the cusp points in the λ_1, λ_2 diagrams for $\lambda_3 < 0$ on the lines $\lambda_2 \equiv \pm 2\sqrt{-\lambda_3}$. We consider the following values of λ_3; (i)$\lambda_3 > 0$,(ii)$-1 < \lambda_3 < 0$, (iii)$\lambda_3 = -1$, (iv)$\lambda_3 < -1$.

(i)$\lambda_3 > 0$. For $\lambda_3 > 0$, (9.3.3),(9.3.4) is only of class $e^{-1}e^1 E^1 M_{1,1}^0 M_{1,1}^0$. The bifurcation diagram is given in Figure 9.33 and the phase portraits in Figure 9.34.

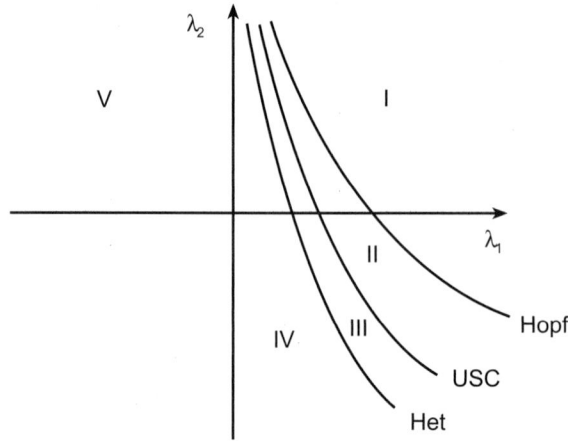

Figure 9.33 Bifurcation diagram for $\alpha \neq 0$, $\lambda_3 > 0$

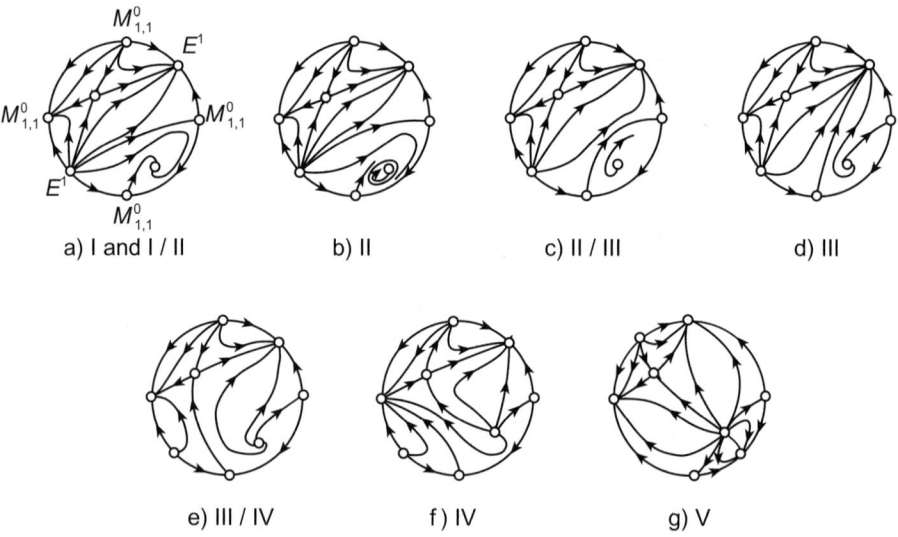

Figure 9.34 Phase portraits for $\alpha \neq 0$, $\lambda_3 > 0$

It should be noticed that λ_1 is a parameter rotating the field. Starting in region I, a weak focus on the curve indicated by Hopf will occur when lowering λ_1, subsequently giving rise to a unique stable hyperbolic limit cycle in region II.

Further decrease of λ_1 yields a connection between the saddle parts of the saddle nodes at infinity $M_{1,1}^0$, indicated by USC, which subsequently breaks open to yield the phase portrait for region III. Still further decrease of λ_1 yields a connection between the finite saddle and the saddle part of the saddle node at the negative end of the y axis, as indicated in the phase portrait corresponding to the curve III/ IV indicated by Het, which subsequently breaks up. To investigate Het further we may consider the line $l \equiv x + \frac{1}{2}(\lambda_2 + \sqrt{\lambda_2^2 + 4\lambda_3}) = 0$, then $\frac{dl}{dt}|_{l=0} = (\lambda_1 + \frac{1}{2}\lambda_2 - \frac{1}{2}\sqrt{\lambda_2^2 + 4\lambda_3})[1 - \frac{1}{2}(\lambda_2 + \sqrt{\lambda_2^2 + 4\lambda_3})y]$ shows that if $\sigma \equiv \lambda_1 + \frac{1}{2}(\lambda_2 - \sqrt{\lambda_2^2 + 4\lambda_3}) = 0$ Het is a straight line saddle connection; for $\sigma \neq 0$ Het is not a straight line. The uniqueness of the curves Hopf, USC and Het follows from the property that λ_1 rotates the vector field.

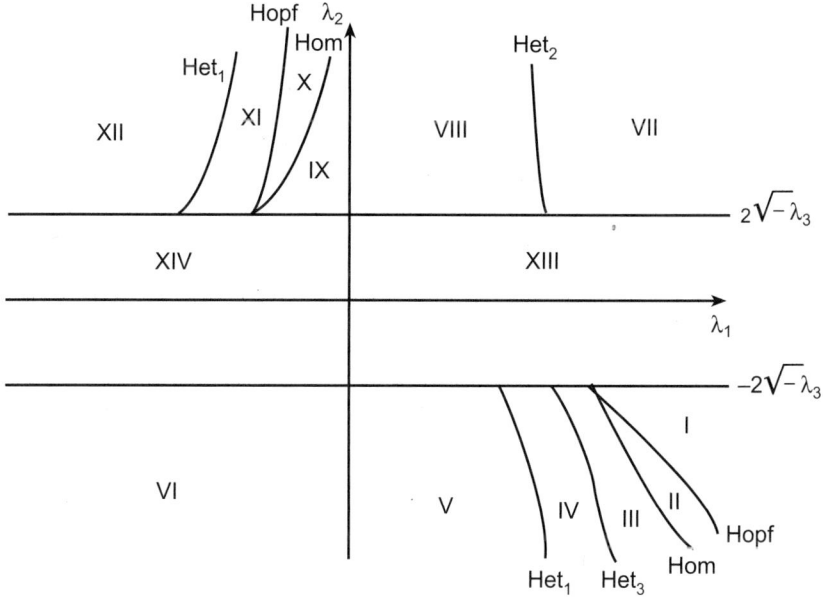

Figure 9.35 Bifurcation diagram for $\alpha \neq 0, -1 < \lambda_3 < 0$

(ii)$-1< \lambda_3 <0$. The bifurcation diagram for $-1< \lambda_3 <0$ is given in Figure 9.35 and the phase portraits in Figure 9.36. There is a cusp point in $\lambda_1 = -\lambda_3 - \sqrt{-\lambda_3} <0$, $\lambda_2 = 2\sqrt{-\lambda_3} >0$ and in $\lambda_1 = -\lambda_3 + \sqrt{-\lambda_3} >0$, $\lambda_2 = -2\sqrt{-\lambda_3} <0$, where, in both cases, a curve "Hopf" starts, corresponding to a stable first order weak focus, extending into the region $|\lambda_2| > 2\sqrt{-\lambda_3}$. If for $\lambda_2 >0$ is started on the Hopf curve X/ XI and λ_1 is increased, a unique stable limit cycle is bifurcated into the region X, which subsequently is adsorbed in a saddle loop indicated by the curve Hom(IX/ X). Lowering λ_1 from the Hopf curve X/ XI into region XI rotates the vector field such that the separatrices change positions and for the curve where $\sigma=0$, indicated by Het, a straight line saddle connection exists, which breaks up again by lowering λ_1 into region XII. As may be derived from the separatrix structure in regions VII and VIII, a curved connection between the finite saddle and the saddle part of the saddle node "at the ends of the y axis", indicated by Het_2 occurs. On the Hopf curve for $\lambda_2 <0$, λ_1 must be decreased to obtain a limit cycle, which also is adsorbed by the saddle loop (homoclinic cycle) Hom. Following the motion of the separatrices while decreasing λ_1 shows that before $\sigma=0$, indicated by Het_1, is reached there should be a curved connection between the finite saddle and the infinite saddle node, indicated by Het_3.

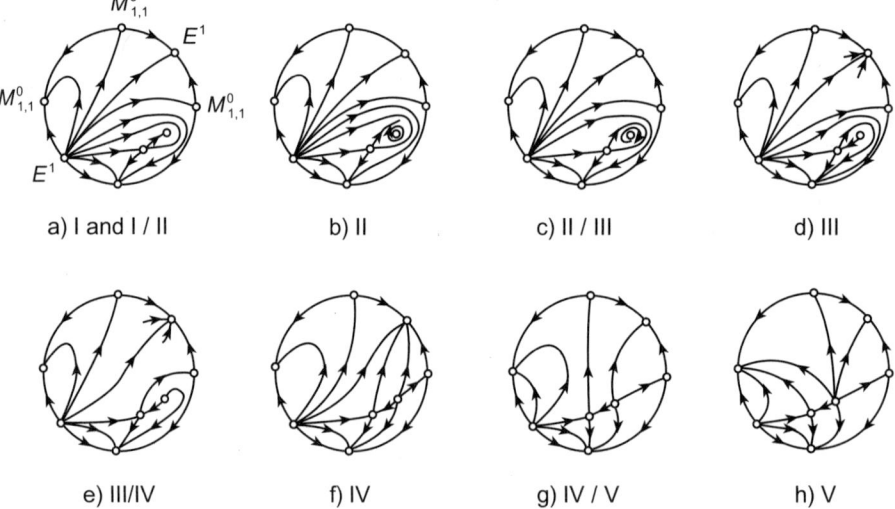

a) I and I / II   b) II   c) II / III   d) III

e) III/IV   f) IV   g) IV / V   h) V

Figure 9.36(i) Phase portraits for $\alpha \neq 0$, $-1< \lambda_3 <0$

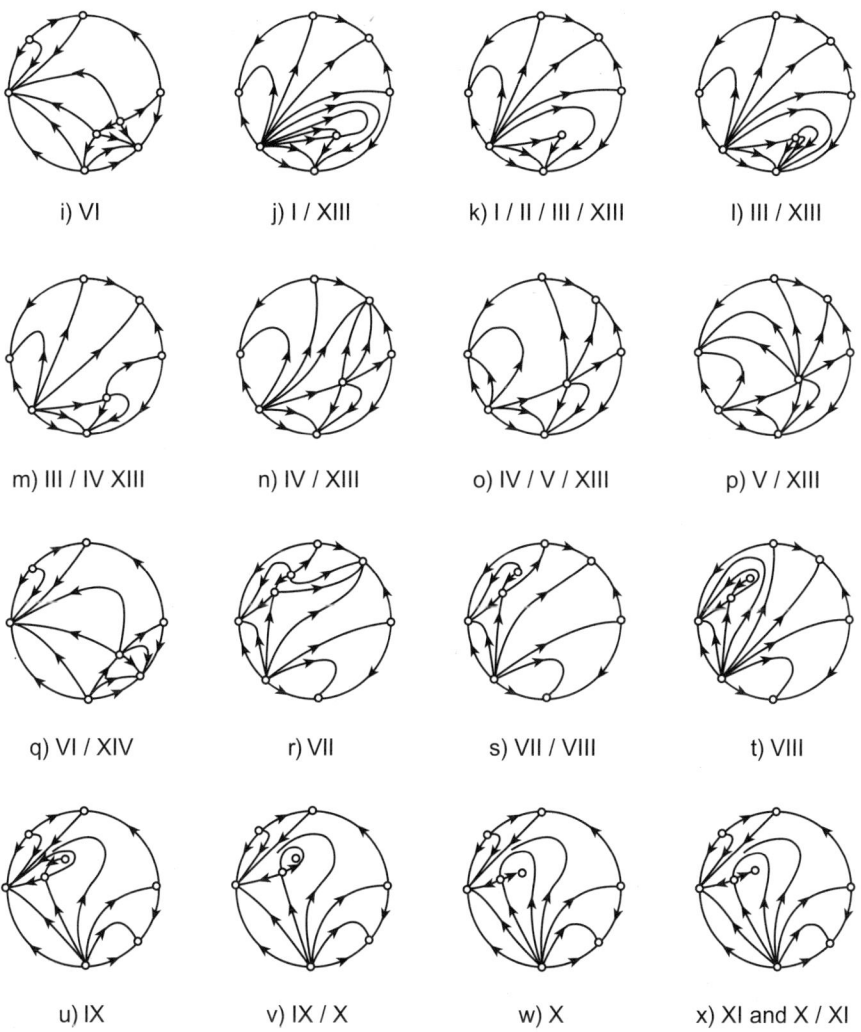

i) VI   j) I / XIII   k) I / II / III / XIII   l) III / XIII

m) III / IV XIII   n) IV / XIII   o) IV / V / XIII   p) V / XIII

q) VI / XIV   r) VII   s) VII / VIII   t) VIII

u) IX   v) IX / X   w) X   x) XI and X / XI

Figure 9.36(ii) Phase portraits for $\alpha \neq 0$, $-1 < \lambda_3 < 0$

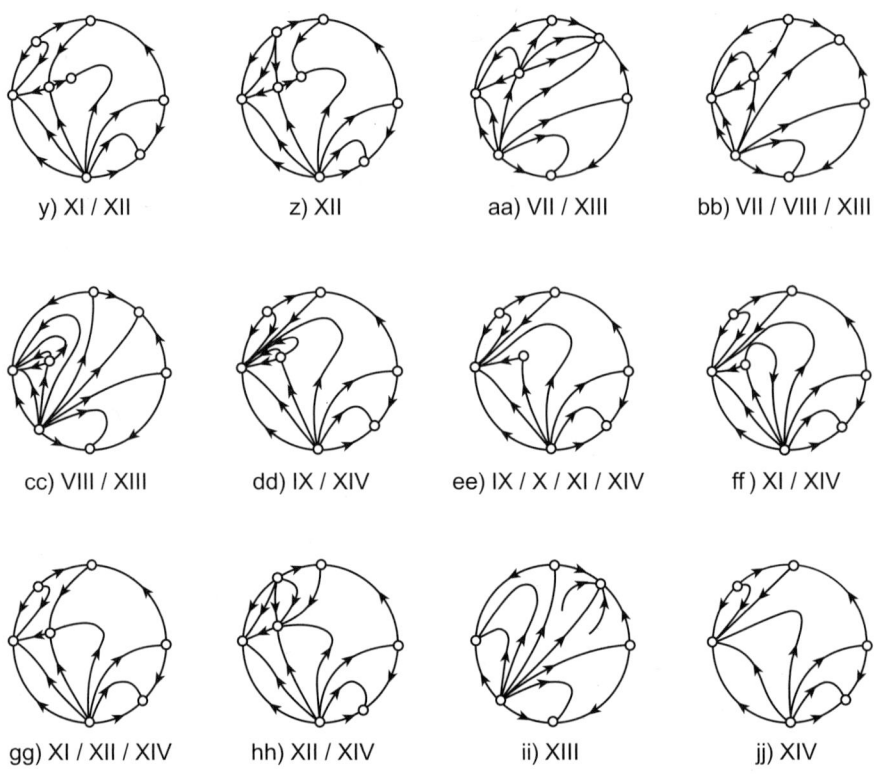

y) XI / XII   z) XII   aa) VII / XIII   bb) VII / VIII / XIII

cc) VIII / XIII   dd) IX / XIV   ee) IX / X / XI / XIV   ff) XI / XIV

gg) XI / XII / XIV   hh) XII / XIV   ii) XIII   jj) XIV

Figure 9.36(iii) Phase portraits for $\alpha \neq 0$, $-1 < \lambda_3 < 0$

(iii) $\lambda_3 = -1$. The bifurcation diagram, given in Figure 9.37 is a limiting case of the previous one in the sense that the cusp for $\lambda_2 > 0$ approaches $\lambda_1 = 0, \lambda_2 = 2\sqrt{-\lambda_3}$ for $\lambda_3 \to -1$, with the result that the phase portraits in Figure 9.36 remain valid for $\lambda_3 = -1$, except that phase portrait ee disappears.

(iv) $\lambda_3 < -1$. The bifurcation diagram for $\lambda_3 < -1$ is given in Figure 9.38. The majority of the phase portraits for $\lambda_3 < -1$ can still be found in Figure 9.36. In addition, new phase portraits exist due to the appearance of the new regions XV and XVI giving rise to the phase portraits given in Figure 9.39.

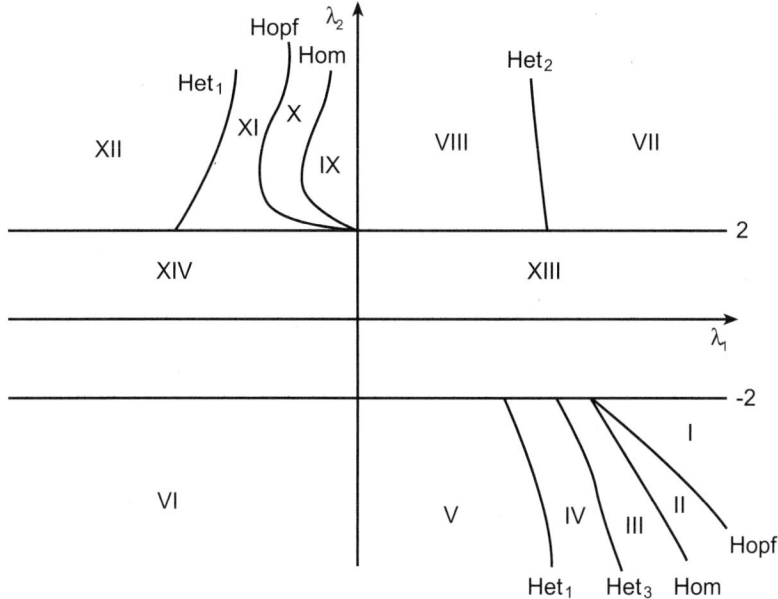

Figure 9.37 Bifurcation diagram for $\alpha \neq 0$, $\lambda_3 = -1$

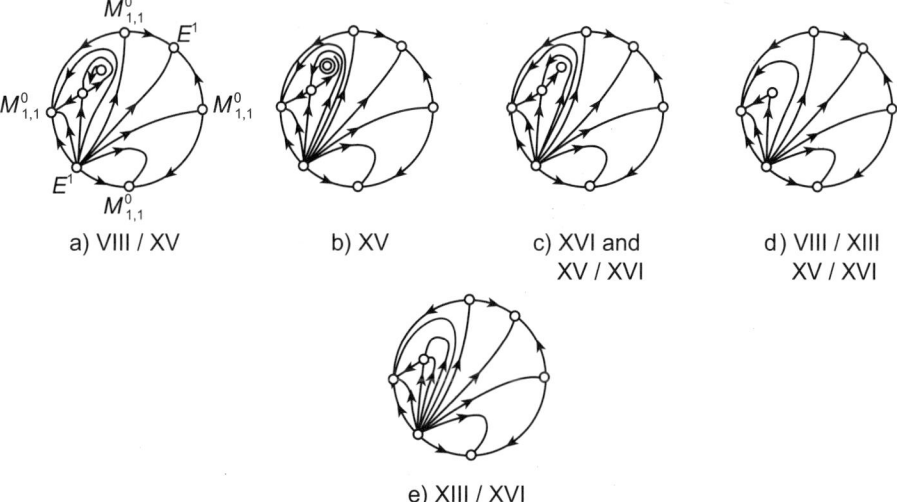

a) VIII / XV   b) XV   c) XVI and   d) VIII / XIII
                                    XV / XVI   XV / XVI

e) XIII / XVI

Figure 9.39 Additional phase portraits for $\alpha \neq 0$, $\lambda_3 < -1$

The location of the curve Hom in Figure 9.35,9.37 and 9.38 for large values of λ_2 has so far only been indicated numerically. Entrance of this curve for larger values of λ_2 into the first quadrant, however, does not increase the number of possible phase portraits.

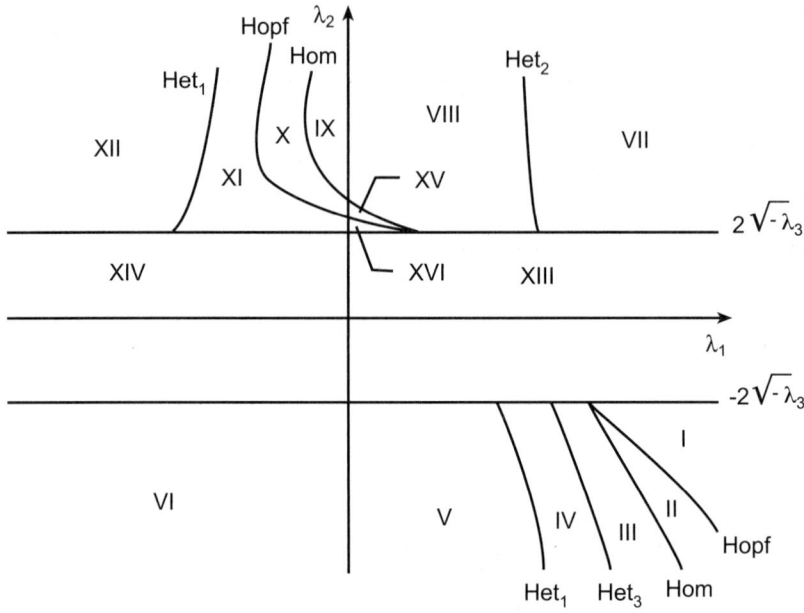

Figure 9.38 Bifurcation diagram for $\alpha \neq 0$, $\lambda_3 < -1$

(ii)$\alpha \equiv b_{00}b_{11} - b_{10}b_{01}{=}0$.
For this case we work with (9.3.5),(9.3.6)

$$\dot{x} \;=\; \lambda_2 + x + y + \lambda_1 xy \equiv P(x,y), \qquad\qquad (9.3.9)$$
$$\dot{y} \;=\; xy \equiv Q(x,y), \qquad\qquad\qquad\qquad\quad (9.3.10)$$

where $\lambda_1 \neq 0, -\infty < \lambda_2 < \infty$. It may be noticed that λ_1 is again a parameter rotating the field.

The finite critical points are located in $(0,\lambda_2)$ and $(-\lambda_2,0)$, their character being determined by $\Omega(x,y) \equiv P_x(x,y)Q_y(x,y) - P_y(x,y)Q_x(x,y)$. Therefore, for $\lambda_2 \neq 0$ there exists a saddle and an antisaddle, and for $\lambda_2{=}0$ a saddle node in $(0,0)$; and $i_f{=}0$. Index theory then tells that the third infinite critical point is a node E^1 at u=$\frac{1}{\lambda_1}$. The system for $\alpha=0$ thus belongs to the classes $e^{-1}e^1E^1M_{1,1}^0M_{1,1}^0$ and $m_2^0\bar{E}^1M_{1,1}^0M_{1,1}^0$.

The bifurcation diagram is given in Figure 9.39 and the phase portraits in Figure 9.40.

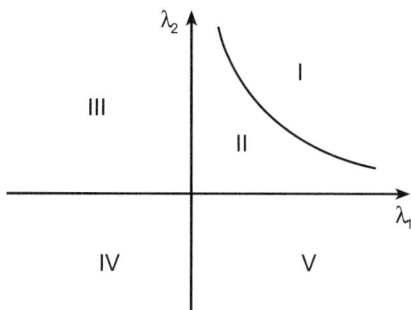

Figure 9.39 Bifurcation diagram for $\alpha=0$

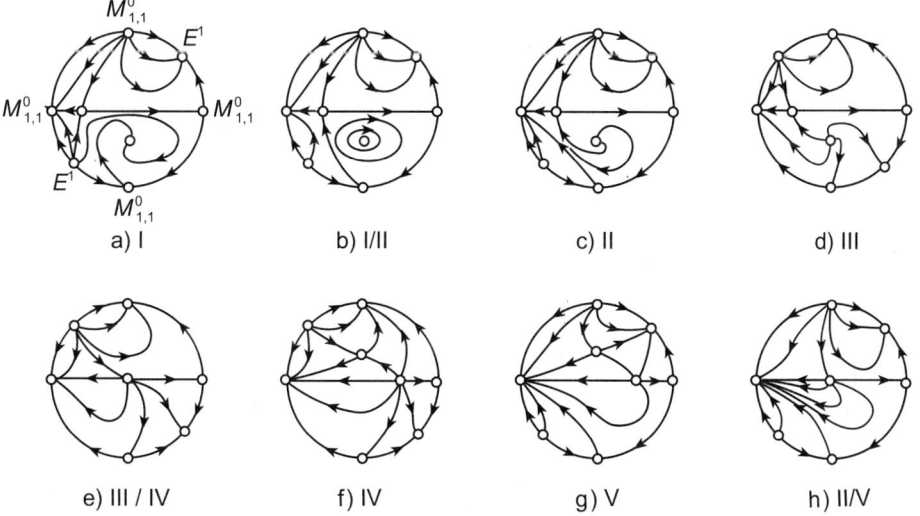

a) I   b) I/II   c) II   d) III

e) III / IV   f) IV   g) V   h) II/V

Figure 9.40 Phase portraits for $\alpha=0$

For $\lambda_2 >0$ the saddle is located in $(-\lambda_2,0)$ and the antisaddle in $(0,-\lambda_2)$, being a center point if $1-\lambda_1\lambda_2=0$, as can be seen using the integrating factor

$\mu(x,y)=\frac{1}{(\lambda_2+x)y}$. Either using the fact that λ_1 is a parameter, rotating the vector field or considering that div $\frac{(P(x,y),Q(x,y))}{(\lambda_2+x)y} \neq 0$ for $\lambda_1\lambda_2-1\neq0$ shows that there exist no limit cycles for $\lambda_2 >0$. For $\lambda_2=0$ there exists an unstable saddle node in $(0,0)$. For $\lambda_2 <0$ the saddle is located in $(0,-\lambda_2)$ and the antisaddle in $(-\lambda_2,0)$, which is a node on the line $y\equiv0$, being covered by straight line orbits. There are no limit cycles.

A saddle node $M_{1,1}^0$ and a third order point $M_{1,2}^i$

We may start with (9.3.1),(9.3.2), being a system with infinite saddle nodes $M_{1,1}^0$ "at the ends of the x axis" and "at the ends of the y axis". Then

$$\dot{x} = \frac{\beta}{b_{11}^2} + \frac{c_{25}}{b_{11}}x + \frac{c_{35}}{b_{11}}y + a_{11}xy, \qquad (9.3.11)$$

$$\dot{y} = \frac{\alpha}{b_{11}} + b_{11}xy, \qquad (9.3.12)$$

with $a_{11}b_{11}c_{25}c_{35} \neq0$, and where $\alpha \equiv b_{00}b_{11} - b_{10}b_{01}$, and $\beta \equiv a_{00}b_{11}^2 - a_{10}b_{01}b_{11} - a_{01}b_{10}b_{11} + a_{11}b_{00}b_{01}$.

The third infinite critical point in u=$\frac{b_{11}}{a_{11}}$ may be sent to "the ends of the y axis" by letting a_{11} vanish, so that then an $M_{1,2}^i$ point exists there. Consulting figure 4 in [95,RK] it follows that $M_{1,2}^i$ is a nilpotent critical point $M_{1,2}^1$ with an elliptic and a hyperbolic sector and the separatrix coinciding with the Poincaré circle.

If $\alpha \neq0$ this system may further be scaled to the system

$$\dot{x} = \lambda_2 + x + \lambda_3 y, \qquad (9.3.13)$$

$$\dot{y} = 1 + xy, \qquad (9.3.14)$$

where $\lambda_2 = \frac{\beta b_{11}}{c_{25}^2}$ and $\lambda_3 = \alpha b_{11}^2 \frac{c_{35}}{c_{25}^3}$.

If $\alpha = 0$, (9.3.11),(9.3.12) may further be scaled to the system

$$\dot{x} = \lambda_3 + x + y, \qquad (9.3.15)$$

$$\dot{y} = xy. \qquad (9.3.16)$$

Both systems may also be obtained by putting $\lambda_1=0$ in (9.3.3),(9.3.4), respectively.

Case1.$\alpha \neq 0$.

Since (9.3.13),(9.3.14) is precisely system (9.3.3),(9.3.4) with $\lambda_1=0$, bifurcation diagram and phase portraits may directly be obtained from the results with $\lambda_1 \neq 0$ by considering the limiting situation as $\lambda_1 \rightarrow 0$.

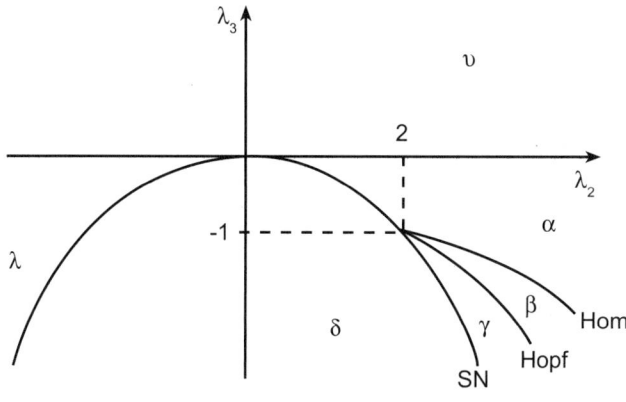

Figure 9.41 Bifurcation diagram for $\alpha \neq 0$

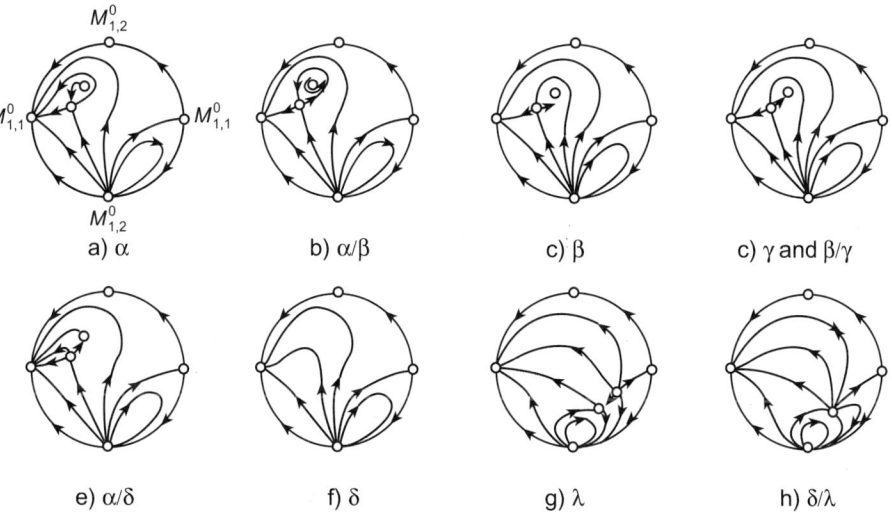

Figure 9.42(i) Phase portraits for $\alpha \neq 0$

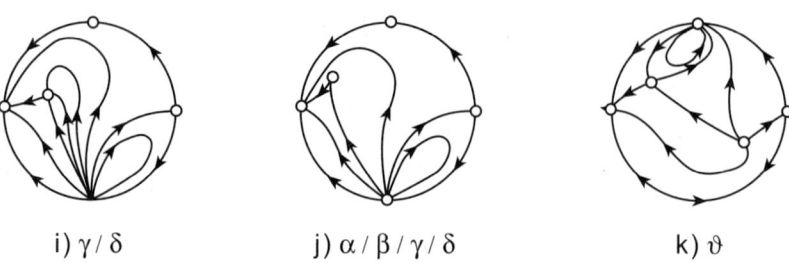

i) $\gamma\,/\,\delta$ j) $\alpha\,/\,\beta\,/\,\gamma\,/\,\delta$ k) ϑ

Figure 9.42(ii) Phase portraits for $\alpha\neq0$

The bifurcation diagram is given in Figure 9.41; "Hopf" and "Hom" have the usual meaning and SN means that the system has a saddle node for $\lambda_2^2+4\lambda_3=0$. There exists precisely one (hyperbolic) limit cycle in the region in between "Hopf" and "Hom"; the uniqueness follows from the fact that (9.3.13),(9.3.14) can be transformed to the system (9.3.7),(9.3.8) with l=0 (indicated by $I_{l=0}$ in the Chinese classification).

The phase portraits are given in Figure 9.42; they correspond also to the bifurcation diagram for (9.3.3),(9.3.4) in the points $\lambda_1=0$ as given in Figures 9.33,9.35,9.37 and 9.38. Case2.$\alpha=0$.

In a similar way the phase portraits may be found for system (9.3.15), (9.3.16); they are given in Figure 9.43.

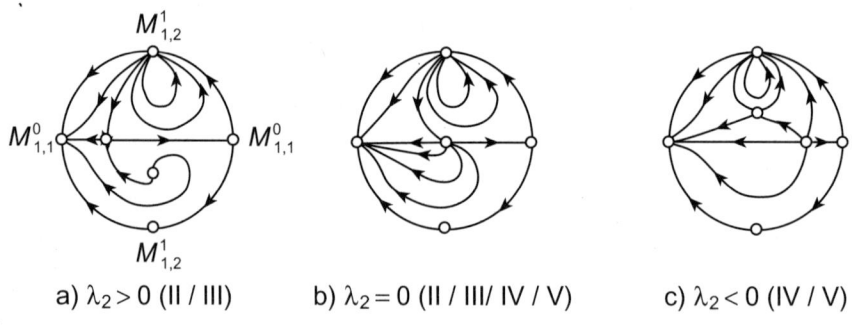

a) $\lambda_2>0$ (II / III)   b) $\lambda_2=0$ (II / III/ IV / V)   c) $\lambda_2<0$ (IV / V)

Figure 9.43 Phase portraits for $\alpha=0$

9.3.2 Two coinciding transversally non-hyperbolic infinite critical points

In order for these points to be "at the ends of the x axis" the quadratic terms in the system should have the common factor y^2 so $a_{20} = a_{11} = b_{20} = b_{11} = 0, a_{02}^2 + b_{02}^2 \neq 0$. The system may then be written as

$$\dot{x} = a_{00} + a_{10}x + a_{01}y + a_{02}y^2, \qquad (9.3.17)$$
$$\dot{y} = b_{00} + b_{10}x + b_{01}y + b_{02}y^2, \qquad (9.3.18)$$

whereas Theorem 2.1 tells that it belongs to the class m_f=2 iff $c_{26} = a_{10}b_{02} - a_{02}b_{10} \neq 0$.

The location of the infinite critical points follows from $(a_{02}u - b_{02})u^2$=0; so for $b_{02} \neq 0$ there is in u=0 an $M_{2,2}^i$ point and for b_{02}=0 an $M_{2,3}^i$ point. The character of these points may be found in [95,RK].

For $b_{02} \neq 0, b_{10} \neq 0$, Figure 5 of [95,RK] shows that $M_{2,2}^i$ is a nilpotent saddle node $M_{2,2}^0$. If $b_{02} \neq 0, b_{10}$=0, (so $a_{10} \neq 0$) the point is highly degenerate and Figure 7 of [95,RK] tells that it is an $M_{2,2}^0$ point with two hyperbolic sectors if $\Delta \equiv b_{01}^2 - 4b_{00}b_{02}$ <0, Figure 8 of [95,RK] that it is an $M_{2,2}^0$ point with two hyperbolic and two parabolic sectors if Δ=0, whereas Figure 9 of [95,RK] that it is also an $M_{2,2}^0$ point with two hyperbolic and two parabolic sectors if Δ >0. Index theory then tells that apart from the $M_{2,2}^0$ point there is an elementary node E^1 in $u = \frac{a_{02}}{b_{02}}$.

If b_{02}=0,$c_{26} \equiv a_{10}b_{02} - a_{02}b_{10} = -a_{02}b_{10} \neq 0$ implies $b_{10} \neq 0$ and $M_{2,3}^i$ is a nilpotent critical point. Figure 6 of [95,RK] then shows that $M_{2,3}^i$ is a nilpotent node $M_{2,3}^1$.

Systems with an $M_{2,2}^0$ point

These systems belong to the classes $e^{-1}e^1E^1M_{2,2}^0, m_2^0E^1M_{2,2}^0$ and $c_1^0c_1^0E^1M_{2,2}^0$, as mentioned in class 4 of Table 3 in Chapter 3.

Using $b_{02}c_{26} \neq 0$, by a linear transformation (9.3.17),(9.3.18) may be transformed to

$$\dot{x} = \frac{c_{26}}{b_{02}}x + (c_{36} + \frac{a_{02}c_{26}}{b_{02}})y, \qquad (9.3.19)$$
$$\dot{y} = \alpha + \frac{b_{10}}{b_{02}}x + b_{02}y^2, \qquad (9.3.20)$$

with

$$\alpha = \frac{1}{4b_{02}^3 c_{26}} (4b_{00}b_{02}^3 c_{26} - 4b_{10}b_{02}^3 c_{16} + a_{02}^2 b_{10}^2 c_{26}$$
$$+ 2b_{10}b_{01}b_{02}^2 c_{36} + 2a_{02}b_{10}^2 b_{02}c_{36} - b_{01}^2 b_{02}c_{26}).$$

If $b_{10} \neq 0$, (9.3.19),(9.3.20) may further be reduced , by scaling, to

$$\dot{x} = x - \lambda_1 y \equiv P(x,y), \qquad\qquad (9.3.21)$$
$$\dot{y} = \lambda_2 + x + y^2 \equiv Q(x,y), \qquad\qquad (9.3.22)$$

where $\lambda_1 = -\frac{b_{10}}{c_{26}^2}(-a_{02}c_{26} + b_{02}c_{36})$, $\lambda_2 = \frac{ab_{02}^2}{c_{26}^2}$.

The finite critical points are located in $(x_\pm, y_\pm)$, where $x_\pm = \lambda_1 y_\pm$ and $y_\pm = \frac{1}{2}(-\lambda_1 \pm \sqrt{\lambda_1^2 - 4\lambda_2})$. There are two elementary critical points for $\lambda_1^2 - 4\lambda_2 > 0$, an m_2^0 point for $\lambda_1^2 - 4\lambda_2 = 0$ and two complex critical points c_1^0 for $\lambda_1^2 - 4\lambda_2 < 0$. Furthermore $\Omega(x,y) \equiv P_x(x,y)Q_y(x,y) - P_y(x,y)Q_x(x,y) = \lambda_1 + 2y$ and $\Omega(x_\pm, y_\pm) = \pm\sqrt{\lambda_1^2 - 4\lambda_2}$. Then, if there are two elementary critical points there is a saddle in (x_-, y_-) and an antisaddle in (x_+, y_+) and $i_f = 0$, as in the remaining cases.

There are weak critical points if $1 - \lambda_1 \pm \sqrt{\lambda_1^2 - 4\lambda_2} = 0$. So for $\lambda_1^2 - 4\lambda_2 > 0$ the $-$ sign corresponds to a weak saddle and the $+$ sign to a weak focus. For $\lambda_1^2 - 4\lambda_2 = 0$ the m_2^0 is a cusp; so for $\lambda_1 = 1$, $\lambda_2 = \frac{1}{4}$ (if $\lambda_1 \neq 1$ or $\lambda_2 \neq \frac{1}{4}$ the m_2^0 point is a saddle node).

It can be shown that for $\lambda_2 = -\frac{1}{16} + \frac{1}{4}\lambda_1$, (9.3.21),(9.3.22) has the parabolic solution $x = \lambda_2 + \frac{1}{2}y - y^2$. For $\lambda_1 > 2$ this implies a connection between the finite saddle and the separatrix entering the saddle node at infinity from the finite part of the plane.

With regard to the limit cycles, it can be seen that for $\lambda_1 \neq 0$ again (9.3.21),(9.3.22) can be transformed to class I of the Chinese classification, where now

$$d = \frac{1 - \lambda_1 + \sqrt{\lambda_1^2 - 4\lambda_2}}{\sqrt[4]{\lambda_1^2 - 4\lambda_2}},$$
$$l = \frac{-1}{\lambda_1}$$
$$m = \frac{2}{\lambda_1 \sqrt[4]{\lambda_1^2 - 4\lambda_2}},$$
$$n = \frac{-1}{\lambda_1 \cdot \sqrt{\lambda_1^2 - 4\lambda_2}}$$

and the result again follows that at most one limit cycle exists, which if it exists is hyperbolic [89,C].

For $\lambda_1 = 0$ the antisaddle is a node and no limit cycle exists.

The bifurcation diagram for (9.3.21),(9.3.22) is given in Figure 9.45. The curves "SN","Hopf" and "Hom" have the same meaning as before; the saddle connection is indicated by "Het".The phase portraits are given in Figure 9.46.

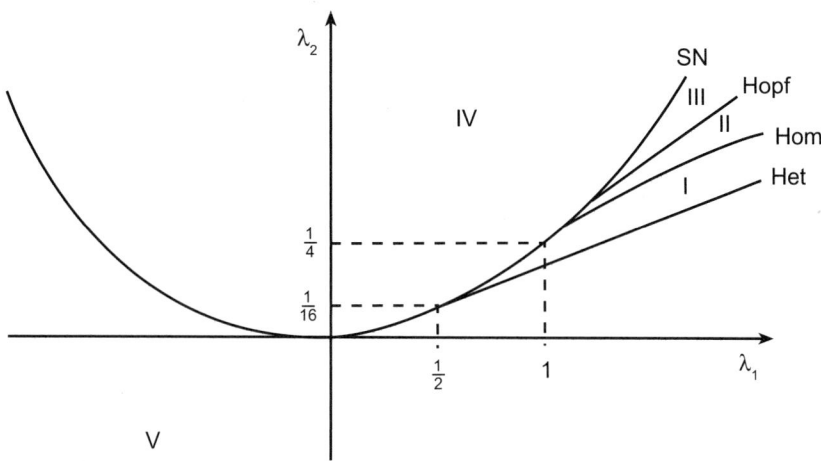

Figure 9.44 Bifurcation diagram for systems with a $M_{2,2}^0$ point:$b_{10} \neq 0$

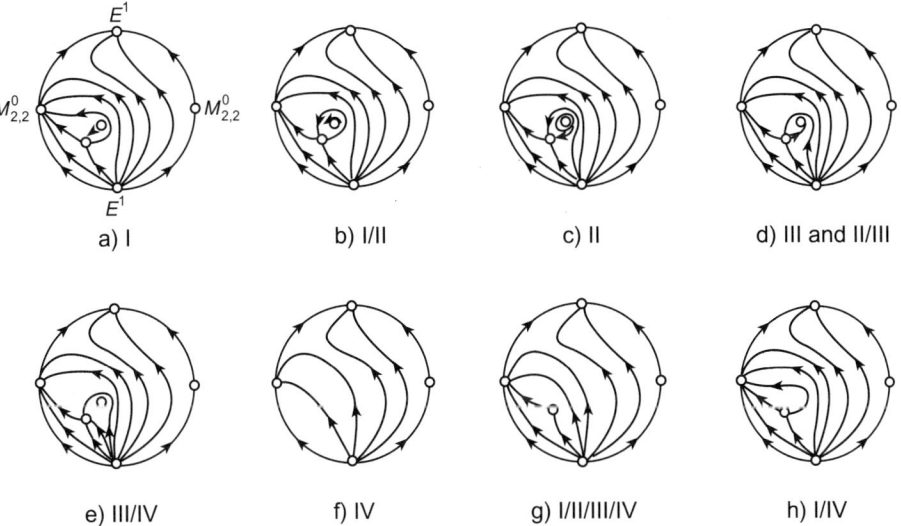

a) I   b) I/II   c) II   d) III and II/III

e) III/IV   f) IV   g) I/II/III/IV   h) I/IV

Figure 9.45(i) Phase portraits for systems with a $M_{2,2}^0$ point:$b_{10} \neq 0$

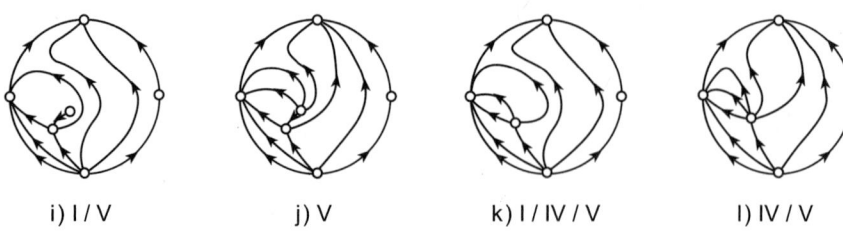

i) I / V   j) V   k) I / IV / V   l) IV / V

Figure 9.45(ii) Phase portraits for systems with a $M_{2,2}^0$ point: $b_{10} \neq 0$

If $b_{10}{=}0$, (9.3.19),(9.3.20) may further be reduced by scaling to

$$\dot{x} = x - \lambda_1 y, \tag{9.3.23}$$
$$\dot{y} = \lambda_2 + y^2, \tag{9.3.24}$$

with $\lambda_1 = \frac{a_{02}c_{26}-b_{02}c_{36}}{b_{02}^2}$, $\lambda_2 = \frac{ab_{02}^3}{c_{26}^2}$.

The phase portraits are given in figure 9.46.

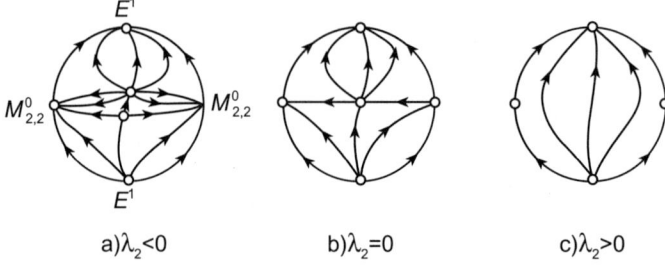

a)$\lambda_2{<}0$ b)$\lambda_2{=}0$ c)$\lambda_2{>}0$

Figure 9.46 Phase portraits for systems with a $M_{2,2}^0$ point: $b_{10}{=}0$

Systems with a nilpotent node $M_{2,3}^1$

These systems belong to the classes $e^{-1}e^1 M_{2,3}^1$, $m_2^0 M_{2,3}^1$ and $c_1^0 c_1^0 M_{2,3}^1$, mentioned in class 4 of Table 3, Chapter3.

Using $b_{02}{=}0$ and $c_{26} \equiv a_{10}b_{02} - a_{02}b_{10} = -a_{02}b_{10} \neq 0$, (9.3.17),(9.3.18) can be transformed to

$$\dot{x} = \lambda + x + y^2 \equiv P(x,y), \tag{9.3.25}$$
$$\dot{y} = x \equiv Q(x,y), \tag{9.3.26}$$

with $\lambda = \frac{4a_{02}b_{10}c_{12} - c_{23}^2}{4a_{10}^4}$.

The phase portraits are given in Figure 9.48. There are no limit cycles since $\text{div}(P(x,y), Q(x,y)) \equiv 1$.

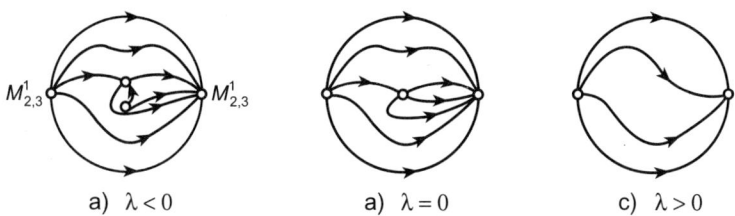

a) $\lambda < 0$ a) $\lambda = 0$ c) $\lambda > 0$

Figure 9.47 Phase portraits for systems with a $M_{2,3}^1$ point: $a_{10} \neq 0$

If $a_{10} = 0$, (9.3.17),(9.3.18) may be reduced to the Hamiltonian system

$$\dot{x} = \lambda + y^2, \tag{9.3.27}$$

$$\dot{y} = x, \tag{9.3.28}$$

where $\lambda = \frac{b_{10}(4a_{02}b_{10}c_{12} - c_{23}^2)}{4b_{11}}$.

The phase portraits are symmetric aroumd the y axis and for $\lambda \leq 0$ given in Figure 9.48. It may be easily seen that for $\lambda < 0$ the antisaddle is a center point and for $\lambda = 0$ a cusp. For $\lambda > 0$ the phase portrait is topologically equivalent to Figure 9.48(c).

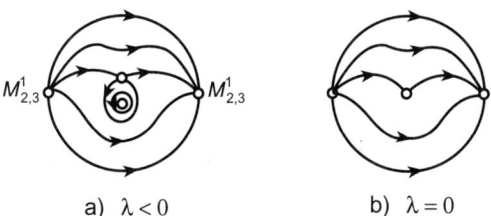

a) $\lambda < 0$ b) $\lambda = 0$

Figure 9.48 Phase portraits for systems with a $M_{2,3}^1$ point: $a_{10} = 0$

9.3.3 Two complex transversally non-hyperbolic critical points

These systems belong to the classes $e^{-1}e^1 E^1 C_{1,1}^0 C_{1,1}^0, m_2^0 E^1 C_{1,1}^0 C_{1,1}^0$ and $c_1^0 c_1^0 E^1 C_{1,1}^0 C_{1,1}^0$, mentioned in class 4 of Table 3, Chapter 3.

If the complex transversally non-hyperbolic infinite critical points are located in u=±i, the system can be written as

$$\dot{x} = a_{00} + a_{10}x + a_{01}y + x^2 + y^2, \qquad (9.3.29)$$
$$\dot{y} = b_{00} + b_{10}x + b_{01}y + x^2 + y^2, \qquad (9.3.30)$$

and the third infinite critical point is located in u=0. Since there is $i_f=0$, as for all systems in this section, this point is an elementary node E^1, using the index argument. Theorem 2.1 (or 2.2) shows furthermore that $b_{10}^2 + b_{01}^2 \neq 0$ in order that (9.3.29),(9.3.30) should be of class $m_f=2$.

By a shift of the origin (9.3.29),(9.3.30) may further be written as

$$\dot{x} = \alpha + x^2 + y^2, \qquad (9.3.31)$$
$$\dot{y} = \beta + b_{10}x + b_{01}y, \qquad (9.3.32)$$

where $\alpha = a_{00} - \frac{1}{4}a_{10}^2 - \frac{1}{4}a_{01}^2$, $\beta = b_{00} - \frac{1}{2}a_{10}b_{10} - \frac{1}{2}a_{01}b_{01}$.

(i)If $\beta \neq 0$ this system can be further reduced to

$$\dot{x} = \lambda_1 + x^2 + y^2 \equiv P(x,y), \qquad (9.3.33)$$
$$\dot{y} = 1 + \lambda_2 x + \lambda_3 y \equiv Q(x,y), \qquad (9.3.34)$$

with $\lambda_1 = \frac{\alpha}{|\beta|}$, $\lambda_2 = \frac{b_{10}}{\sqrt{|\beta|}}sgn\beta$, $\lambda_3 = \frac{b_{01}}{\sqrt{|\beta|}}$, as a result there is $\lambda_2^2 + \lambda_3^2 \neq 0$ in order to yield $m_f=2$.

Moreover we may assume $\lambda_2 \geq 0$. If not apply the transformation $(x,y,t;\lambda_3) \rightarrow (-x,-y,-t;-\lambda_3)$. The finite critical points are located at $(x_\pm, y_\pm)$, where $x_\pm = \frac{-(\lambda_2 \pm \lambda_3\sqrt{D})}{\lambda_2^2+\lambda_3^2}$, $y_\pm = \frac{-\lambda_3 \pm \lambda_2\sqrt{D}}{\lambda_2^2+\lambda_3^2}$ and $D \equiv -1-\lambda_1(\lambda_2^2+\lambda_3^2)$. For D>0 there are two finite elementary critical points, for D=0 an m_2^0 point, and for D<0 two complex critical points. As $\Omega(x,y) \equiv P_x(x,y)Q_y(x,y) - P_y(x,y)Q_x(x.y) = 2(\lambda_3 x - \lambda_2 y)$, it follows that $\Omega(x_\pm, y_\pm) = \mp 2\sqrt{D}$, so if D>0 the + sign corresponds to a saddle and the − sign to an antisaddle.

There exist weak critical points if $2(\lambda_2 \pm \lambda_3\sqrt{D}) - \lambda_3(\lambda_2^2 + \lambda_3^2)=0$, where the + sign corresponds to a weak saddle, the − sign to a first order weak focus, unless there is also $\lambda_2^2 + \lambda_3^2 = -\frac{1}{\lambda_1}$, $\lambda_3 = -2\lambda_1\lambda_2$; then a cusp prevails.

With respect to the limit cycles , for $\lambda_2 \neq 0$, it can be seen that (9.3.3),(9.34) also transforms into class I of the Chinese classification, where now

$$d = \frac{-2\lambda_2 + \lambda_3(\lambda_2^2 + \lambda_3^2) + 2\lambda_3\sqrt{D}}{\sqrt[4]{4D}},$$

$$l = \frac{1}{\lambda_2}, \; m = \frac{2\lambda_3}{\lambda_2 \sqrt[4]{4D}}, \; n = \frac{\lambda_2^2 + \lambda_3^2}{2\lambda_2 \sqrt{D}}.$$

Uniqueness and hyperbolicity of possible limit cycles again follows from [89,Co]. If $\lambda_2 = 0$ the finite critical points are located in $y \equiv -\frac{1}{\lambda_3}$ and the antisaddle is node; so there exists no limit cycle.

The bifurcation diagram for $\lambda_1 < 0, \lambda_2 > 0$ is given in Figure 9.49.

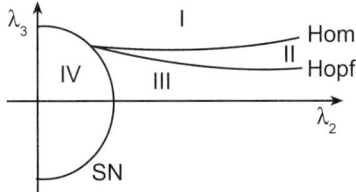

Figure 9.49 Bifurcation diagram for systems with $C_{1,1}^0$ points: $\beta \neq 0$

There exists a limit cycle only in region II, bounded by the bifurcation curves "Hopf" and "Hom". On the circle $\lambda_2^2 + \lambda_3^2 = -\frac{1}{\lambda_1}, \lambda_3 \neq -2\lambda_1\lambda_2$ there exists in the corresponding system a finite saddle node m_2^0. In the interior of the circle the finite critical points are complex . This region expands without bounds if $\lambda_1 \to 0$; for $\lambda_1 > 0$ only the phase portrait corresponding to region IV prevails. The phase portraits for (9.33),(9.34) are given in Figure 9.50.

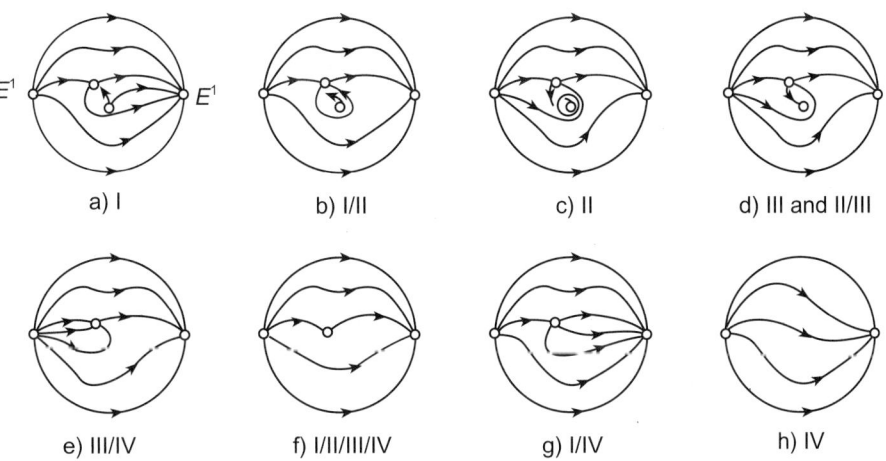

Figure 9.50 Phase portraits for systems with $C_{1,1}^0$ points: $\beta \neq 0$

(ii) If $\beta=0$, system (9.3.31),(9.3.32) reads in the notation used above

$$\dot{x} = \lambda_1 + x^2 + y^2, \qquad\qquad (9.3.35)$$
$$\dot{y} = \lambda_2 x + \lambda_3 y, \qquad\qquad (9.3.36)$$

where $\lambda_1 = \alpha$, $\lambda_2 = b_{10}$, $\lambda_3 = b_{01}$ and $\lambda_2^2 + \lambda_3^2 \neq 0$. We may assume $\lambda_3 \geq 0$; if not apply $(x,t) \rightarrow (-x,-t)$.

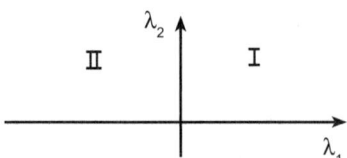

Figure 9.51 Bifurcation diagram for systems with $C_{1,1}^0$ points:$\beta=0$

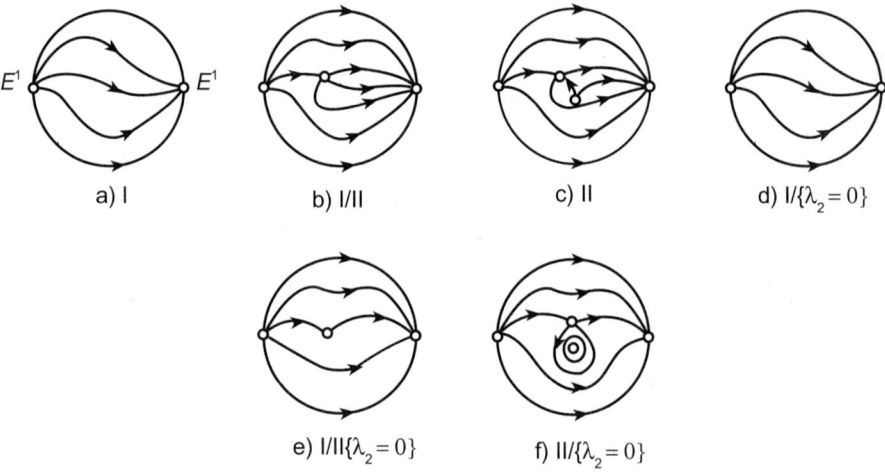

a) I   b) I/II   c) II   d) I/$\{\lambda_2 = 0\}$

e) I/II$\{\lambda_2 = 0\}$ f) II/$\{\lambda_2 = 0\}$

Figure 9.52 Phase portraits for systems with $C_{1,1}^0$ points:$\beta=0$

The finite critical points are located in $(x_\pm, y_\pm)$, where $x_\pm = \pm\lambda_3\sqrt{\frac{-\lambda_1}{\lambda_2^2+\lambda_3^2}}$, $y_\pm = \mp\lambda_2\sqrt{\frac{-\lambda_1}{\lambda_2^2+\lambda_3^2}}$. So for $\lambda_1 < 0$ there are two finite elementary critical points, for $\lambda_1 = 0$ an m_2^0 point and for $\lambda_1 > 0$ two complex critical points c_1^0. As

$\Omega(x, y) \equiv P_x(x, y)Q_y(x, y) - P_y(x, y)Q_x(x, y) = 2(\lambda_3 x - \lambda_2 y)$, it follows that $\Omega(x_\pm, y_\pm) = \pm 2(\lambda_2^2 + \lambda_3^2)\sqrt{\frac{-\lambda_1}{\lambda_2^2 + \lambda_3^2}}$, so for $\lambda_1 < 0$, the $-$ sign corresponds to a saddle and the $+$ sign to an antisaddle.   There are weak critical points if $\lambda_3(1 \pm 2\sqrt{\frac{-\lambda_1}{\lambda_2^2 + \lambda_3^2}}) = 0$, a weak saddle for the $-$ sign and a weak focus for the $+$ sign, which implies $\lambda_3 = 0$. Further analysis shows that if $\lambda_1 < 0$, the antisaddle is a node or a focus. In fact the latter is true if also $\lambda_3 > 0$ and becomes a center point if $\lambda_3 = 0$, as follows from the symmetry of the phase portrait around the y axis. If $\lambda_1 = 0$ the m_2^0 point is a saddle node for $\lambda_3 \neq 0$ and a cusp for $\lambda_3 = 0$. If $\lambda_1 > 0$ there exist two finite complex critical points.

The bifurcation diagram for (9.3.31),(9.3.32) is given in Figure 9.51, wherein region I corresponds to $\lambda_1 > 0$ and II to $\lambda_1 < 0$; the phase portraits are given in Figure 9.52. It can be shown that system (9.3.31),(9.3.32) has no limit cycles [97,RK].

9.3.4   Conclusion

Theorem 9.2 *There are 99 possible phase portraits for quadratic systems with $m_f = 2$ and two transversally hyperbolic infinite critical points. They are given in Figures 9.33–9.53.   There exists at most one limit cycle in these systems.*

9.4   Conclusion

Theorem 9.3 *There exist 230 possible phase portraits for quadratic systems with $m_f = 2$. They are indicated in Theorems 9.1 and 9.2.*

Chapter 10

Phase portraits of quadratic systems in the class $m_f=3$

10.1 Introduction

The tendency that increase of the finite multiplicity from $m_f=0$ and 1 to $m_f=2$, causing an increase of the number of parameters in the system , results in a drastic increase of the number of phase portraits and continues to prevail when $m_f=3$ is considered. Almost a doubling occurs compared with $m_f=2$. Moreover, not all phase portraits are known, due to the fact that the limit cycle problem has only partly been solved for $m_f=3$. This effectively means that, although all possible separatrix structures are known, it is not known in all cases what the limit cycle distribution is. From this results that the statement that a quadratic system with $m_f=3$ has at most three limit cycles is still an unproved conjecture. It is known for a system in $m_f=3$ that, if there exist two nests of limit cycles, then there exists in each nest precisely one (hyperbolic) limit cycle [99,ZK], this verifies the above conjecture and means we still have to show that, if precisely one nest of limit cycles exists, it contains at most three limit cycles.

In this chapter the discussion on the possible phase portraits for $m_f=3$ will be less detailed than is given for $m_f=0,1$ and 2. Instead of repeating all the arguments used in the previous cases to arrive at the possible phase portraits, results will be indicated in a more global manner and open questions will be stressed.

Since for $m_f=3$ there exists precisely one transversally non-hyperbolic infinite critical point, this point is an $M^i_{1,q}$ point, with q=1,2 or 3. The $M^i_{1,1}$ point is a semi-elementary critical point and the $M^i_{1,2}$ and $M^i_{1,3}$ points are degenerate infinite critical points. The number of infinite critical points can be infinite as well.

10.2   Phase portraits with a degenerate infinite critical point

10.2.1   Phase portraits of systems with a fourth order infinite critical point

Without loss of generality we may take the $M^i_{1,3}$ point "at the end of the x axis" so that the common linear factor in the quadratic terms equals y, so $a_{20} = b_{20}{=}0$, whereas $c_{56} \equiv a_{11}b_{02} - a_{02}b_{11} \neq 0$ to ensure that there is only one such factor. Since q=3 there is $b_{11} = a_{11} - b_{02} {=}0$ with $a_{11} = b_{02} \neq 0$ as there should be $c_{56} = a^2_{11} \neq 0$ in order to ensure that $M^i_{1,3}$ is the only transversally non-hyperbolic infinite critical point. Also, it may be seen that for $m_f{=}3$ there exists at least one real finite critical point, which then may be thought to be located at the origin, thus $a_{00} = b_{00}{=}0$.
    For (1.0.1),(1.0.2) we may then write

$$\dot{x} \;=\; a_{10}x + a_{01}y + a_{11}xy + a_{02}y^2 \equiv P(x, y), \qquad (10.2.1)$$
$$\dot{y} \;=\; b_{10}x + b_{01}y + b_{02}y^2 \equiv Q(x, y), \qquad (10.2.2)$$

with $a_{11} \neq 0$, $a_{02} \neq 0$ and we may take $a_{11} = a_{02} {=}1$, if needed by replacing x by $a_{02}x$, y by $a_{11}y$ and $a^2_{11}t$ by t. Furthermore if $b_{10}{=}0$, there are two roots of $Q(x, y){=}0$ and, correspondingly two roots of $P(x, y) = 0$, thus $m_f \leq 2$. Thus $b_{10} \neq 0$ and we may take $b_{10}{=}1$, if needed by replacing x by $b_{10}x$, y by $b_{10}y$ and $b_{10}t$ by t. Finally, by replacing $x + b_{01}y$ by x, (10.2.1),(10.2.2) takes the form

$$\dot{x} \;=\; \lambda x + \mu y + xy + y^2, \qquad (10.2.3)$$
$$\dot{y} \;=\; x + y^2, \qquad (10.2.4)$$

with $\lambda, \mu \in \mathbb{R}$, which is in the class $m_f{=}3$ and has an infinite critical point $M^i_{1,3}$ "at the ends of the x axis".
    Conditions on the coefficients in (1.0.1),(1.0.2) such that the system is in $m_f{=}3$ and has an $M^i_{1,3}$ point are given in [97,RH].

Since $M_{1,3}^i$ is a degenerate critical point it follows that (10.2.3),(10.2.4) has at most one limit cycle and if it exists is hyperbolic [89,C].

It may further be seen [95,RK],[97,RH], that $M_{1,3}^i$ is a fourth order nilpotent saddle node $M_{1,3}^0$, having hyperbolic sectors with opening angles 0 and π and a parabolic sector with orbits tangent to the Poincaré circle. A special feature of (10.2.3),(10.2.4) is that it has parabolic orbits; they are given by

$$x = -\frac{1}{2}\mu - y - \frac{1}{2}y^2, \lambda \equiv -1, -\infty < \mu < \infty, \quad (10.2.5)$$

$$x = \frac{1}{3}(\lambda - 2)y - \frac{1}{2}y^2, -\infty < \lambda < \infty, \mu = -\frac{2}{9}(\lambda + 1)(\lambda - 2). \quad (10.2.6)$$

As $c_{45} + c_{56} = a_{11}^2 > 0$, Theorem 3.14 (Chapter3) says that $i_f=1$. Analysis of (10.2.3),(10.2.4) shows that it represents the classes $e^{-1}e^1e^1M_{1,3}^0$, $c_1^0c_1^0e^1M_{1,3}^0$, $e^1m_2^0M_{1,3}^0$ and $m_3^1M_{1,3}^0$, all belonging to class 2 in Table 2 of Chapter 3. It appears that there exist 22 possible phase portraits for quadratic systems in $m_f=3$ with an $M_{1,3}^0$ point; they are given in [97,RH]. Hopf bifurcation of a (unique) hyperbolic limit cycle takes place from a first order weak focus. Bifurcation from a separatrix cycle takes place either from a finite saddle loop or an unbounded separatrix loop containing the point $M_{1,3}^0$. For $\lambda = \mu=0$ a Bogdanov−Takens cusp bifurcation takes place.

The theory of rotated vector fields is used to follow the behaviour of the limit cycles.

10.2.2   Phase portraits with a third order infinite critical point

The derivation of (10.2.3),(10.2.4) in the previous section may be partly repeated to obtain a similar result if there exists an $M_{1,2}^i$ point, now also thought to be located "at the ends of the x axis". Starting again with (1.0.1),(1.0.2), imposing the same conditions, with the difference that now "at the ends of the x axis" there is an $M_{1,2}^i$ point, thus $b_{11} = 0, a_{11} \neq b_{02}$, leads to

$$\dot{x} = a_{10}x + a_{01}y + a_{11}xy + a_{02}y^2, \quad (10.2.7)$$
$$\dot{y} = x + y^2, \quad (10.2.8)$$

with $a_{10}, a_{01}, a_{11}, a_{02} \in \mathbb{R}$ and $a_{11} \neq 0$ or 1.

In order to be able to use the theory of rotated vector fields, (10.2.7),(10.2.8) may be written as

$$\dot{x} = \lambda x + \mu y + \gamma xy + \delta(x + y^2), \qquad (10.2.9)$$
$$\dot{y} = x + y^2, \qquad (10.2.10)$$

with λ, μ, γ and $\delta \in \mathbb{R}$ and $\gamma \neq 0$ or 1. Also, we may take $\lambda \geq 0$; if necessary, we can replace x, y and t by $x, -y, -t$, respectively. In addition, without loss of generality, we can put $\lambda=0$ or 1.

Conditions on the coefficients of (1.0.1),(1.0.2) such that the system is in $m_f=3$ and has an $M_{1,2}^i$ point are given in [97,RH].

Since $M_{1,2}^i$ is a degenerate critical point it follows [89,C] that (10.2.9), (10.2.10) has at most one limit cycle and if it exists is hyperbolic.

A study of the infinite critical points of (10.2.9),(10.2.10) reveals that the $M_{1,2}^i$ point is a nilpotent third order saddle $M_{1,2}^{-1}$ if $0 < \gamma < 1$ and a nilpotent $M_{1,2}^1$ point with an elliptic and a hyperbolic sector (possibly also parabolic sectors) if $\gamma < 0$ or $\gamma > 1$ [95,RK],[97,RH]. The remaining elementary infinite critical point is a saddle E^{-1} for $\gamma > 1$ and a node for $\gamma < 1$. This follows from the index argument and that Theorem 3.14 in Chapter 3 yields that $i_f=-1(1)$ for $\gamma < 0(>0)$, as $c_{45}^2 + c_{56}^2 = \gamma$.

Analysis of (10.2.9),(10.2.10) shows that it represents for $\gamma < 0$ the classes $e^{-1}e^{-1}e^1E^1M_{1,2}^1, e^{-1}c_1^0c_1^0E^1M_{1,2}^1, e^{-1}m_2^0E^1M_{1,2}^1$ and $m_3^{-1}E^1M_{1,2}^1$, all belonging to class 1 in Table 2 of Chapter 3 and for $\gamma > 0$ the classes $e^{-1}e^1e^1E^{-1}M_{1,2}^1$, $c_1^0c_1^0e^1E^{-1}M_{1,2}^1, e^1m_2^0E^{-1}M_{1,2}^1, m_3^1E^{-1}M_{1,2}^1$ and $e^{-1}e^1e^1E^1M_{1,2}^{-1}, c_1^0c_1^0e^1E^1M_{1,2}^{-1}$, $e^1m_2^0E^1M_{1,2}^{-1}, m_3^1E^1M_{1,2}^{-1}$, all belonging to class 2 in Table 2 of Chapter 3.

It appears that for quadratic systems in $m_f=3$ with an $M_{1,2}^i$ point there exist 119 possible phase portraits; they are given in [97,RH].

(i)For $\lambda=0$ there exist 39 possible phase portraits.

For $\lambda = \delta = 0$, system (10.2.9),(10.2.10) reads

$$\dot{x} = \mu y + \gamma xy, \qquad (10.2.11)$$
$$\dot{y} = x + y^2, \qquad (10.2.12)$$

which may be integrated as a linear system in y^2. Its phase portraits contain for $\mu < 0$ a center point, for $\mu=0$ an m_3^1 point with an elliptic and a hyperbolic sector if $\gamma > 0$ and a nilpotent saddle m_3^{-1} if $\gamma < 0$, whereas for $\mu > 0$ there exist only elementary finite critical points.

Phase portraits for $\lambda=0$, $\delta \neq 0$, as given in [97,RH], may be seen as bifurcations from those for $\lambda = \delta=0$. In fact all phase portraits can be bifurcated from $\lambda = \delta = \mu = 0$ as they are affine equivalent for fixed κ, where $\mu = \kappa\delta^2$. Hopf bifurcation of a (unique) hyperbolic limit cycle occurs from a first order weak critical point, whereas bifurcation from a separatrix cycle takes place from a finite saddle loop.

(ii)For $\lambda=1$ there exist 80 possible phase portraits not yet contained in the classification for $\lambda=0$. Bifurcation of a (unique) hyperbolic limit cycle occurs again from a Hopf bifurcation out of a first order weak focus, from a cusp bifurcation, a finite saddle loop or a saddle node loop and also from an unbounded separatrix loop. Various types of saddle connections occur and bifurcations from them.

10.2.3 Conclusion

Theorem 10.1 *For a quadratic system of finite multiplicity $m_f=3$ and a degenerate infinite critical point (either an $M_{1,3}^0$, $M_{1,2}^{-1}$ or an $M_{1,2}^1$ point) there exist 141 possible phase portraits. For such a system there exists at most one limit cycle and, if it exists, it is hyperbolic.*

10.3 Phase portraits with a saddle node at infinity

If the transversally non-hyperbolic infinite critical point $M_{1,1}^i$ is thought to be located "at the ends of the x axis" it follows that $a_{20} = b_{20} = 0$ should be satisfied. If $M_{1,1}^i$ is the only transversally non-hyperbolic infinite critical point, it follows then that $a_{02}^2 + b_{02}^2 \neq 0$ and we may apply Theorem 1 of Chapter 3. The requirement that $m_f=3$ then leads to $c_{25}c_{56} \neq 0$. From $c_{56} \neq 0$ it follows that p=1 in $M_{p,q}^i$. From $c_{25} \neq 0$ it follows with the classification of infinite critical points in [95,RK] and taking q=1, that $M_{1,1}^i$ is a second order semi-elementary saddle node $M_{1,1}^0$ having (a) center manifold(s) transversal to the Poincaré circle.

The separatrix structure for systems in $m_f=3$ having an $M_{1,1}$ saddle node were determined by Huang and Reyn [95,HR2],[96,H] leaving the limit cycle problem untouched and about which only scattered information can be found in the literature [91,C],[89,H],[98,Z],[79,CW].

Use was made of a representation of these systems in the form

$$\dot{x} = x + \lambda y + \varepsilon y^2 + \delta(\mu x + \gamma y + xy), \tag{10.3.1}$$
$$\dot{y} = \mu x + \delta y + xy, \tag{10.3.2}$$

where $\mu \geq 0$, $\varepsilon = \pm 1$, $\lambda, \gamma, \delta \in \mathbb{R}$.

This system may be derived from $(1.0.1),(1.0.2)$ by taking $a_{20} = b_{20}=0$, $b_{11} \neq 0$ to create an $M^i_{p,1}(p \geq 1)$ "at the ends of the x axis". From Theorem 2.1 it follows that for $m_f=3$ it is required that $c_{25}c_{56} \neq 0$, then p=1 and $M^i_{p,1} = M^0_{1,1}$. Since there exists at least one real finite critical point we put $a_{00} = b_{00}=0$. If, furthermore $b_{11}x + b_{02}y$ is replaced by x we arrive at

$$\dot{x} = a_{10}x + a_{01}y + a_{11}xy + a_{02}y^2, \tag{10.3.3}$$
$$\dot{y} = b_{10}x + b_{01}y + xy, \tag{10.3.4}$$

with $a_{02} \neq 0$ since $c_{56} = -a_{02} \neq 0$, $c_{25} \neq 0$.

In order to use the theory of rotated vector fields, and also restrict the number of parameters that determine the location of the finite critical points, we write this system as

$$\dot{x} = a_{10}x + a_{01}y + a_{02}y^2 + \delta(\mu x + \gamma y + xy), \tag{10.3.5}$$
$$\dot{y} = \mu x + \gamma y + xy, \tag{10.3.6}$$

with $a_{02} \neq 0$, and $a_{10} \neq 0$ since $c_{25} = a_{10} \neq 0$.

Replacing x by $a_{10}x$, y by $\frac{a_{10}}{\sqrt{|a_{02}|}}y$ and t by $\frac{t}{a_{10}}$ yields

$$\dot{x} = x + \lambda y \pm y^2 + \delta(\mu x + \gamma y + xy), \tag{10.3.7}$$
$$\dot{y} = \mu x + \gamma y + xy, \tag{10.3.8}$$

where $\lambda, \delta, \gamma \in \mathbb{R}$ and where we can set $\mu \geq 0$, if needed, replacing y by $-y$ (then also λ and δ change their sign).

For $(10.3.1),(10.3.2)$ there follows $c_{45} + c_{56} = -\varepsilon$. As a result it may be seen that for $\varepsilon = -1$ $(\varepsilon=1)$ there is $i_f=1(i_f = -1)$ and the system belongs to class 2 (class 1), mentioned in Table 2 of Chapter 3.

10.3.1   Systems with finite index $i_f{=}1$

System (10.3.1),(10.3.2) now reads

$$\dot{x} = x + \lambda y - y^2 + \delta(\mu x + \gamma y + xy), \qquad (10.3.9)$$
$$\dot{y} = \mu x + \gamma y + xy, \qquad (10.3.10)$$

where $\mu \geq 0$, $\lambda, \gamma, \delta \in \mathbb{R}$. The coordinates of the finite critical points are given by $(0,0)$, (x_+, y_+) and (x_-, y_-), where

$$x_\pm = -\gamma + \frac{1}{2}(\lambda + \mu \pm \sqrt{(\lambda + \mu)^2 - 4\gamma}), \qquad (10.3.11)$$

$$y_\pm = \frac{1}{2}(\lambda - \mu \mp \sqrt{(\lambda + \mu)^2 - 4\gamma}). \qquad (10.3.12)$$

All combinations of finite critical points listed in class 2 of Table 2 in Chapter3 are possible: $e^{-1}e^1e^1, c_1^0c_1^0e^1, e^1m_2^0$ and m_3^1. Here e^{-1} indicates a saddle and e^1 an antisaddle, being either a node, a (rough) focus or of first, second or third order. A second order point m_2^0 is either a saddle node or a cusp. A third order point m_3^1 is either a semi-elementary node or a nilpotent point with an elliptic and a hyperbolic sector.

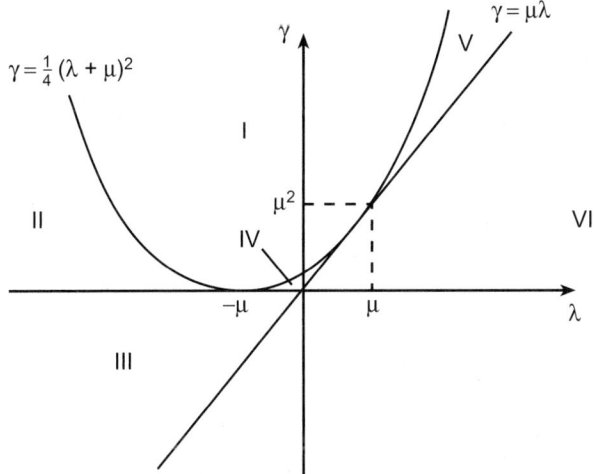

Figure 10.1 Parameter plane for (10.3.9),(10.3.10)($\mu > 0$

For a given value of μ we may use a λ, γ plane wherein the location of the possible combinations of finite critical points are indicated, as done in Figure

10.1 for $\mu > 0$. Systems with an m_3^1 point are marked by $\lambda = \mu, \gamma = \mu^2$, those with the $e^1 m_2^0$ combination by $\gamma = \frac{1}{4}(\lambda + \mu)^2$ and by $\gamma = \mu\lambda(\lambda \neq \mu)$.

Three critical points are present in the region I:$-\infty < \lambda < \infty, \gamma > \frac{1}{4}(\lambda + \mu)^2$; with the $c_1^0 c_1^0 e^1$ combination , the regions II:$-\infty < \lambda < -\mu, 0 < \gamma < \frac{1}{4}(\lambda + \mu)^2$, III: $-\infty < \lambda < 0, \lambda\mu < \gamma \leq 0, (\lambda, \mu) \neq (-\mu, 0)$, IV: $-\mu < \lambda < \mu$, max of 0 and $\lambda\mu < \gamma < \frac{1}{4}(\lambda + \mu)^2$, V: $\mu < \lambda < \infty, \lambda\mu < \gamma < \frac{1}{4}(\lambda + \mu)^2$ and region VI: $-\infty < \lambda < \infty, \gamma < \lambda\mu$; with the $e^{-1} e^1 e^1$ combination.

If the $e^{-1} e^1 e^1$ combination prevails the situation is as indicated in Figures 3.5 and 3.6. The central conic of (10.3.9),(10.3.10) is given by

$$\Omega(x, y) = \gamma - \lambda\mu + x - (\lambda - 2\mu)y + 2y^2 = 0, \qquad (10.3.13)$$

which is illustrated in Figure 10.2 together with the degenerate isoclines and the $e^{-1} e^1 e^1$ configuration.

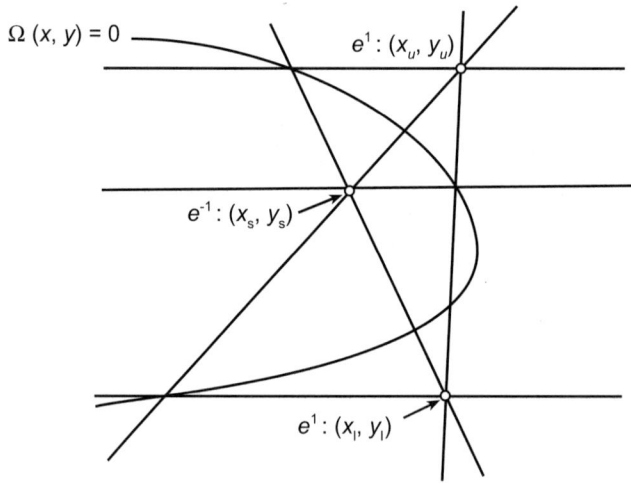

Figure 10.2 Central conic and degenerate isoclines for
(10.3.9),(10.3.10):$e^{-1} e^1 e^1$ configuration

The upper (lower) antisaddle is indicated by $(x_u, y_u)((x_l, y_l))$ and the saddle by (x_s, y_s). If $\gamma < \lambda\mu(\gamma > \lambda\mu)$ the origin is a saddle (antisaddle). If the origin is transferred from one critical point to another (10.3.9),(10.3.10) is still valid for different values of the parameters. In fact $\bar{x} = x - x_c, \bar{y} = y - y_c$, where (x_c, y_c) are the coordinates of a critical point of (10.3.9),(10.3.10), yields

$$\dot{\bar{x}} = \bar{x} + \bar{\lambda}\bar{y} - \bar{y}^2 + \bar{\delta}(\bar{\mu}\bar{x} + \bar{\gamma}\bar{y} + \bar{x}\bar{y}), \qquad (10.3.14)$$

$$\dot{\bar{y}} = \bar{\mu}\bar{x} + \bar{\gamma}\bar{y} + \bar{x}\bar{y}, \qquad (10.3.15)$$

where $\bar{\lambda} = \lambda - 2\varepsilon y_c, \bar{\delta} = \delta, \bar{\mu} = \mu + y_c$ and $\bar{\gamma} = \gamma + x_c$.

If the lower antisaddle is located at the origin, there is $0 < y_s < y_u$ so $0 < y_+ < y_-$ and $\lambda > \mu, \frac{1}{4}(\lambda + \mu)^2 > \gamma > \lambda\mu$ which corresponds to region V.

If the saddle is located at the origin, parameters are in the region VI and if the upper antisaddle is at the origin, parameters are in the regions II, III and IV.

Apart from the saddle node $M_{1,1}^0$, for $0 < |\delta| < 2$ there exist two complex critical points $C_{0,1}^0$, for $|\delta| = 2$ a tangentially saddle node $M_{0,2}^0$ and for $|\delta| > 2$ a saddle E^{-1} and a node E^1.

Systems with a $C_{0,1}^0 C_{0,1}^0 M_{1,1}^0$ combination

The classes $e^{-1}e^1e^1C_{0,1}^0C_{0,1}^0M_{1,1}^0$, $c_1^0c_1^0e^1C_{0,1}^0C_{0,1}^0M_{1,1}^0$, $e^1m_2^0C_{0,1}^0C_{0,1}^0M_{1,1}^0$ and $m_3^1C_{0,1}^0C_{0,1}^0M_{1,1}^0$ occur as listed in class 2 of Table 2 of Chapter 3. These classes constitute the bounded quadratic systems in $m_f = 3$ if (10.3.9),(10.3.10) is used with t replaced by $-t$. As mentioned before the solutions of these systems remain in the finite part of the plane for $t \geq 0$; their study was introduced in 1970 by Dickson and Perko [70,DP]. In that paper necessary and sufficient conditions were given for a quadratic system to be bounded and a start was made to determine for this class all possible separatrix configurations not yet considering the limit cycle problem. The study of the latter problem was subsequently initiated by Perko in 1975 [75,P], using the theory of rotated vector fields. Non-existence, existence and uniqueness results on limit cycles in bounded quadratic systems were thereafter obtained by Perko and Shü[84,PS], Yang Xinan [82,Y],[83,Y] and [88,Y], Coll,Gasull and Llibre [87,CGL], Xie Xiangdong [92,X], Salam, van Gils and Zhang Zhifen [91,SGZ], Huang Xianhua and Reyn [97,HR],[95,HR3], Huang[96,H] and Coppel[96,C].

An extensive exploration of the possible phase portraits of bounded quadratic systems (including those in $m_f = 1$ and $m_f = \infty$) was given by Perko [94,P]. It is conjectured that there exist at most two limit cycles in a bounded quadratic system. Assuming this to be true, all possible phase portraits (including the limit cycles) were determined. A partition of the $\mathbb{R}^4$ parameter space is constructed by numerical and analytical methods to the extent that for a given particular bounded system the corresponding phase portrait can be determined from this partition.

As a result it may be concluded that there exists precisely one phase portrait in class $m_3^1 C_{0,1}^0 C_{0,1}^0 M_{1,1}^0$; the m_3^1 point being a third order semi-elementary node and, trivially, no limit cycle occurs in the phase portrait which can now easily be constructed.

Similarly, in the class $c_1^0 c_1^0 e^1 C_{0,1}^0 C_{0,1}^0 M_{1,1}^0$ there exists a simple phase portrait with an elementary node or focus; the latter also possibly being a first order weak focus from which a unique (hyperbolic) limit cycle bifurcates within this class without meeting another limit cycle in this class.

In the class $e^1 m_2^0 C_{0,1}^0 C_{0,1}^0 M_{1,1}^0$, the point m_2^0 is either a saddle node or a cusp. Moreover, phase portraits with a single (hyperbolic) limit cycle occur as well as homoclinic cycles running from and towards a saddle node. As a result 12 phase portraits exist.

In the class $e^{-1} e^1 e^1 C_{0,1}^0 C_{0,1}^0 M_{1,1}^0$ the assumption that there exist at most two limit cycles in a bounded quadratic system is used to arrive at the conclusion that there exist 15 possible phase portraits in this class. Apart from Hopf bifurcations from a first order weak focus, now they also occur from a second order weak focus (in such a system no limit cycles occur), yielding a semi-stable second order limit cycle or two hyperbolic limit cycles of opposite stability. Also stable or unstable homoclinic saddle loops occur, the latter one also with a stable limit cycle in its interior. Phase portraits with two nests of limit cycles occur ; they have precisely one hyperbolic limit cycle in each nest, as indeed is the only distribution of limit cycles in $m_f=3$ over two nests [99,ZK].

Proving Perko's conjecture, stating that two is the maximum number of limit cycles in a bounded quadratic system, appears to be the one, and most difficult, remaining problem in the class $e^{-1} e^1 e^1 C_{0,1}^0 C_{0,1}^0 M_{1,1}^0$. A step in this direction was taken by Li Chengzhi, Llibre and Zhang Zhifeng, who considered bounded quadratic systems near systems with a center point [95,LLZ1]. A survey of bounded quadratic systems was given by Dumortier, Herssens and Perko; they also studied local bifurcations within these systems [00,DHP],[94,P,p.487-538].

Systems with an $M_{0,2}^0 M_{1,1}^0$ combination

Involved are the classes $e^{-1}e^1e^1 M_{0,2}^0 M_{1,1}^0$, $c_1^0 c_1^0 e^1 M_{0,2}^0 M_{1,1}^0$, $e^1 m_2^0 M_{0,2}^0 M_{1,1}^0$, and $m_3^1 M_{0,2}^0 M_{1,1}^0$ as listed in class 2 of table 2 of Chapter 3. They can be represented by (10.3.14),(10.3.15) with $|\delta|=2$. We will discuss the separatrix structures for these classes following [95,HR2],[96,H]. Information about limit cycle behaviour would probably be obtainable when considering these classes as limits of the bounded quadratic systems discussed in the previous section. Moreover, partial results may be found scattered in the literature, as in [96,H],[97,HR] and [95,HR3]. Use might also be made by writing these systems in the Ye normal form (Table 5 of Chapter 3) and collecting results found for these classes.

For $\delta = -2$, the determination of the possible separatrix structures in these classes, as given in [96,H], yields 73 classes.

In the class $m_3^1 M_{0,2}^0 M_{1,1}^0$, three possible phase portraits are found. The m_3^1 point is either a nilpotent critical point with an elliptic and a hyperbolic sector ($\mu=1$) or an unstable semi-elementary node ($\mu \neq 1$).Trivially, there exists no limit cycle in this class.

In class $e^1 m_2^0 M_{0,2}^0 M_{1,1}^0$, 33 possible separatrix structures are found. The m_2^0 point is either a saddle node or a cusp. Finite saddle loops and unbounded separatrix cycles without a finite critical point occur. For $(\gamma, \lambda)=(1,-4),\mu=2$ there exists the hyperbolic orbit $(2 + y)(4 + x + y) = 2$, one branch being an unbounded separatrix cycle. Seen as a limit of the class $e^1 m_2^0 C_{0,1}^0 C_{0,1}^0 M_{1,1}^0$ leads to the conjecture that in the class $e^1 m_2^0 M_{0,2}^0 M_{1,1}^0$ there exists at most one (hyperbolic) limit cycle. Correspondingly only weak foci of first order occur.

In the class $c_1^0 c_1^0 e^1 M_{0,2}^0 M_{1,1}^0$, five possible separatrix sructures are found. An unbounded separatrix cycle and a first order weak focus occurs. Seen as a limiting case of the class $c_1^0 c_1^0 e^1 C_{0,1}^0 C_{0,1}^0 M_{1,1}^0$ it may again be conjectured that there exists at most one (hyperbolic) limit cycle.

In the class $e^{-1}e^1e^1 M_{0,2}^0 M_{1,1}^0$, 32 possible phase portraits are found. Finite saddle loops and unbounded separatrix cycles with and without finite critical points occur. The limit cycle problem needs further study [96,H],[97,HR].

For $\delta=2$, the determination of the possible separatrix structures in the classes with an $M_{0,2}^0 M_{1,1}^0$ combination at infinity as given in [96,H],[97,HR] yields 18 cases.

In the class $m_3^1 M_{0,2}^0 M_{1,1}^0$ there exists one phase portrait and the m_3^1 point is an unstable semi-elementary node. Trivially, there exists no limit cycle in this class.

In the class $e^1 m_2^0 M_{0,2}^0 M_{1,1}^0$ there are eight possible separatrix structures found. The m_2^0 point is either a saddle node or a cusp. The limit cycle problem needs further study as is also needed for the class $c_1^0 c_1^0 e^1 M_{0,2}^0 M_{1,1}^0$ for which there exists only one possible separatrix structure.

In the class $e^{-1} e^1 e^1 M_{0,2}^0 M_{1,1}^0$ there are found eight possible separatrix structures.

Systems with a $E^{-1} E^1 M_{1,1}^0$ combination

Involved are the classes $e^{-1} e^1 e^1 E^{-1} E^1 M_{1,1}^0$, $c_1^0 c_1^0 e^1 E^{-1} E^1 M_{1,1}^0$, $e^1 m_2^0 E^{-1} E^1 M_{1,1}^0$ and $m_3^1 E^{-1} E^1 M_{1,1}^0$ as listed in class 2 of Table 2 of Chapter 3. They can be represented by (10.3.14),(10.3.15) with $|\delta| > 2$. Compared to the classes in $i_f = 1$, discussed in the previous sections, there is a richer variety in the classes with an $E^{-1} E^1 M_{1,1}^0$ combination at infinity. Apart from a second order weak focus (also occurring in the bounded case) around which at most one (hyperbolic) limit cycle can exist as the only limit cycle in such a system, also systems with a third order weak focus may occur which cannot contain any limit cycles at all.

For $\delta < -2$ the determination of possible separatrix structures in the classes with an $E^{-1} E^1 M_{1,1}^0$ combination as given in [96,H] yields 75 cases. In the class $m_3^1 E^{-1} E^1 M_{1,1}^0$ 4 possible phase portraits are found. The m_3^1 point is either a nilpotent critical point with an elliptic and a hyperbolic sector or a semi-elementary node. Trivially there exists no limit cycle. In the class $e^1 m_2^0 E^{-1} E^1 M_{1,1}^0$ 34 possible separatrix structures are found. The m_2^0 point is either a saddle node or a cusp. Finite saddle loops and unbounded separatrix loops without and with finite critical points occur as there are also first order and second order weak foci. In the class $c_1^0 c_1^0 e^1 E^{-1} E^1 M_{1,1}^0$ 32 possible separatrix structures are found. Finite saddle loops and unbounded separatrix loops without finite critical points occur. Weak foci of first and second order also occur.

For $\delta > 2$ the determination of possible separatrix structures in the classes with a $E^{-1} E^1 M_{1,1}^0$ combination as given in [96,H] yields 20 cases. In the class $m_3^1 E^{-1} E^1 M_{1,1}^0$ there is one possible phase portrait; it contains an unstable semi-elementary node. In the class $e^1 m_2^0 E^{-1} E^1 M_{1,1}^0$ eight possible separatrix structures are found. The m_2^0 point is either a saddle node or a cusp. Only first order weak foci occur. In the class $c_1^0 c_1^0 e^1 E^{-1} E^1 M_{1,1}^0$ only one separatrix structure occurs. In the class $e^{-1} e^1 e^1 E^{-1} E^1 M_{1,1}^0$, 10 possible separatrix structures are found. Finite saddle loops and unbounded loops with a finite critical point occur. Weak foci are of the first order.

10.3.2   Systems with finite index $i_f = -1$

System (10.3.2),(10.3.3) now reads

$$\dot{x} = x + \lambda y + y^2 + \delta(\mu x + \gamma y + xy), \qquad (10.3.16)$$
$$\dot{y} = \mu x + \gamma y + xy, \qquad (10.3.17)$$

where $\mu \geq 0$ and $\lambda, \gamma, \delta \in \mathbb{R}$. The coordinates of the finite critical points are given by $(0,0),(x_+, y_+)$ and (x_-, y_-), where

$$x_\pm = -\gamma + \frac{1}{2}\mu(\lambda - \mu \pm \sqrt{(\lambda - \mu)^2 + 4\gamma}), \qquad (10.3.18)$$

$$y_\pm = \frac{1}{2}(-\lambda - \mu \pm \sqrt{(\lambda - \mu)^2 + 4\gamma}). \qquad (10.3.19)$$

All combinations of finite critical points listed in class 1 of Table 2 of Chapter 3 are possible: $e^{-1}e^{-1}e^1, e^{-1}c_1^0c_1^0, e^{-1}m_2^0$ and m_3^{-1}. Here e^{-1} indicates a saddle, e^1 indicates an antisaddle, being either a node, a (rough)focus or a weak focus of first, second or third order. A second order point m_2^0 is either a saddle node or a cusp. A third order point m_3^{-1} is either a semi-linear or a nilpotent saddle.

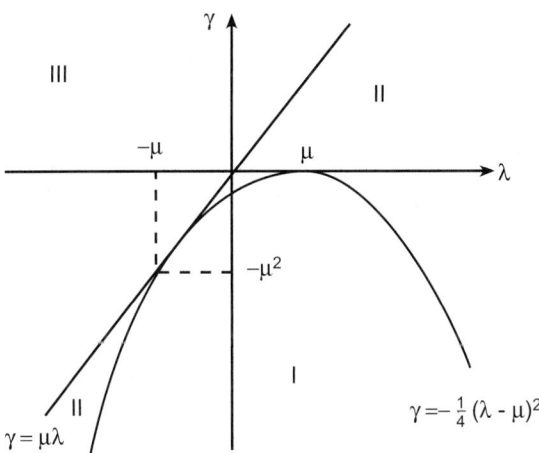

Figure 10.3 Parameter plane for (10.3.16),(10.3.17)

For a given value of μ we may use a λ, γ plane, wherein the location of the possible combinations of finite critical points are indicated, as done for $\mu > 0$ in Figure 10.3. Systems with an m_3^{-1} point are given by $\lambda = -\mu, \gamma = -\mu^2$, those with the $e^{-1}m_2^0$ combination by $\gamma = -\frac{1}{4}(\lambda - \mu)^2$ and by $\gamma = \lambda\mu(\lambda \neq -\mu)$. Three critical points are present in the region I: $-\infty < \lambda < \infty, \gamma < -\frac{1}{4}(\lambda - \mu)^2$ with the $e^{-1}c_1^0c_1^0$ combination and the regions II:$-\infty < \lambda < \infty, -\frac{1}{4}(\lambda - \mu)^2 < \gamma < \lambda\mu, (\lambda, \gamma) \neq (-\mu, -\mu^2)$, and III:$-\infty < \lambda < \infty, \gamma > \mu\lambda$ with the $e^{-1}e^{-1}e^1$ combination.

If the $e^{-1}e^{-1}e^1$ combination prevails the situation is indicated geometrically in Figures 3.5 and 3.6. The central conic of (10.3.16),(10.3.17) is given by

$$\Omega(x, y) = \gamma - \lambda\mu + x - (\lambda + 2\mu)y - 2y^2 = 0, \qquad (10.3.20)$$

which is illustrated in figure 10.4, together with the degenerate isoclines and the $e^{-1}e^{-1}e^1$ configuration.

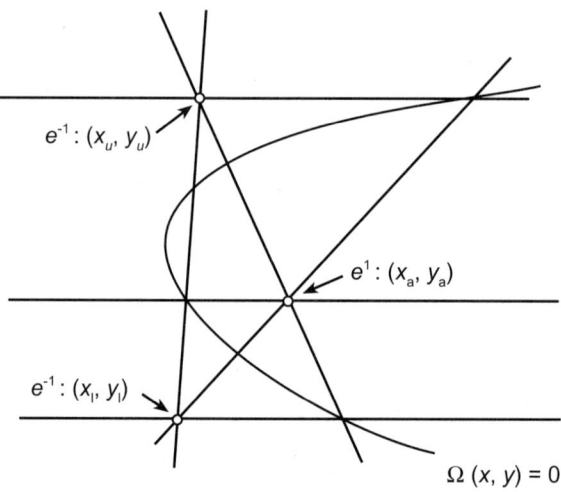

$e^{-1}: (x_u, y_u)$

$e^1: (x_a, y_a)$

$e^{-1}: (x_l, y_l)$

$\Omega(x, y) = 0$

Figure 10.4 Central conic and degenerate isoclines for (10.3.16),(10.3.17): $e^{-1}e^{-1}e^1$ configuration

The upper(lower) saddle is indicated by $(x_u, y_u)((x_l, y_l))$ and the antisaddle by (x_a, y_a). If $\gamma < \lambda\mu(\gamma > \lambda\mu)$ the origin is a saddle (antisaddle). Transfer of the origin from one critical point to another leaves (10.3.16),(10.3.17) invariant in form yielding different values for the parameters.

If the lower(upper) saddle is located in the origin, then $(x_l, y_l)=(0,0)$, $((x_u, y_u)=(0,0))$ and there is $0< y_a < y_u(y_l < y_a < 0)$ so $0< y_- < y_+$ $(y_- < y_+ < 0)$ and $\lambda < \mu(\lambda > \mu)$, $-\frac{1}{4}(\lambda - \mu)^2 < \gamma < \lambda\mu$, which is region II. If the antisaddle is located in the origin, there is $-\infty < \lambda < \infty, \gamma > \lambda\mu$ (region III).

Apart from the saddle node $M_{1,1}^0$ there exist two elementary nodes E^1 at infinity. So, involved are the classes $e^{-1}e^{-1}e^1E^1E^1M_{1,1}^0$, $e^{-1}c_1^0c_1^0E^1E^1M_{1,1}^0$, $e^{-1}m_2^0E^1E^1M_{1,1}^0$ and $m_3^{-1}E^1E^1M_{1,1}^0$ as listed in class 1 of Table 2 of Chapter 3. We will again discuss the possible separatrix structures for these classes, following [96,H], where 54 classes are presented. Clearly at most one nest of limit cycles is possible.

In the class $m_3^{-1}E^1E^1M_{1,1}^0$ two possible separatrix structures are found. The m_3^{-1} point is a semi-elementary saddle for $1+\delta\mu - \mu^2 \neq 0$ or a nilpotent saddle for $1+\delta\mu - \mu^2 =0$; obviously there are no limit cycles. In the class $e^{-1}m_2^0E^1E^1M_{1,1}^0$, 26 separatrix structures are presented; m_2^0 is a saddle node or a cusp. Saddle connections occur involving finite or infinite separatrices. Trivially, no limit cycles exist. In the class $e^{-1}c_1^0c_1^0E^1E^1M_{1,1}^0$ there appears one phase portrait; no limit cycles are possible. In the class $e^{-1}e^{-1}e^1E^1E^1M_{1,1}^0$, 25 separatrix structures are presented. There exist finite saddle loops and separatrix cycles with two saddles. Weak foci of order 1, 2 and 3 occur. The limit cycle problem seems to be unsolved.

10.3.3 Conclusion

There exist 269 separatrix structures in a quadratic system of the class $m_f=3$ and an $M_{1,1}^0$ saddle node at infinity. The limit cycle problem is still open.

10.4 Phase portraits with infinite critical points at infinity

We may start with system $(3.5.21),(3.5.22)$

$$\dot{x} = a_{00} + a_{10}x + a_{01}y + a_{20}x^2 + a_{11}xy, \qquad (10.4.1)$$
$$\dot{y} = b_{00} + b_{10}x + b_{01}y + a_{20}xy + a_{11}y^2, \qquad (10.4.2)$$

where $c_{45} = a_{20}^2, c_{46} = a_{20}a_{11}, c_{56} = a_{11}^2$ so $A \equiv c_{46}^2 - c_{45}c_{56}=0$. Moreover $c_{45} + c_{56} = a_{20}^2+a_{11}^2 >0$ so $i_f=1$. In order that $(10.4.1),(10.4.2)$ belongs to the class

$m_f=3$, there should be H $\equiv -2[a_{20}(c_{26} - \frac{1}{2}c_{35}) + a_{11}(c_{34} - \frac{1}{2}c_{25})]^2 \neq 0$, so that the central conic is a non-degenerate parabola. If this condition is satisfied the axis of the parabola is in the direction given by the common linear factor in the quadratic terms being $a_{20}x + a_{11}y$. Rotating the axes such that the axis of the parabola is aligned along the x axis amounts to putting $a_{20}=0$, whereas since $a_{11} \neq 0$, scaling yields $a_{11}=1$. As a result (10.4.1),(10.4.2) may be written as

$$\dot{x} = a_{00} + a_{10}x + a_{01}y + xy, \tag{10.4.3}$$
$$\dot{y} = b_{00} + b_{10}x + b_{01}y + y^2, \tag{10.4.4}$$

where now H$=-\frac{1}{2}b_{10}^2 \neq 0$ as $b_{10} \neq 0$ should be satisfied.

It should be noted that there exists a straight line solution given by $x = \alpha+\beta y$, where α, β are given by the solutions of $\alpha = -a_{01}-(a_{10}-b_{01})\beta+b_{10}\beta^2$, and $a_{00}-a_{10}a_{01}+[-b_{00}-a_{10}(a_{10}-b_{01})+a_{01}b_{10}]\beta+[2a_{10}b_{10}-b_{10}a_{01}]\beta^2-b_{10}^2\beta^3=0$. Since $b_{10} \neq 0$ there exists at least one real solution $\beta=\bar{\beta}$ and correspondingly $\alpha = \bar{\alpha}$. Using the transformation $\bar{x} = x - \bar{\alpha} - \bar{\beta}y, \bar{y} = a_{10} - \bar{\beta}b_{10} + y$ then yields from (10.4.3),(10.4.4), changing the notation, the system

$$\dot{x} = xy, \tag{10.4.5}$$
$$\dot{y} = \alpha + \beta x + \gamma y + y^2, \tag{10.4.6}$$

where $\beta \neq 0$, and by a scaling we may put $\beta=1$ and $\gamma \geq 0$. Also x$\equiv$0 is a straight line solution. This equation was also studied by Gasull and Prohens [96,GP1]. It may be seen that at infinity the system may be written as

$$z' = -u, \tag{10.4.7}$$
$$u' = 1 + \alpha z + \gamma u. \tag{10.4.8}$$

This system has only ordinary points for $z=0$ and can be integrated as a linear system in the variables $z = \frac{1}{x}$ and $u = \frac{y}{x}$, yielding also the solutions of (10.4.5),(10.4.6). The possible phase portraits can now easily be determined and are given in Figure 10.5. They are topologically equivalent to those given by Gasull and Prohens [96,GP1]. There are no limit cycles.

System (10.4.5),(10.4.6) belongs to class 2 of Table 2 in Chapter 3. Class $m_3^1\infty$ occurs for $\alpha = \gamma =0$, where m_3^1 is a nilpotent critical point with a hyperbolic and an elliptic sector. The class $e^1m_2^0\infty$ occurs for $\alpha=0,\gamma >0$ and $\alpha = \frac{1}{4}\gamma^2$, $\gamma >0$; m_2^0 is a saddle node and e^1 a node. The class $e^{-1}e^1e^1\infty$ occurs for $\alpha <0,\gamma \geq 0$ and $0< \alpha < \frac{1}{4}\gamma^2,\gamma >0$; the e^1 points are nodes. The class $c_1^0 c_1^0 e^1 \infty$ occurs for $\alpha > \frac{1}{4}\gamma^2, \gamma > 0$; then the e^1 point is an unstable focus. For $\alpha > 0$, $\gamma = 0$, then e^1 is a center point; the phase portrait can also be found in Figure 6.2(d)(portrait 27).

The non-existence of limit cycles can be shown by noting that $x \equiv 0$ can not be crossed by a limit cycle and using the solutions of (10.4.7),(10.4.8).

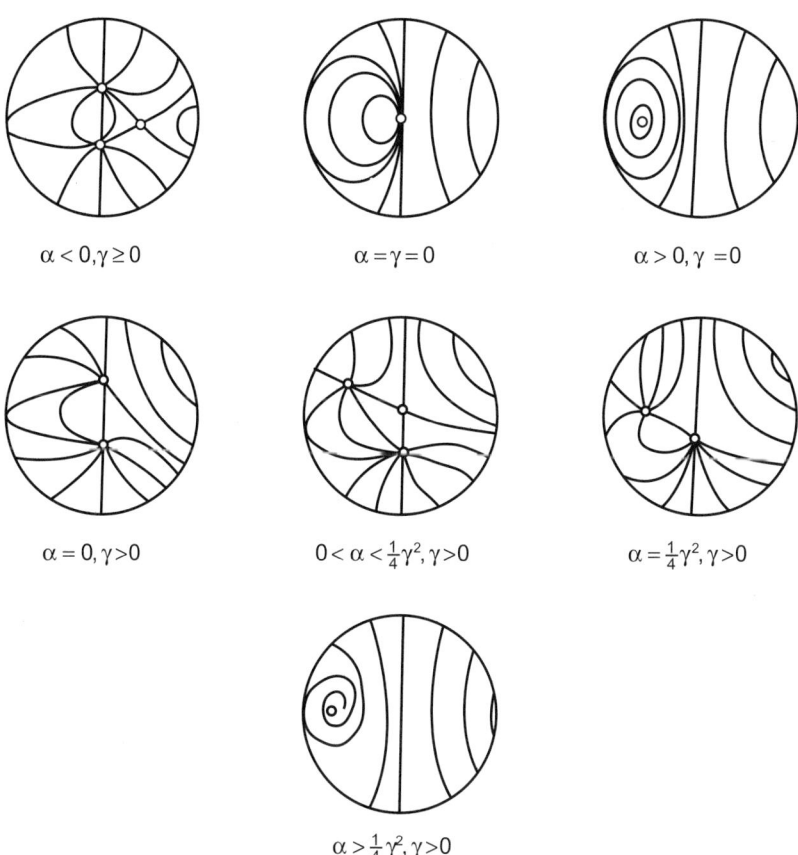

$\alpha < 0, \gamma \geq 0$ $\alpha = \gamma = 0$ $\alpha > 0, \gamma = 0$

$\alpha = 0, \gamma > 0$ $0 < \alpha < \frac{1}{4}\gamma^2, \gamma > 0$ $\alpha = \frac{1}{4}\gamma^2, \gamma > 0$

$\alpha > \frac{1}{4}\gamma^2, \gamma > 0$

Figure 10.5 Phase portraits in $m_f = 3$ with an infinite number of infinite critical points

Chapter 11

Phase portraits of quadratic systems in the class $m_f=4$

11.1 Introduction

What has been said about the drastic increase of the number of possible phase portraits when the finite multiplicity m_f is increased to $m_f=3$ can now equally be repeated when $m_f=4$ is considered. Giving an estimate of this number, moreover, is hindered by the circumstance that at variance with the case of $m_f=3$, a classification of phase portraits using the definition of classes in Chapter 3 is not given yet. Various significant results that can be found in the literature may be used, however, to perform such a classification to a certain degree of completeness.

In this chapter we indicate a beginning of this classification and at least inform the reader where useful information can be found in the literature.

Since for $m_f=4$ no transversally non-hyperbolic infinite critical points occur, the quadratic terms are dominant at infinity and the behavior at infinity is the same as for a non-degenerate homogeneous system. In fact, by adding constant and linear terms, all phase portraits for $m_f=4$ can be thought to be generated from these systems.

In Theorem 3.9 of Chapter 3, the topologically equivalent phase portraits of the non-degenerate homogeneous systems are described and illustrated in Figure 3.3. Further details on the geometric behaviour of the orbits, in particular near critical points, can be found at various places in the literature.

Using the method of Shilov, Ljagina [51,L] studies the orbits near the finite critical point, distinguishing 16 equivalent classes, whereas using the same method and expressing their results in terms of affine invariants, as obtained in an earlier paper [74,VS], Vulpe an Sibirskii [77,VS] gave a geometric classification of the phase portraits on the Poincaré disk. In [78,N] Newton presents an analysis, avoiding the use of affine transformations, in giving a classification and sketches of the orbits near the finite critical point of an arbitrarily given homogeneous quadratic system. Using invariant theory, a conclusive account of the classification of orbits near the finite critical point of a homogeneous quadratic system is presented by Date[79,D].

11.2 Systems with finite index $i_f=-2$

These systems are characterized by the conditions A$\equiv c_{46}^2 - c_{45}c_{56} <0, c_{45} + c_{56} <0$ and involves the classes $e^{-1}e^{-1}e^{-1}e^{1}E^{1}E^{1}E^{1}, e^{-1}e^{-1}c_1^0c_1^0E^{1}E^{1}E^{1}, e^{-1}e^{-1}m_2^0E^{1}E^{1}E^{1}, e^{-1}m_3^{-1}E^{1}E^{1}E^{1}$ and $m_4^{-2}E^{1}E^{1}E^{1}$.

11.2.1 Systems with less than four real finite critical points

Since no antisaddles are present in these classes, no limit cycles exist. The number of possible phase portraits is very limited.

The class $m_4^{-2}E^{1}E^{1}E^{1}$ consists of the translated homogeneous systems discussed above, containing a finite critical point with six hyperbolic sectors.

In the class $e^{-1}m_3^{-1}E^{1}E^{1}E^{1}$ the third order critical point m_3^{-1} may be either a semi-elementary or a nilpotent saddle. Phase portraits in this class were classified by de Jager [89,J],[90J] and later by Gasull and Prohens [96,GP1]. Invariant theory is used by Vulpe and Nikolaev to locate this class in coefficient space [92,VN1]; this work was later extended by Voldman [96,V].

If m_3^{-1} is semi-elementary we may work with (2.3.3),(2.3.4)

$$\dot{x} = a_{20}x^2 + a_{11}xy + a_{02}y^2, \tag{11.2.1}$$
$$\dot{y} = y + b_{20}x^2 + b_{11}xy + b_{02}y^2, \tag{11.2.2}$$

with $a_{20} = 0, a_{11}b_{20} > 0$. Scaling then yields

$$\dot{x} = xy + \lambda_3 y^2, \tag{11.2.3}$$
$$\dot{y} = y + x^2 + \lambda_1 xy + \lambda_2 y^2, \tag{11.2.4}$$

where $\lambda_1, \lambda_2, \lambda_3 \in \mathbb{R}$ and the requirement that the system belongs to the class $e^{-1}m_3^{-1}E^1E^1E^1$ leads to $\lambda_2 - \lambda_1\lambda_3 < -\lambda_3^2$.

For $\lambda_3=0$ the system is symmetric around the y axis; apart from the m_3^{-1} point in (0,0) there exists a saddle in $(0,-\frac{1}{\lambda_2})$; they are connected through a straight line orbit on $x\equiv0$ as in Figure 11.1 (a). For $\lambda_3 \neq 0$ this saddle connection is broken as shown in Figure 11.1 (b).

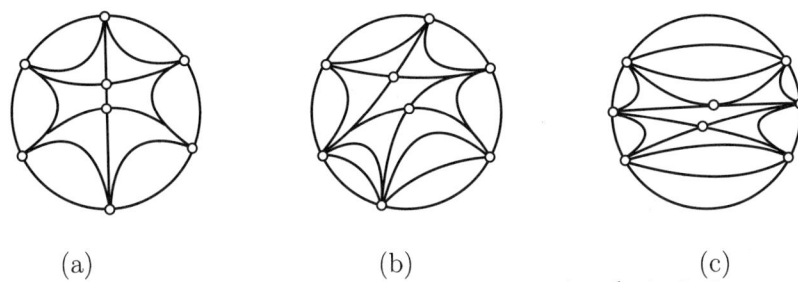

(a)   (b)   (c)
Figure 11.1 Phase portraits in the class $e^{-1}m_3^{-1}E^1E^1E^1$

If m_3^{-1} is nilpotent we may work with

$$\dot{x} = y + a_{20}x^2 + a_{11}xy + a_{02}y^2, \tag{11.2.5}$$
$$\dot{y} = b_{20}x^2 + b_{11}xy + b_{02}y^2, \tag{11.2.6}$$

with $b_{20}=0, a_{20}b_{11} <0$. Scaling then yields

$$\dot{x} = y + \lambda_1x^2 + \lambda_2xy + \lambda_3y^2, \tag{11.2.7}$$
$$\dot{y} = xy + \lambda_4y^2, \lambda_1 < 0. \tag{11.2.8}$$

The phase portraits are illustrated by Figure 11.1 (c); they are topologically equivalent to those shown in Figure 11.1 (b).

The remaining classes in this section are $e^{-1}e^{-1}m_2^0E^1E^1E^1$ and $e^{-1}e^{-1}c_1^0c_1^0E^1E^1E^1$, both containing two saddle points. Rotating the axes and stretching them will establish that there are critical points in $(-1,0)$ and $(1,0)$ and we can work with

$$\dot{x} = a_{01}y + a_{20}(x^2 - 1) + a_{11}xy + a_{02}y^2, \tag{11.2.9}$$
$$\dot{y} = b_{01}y + b_{20}(x^2 - 1) + b_{11}xy + b_{02}y^2. \tag{11.2.10}$$

The degenerate isocline l through $(-1,0)$ and $(1,0)$ is given by $y(c_{34} - c_{45}x - c_{46}y) - 0$, where we may put $c_{46} - 0$, if needed by replacing $c_{45}x$ by

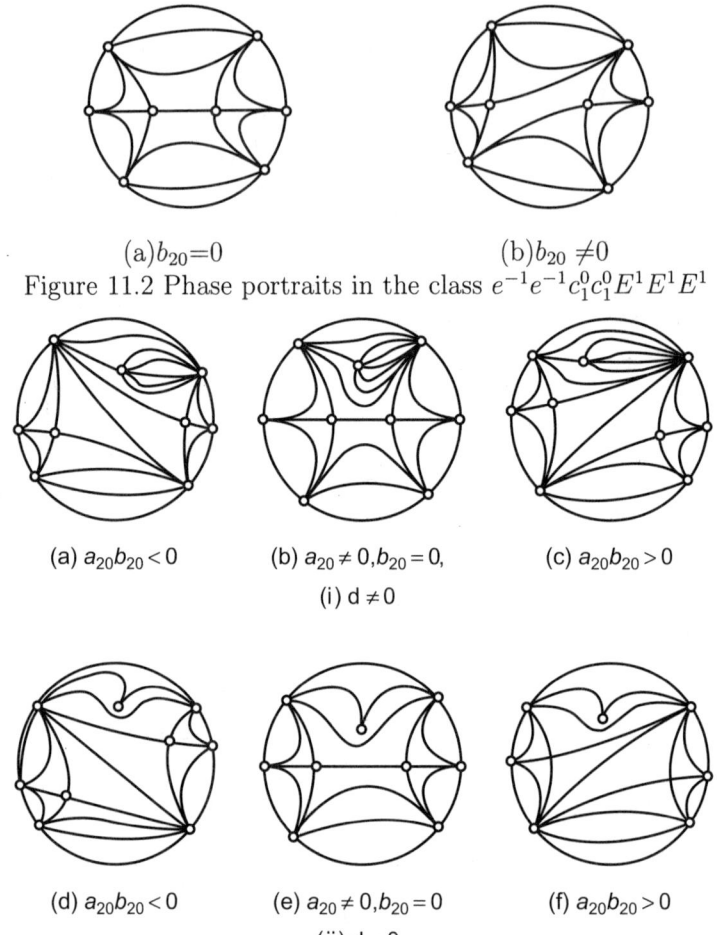

(a)$b_{20}=0$ (b)$b_{20} \neq 0$

Figure 11.2 Phase portraits in the class $e^{-1}e^{-1}c_1^0c_1^0E^1E^1E^1$

(a) $a_{20}b_{20} < 0$ (b) $a_{20} \neq 0, b_{20} = 0,$ (c) $a_{20}b_{20} > 0$

(i) d $\neq 0$

(d) $a_{20}b_{20} < 0$ (e) $a_{20} \neq 0, b_{20} = 0$ (f) $a_{20}b_{20} > 0$

(ii) d $= 0$

Figure 11.3 Phase portraits in the class $e^{-1}e^{-1}m_2^0E^1E^1E^1$

$c_{45}x - c_{46}y$. Since $A \equiv c_{46}^2 - c_{45}c_{56} < 0$, $c_{45} + c_{56} < 0$ there is $c_{45} < 0$, $c_{56} < 0$. Furthermore, $\Omega(x,y) \equiv P_x(x,y)Q_y(x,y) - P_y(x,y)Q_x(x,y) = -2c_{34}x - c_{35}y + 2c_{45}x^2 + 2c_{56}y^2$, and $(-1,0)$ and $(1,0)$ are saddle points if $c_{45} < c_{34} < -c_{45}$. The other critical point(s) are on $c_{34} - c_{45}x = 0$, where $y_{\pm} = \frac{1}{2c_{45}c_{56}}[c_{35}c_{45} \pm \sqrt{\Delta}]$ and $\Delta = 4c_{34}^2c_{45}c_{56} + c_{35}^2c_{45}^2 - 4c_{45}^3c_{56}$. So, for $\Delta < 0$ there are two complex critical points c_1^0, for $\Delta=0$ an m_2^0 point and for $\Delta>0$ a saddle and an antisaddle on l. If $\Delta<0$, so for the class $e^{-1}e^{-1}c_1^0c_1^0E^1E^1E^1$, there exist two phase portraits; they are given in Figure 11.2 (a) and (b). If $\Delta=0$, so for class $e^{-1}e^{-1}m_2^0E^1E^1E^1$ there are six phase portraits, as given in Figure 11.3. The m_2^0 point, located in $(\frac{c_{34}}{c_{45}}, \frac{c_{35}}{2c_{56}})$, is a cusp if $d\equiv 2b_{01}c_{45}c_{56}+2(2a_{20}+b_{11})c_{34}c_{56}+(1+2b_{02})c_{35}c_{45}=0$ and a saddle node if $d\neq0$.

11.2.2 The system $e^{-1}e^{-1}e^{-1}e^{1}E^1E^1E^1$, having four real finite critical points

As follows from the discussion in section 3.4.1 we may work with

$$\dot{x} = a_1x(1-x) + b_1y(1-y) + (\frac{\alpha-1}{\beta}a_1 + \frac{\beta-1}{\alpha}b_1)xy, \quad (11.2.11)$$

$$\dot{y} = b_2x(1-x) + a_2y(1-y) + (\frac{\beta-1}{\alpha}a_2 + \frac{\alpha-1}{\beta}b_2)xy, \quad (11.2.12)$$

with $a_1a_2 - b_1b_2 >0$, $\alpha<0$, $\beta<0$. Then, there is an antisaddle in $(0,0)$ and there are saddle points in $(1,0),(0,1)$ and (α,β).

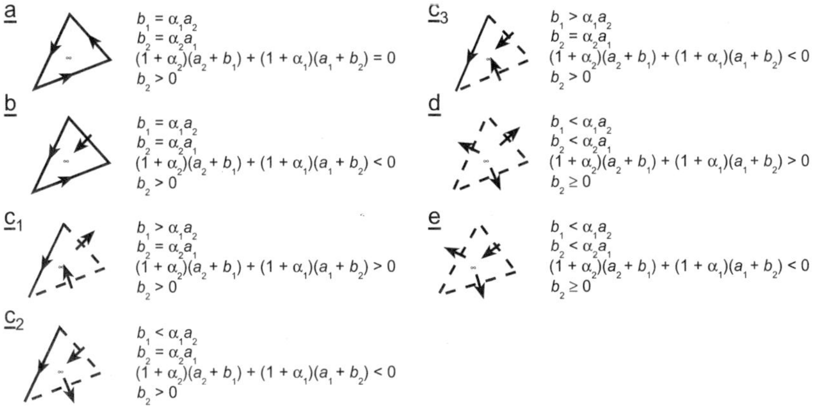

a	$b_1 = \alpha_1 a_2$ $b_2 = \alpha_2 a_1$ $(1+\alpha_2)(a_2+b_1) + (1+\alpha_1)(a_1+b_2) = 0$ $b_2 > 0$
b	$b_1 = \alpha_1 a_2$ $b_2 = \alpha_2 a_1$ $(1+\alpha_2)(a_2+b_1) + (1+\alpha_1)(a_1+b_2) < 0$ $b_2 > 0$
c_1	$b_1 > \alpha_1 a_2$ $b_2 = \alpha_2 a_1$ $(1+\alpha_2)(a_2+b_1) + (1+\alpha_1)(a_1+b_2) > 0$ $b_2 > 0$
c_2	$b_1 < \alpha_1 a_2$ $b_2 = \alpha_2 a_1$ $(1+\alpha_2)(a_2+b_1) + (1+\alpha_1)(a_1+b_2) < 0$ $b_2 > 0$
c_3	$b_1 > \alpha_1 a_2$ $b_2 = \alpha_2 a_1$ $(1+\alpha_2)(a_2+b_1) + (1+\alpha_1)(a_1+b_2) < 0$ $b_2 > 0$
d	$b_1 < \alpha_1 a_2$ $b_2 < \alpha_2 a_1$ $(1+\alpha_2)(a_2+b_1) + (1+\alpha_1)(a_1+b_2) > 0$ $b_2 \geq 0$
e	$b_1 < \alpha_1 a_2$ $b_2 < \alpha_2 a_1$ $(1+\alpha_2)(a_2+b_1) + (1+\alpha_1)(a_1+b_2) < 0$ $b_2 \geq 0$

Figure 11.4 Possible cross flow over the sides of the triangle formed by the saddle points

Possible cross flow over the sides of the triangle formed by the saddle points arc given in Figure 11.4, where $\alpha_1 = \frac{\alpha-1}{\beta}, \alpha_2 = \frac{\beta-1}{\alpha}$

Using (11.2.11),(11.2.12) the phase portraits of class $e^{-1}e^{-1}e^{-1}e^1E^1E^1E^1$ were determined by Zegeling [91,Z]. Proper choices of the parameters allow, on each side of the triangle formed by the saddle points as vertices, flow into, and out of the triangle or along a straight line orbit, coinciding with such a side. A first classification of the phase portraits as allowed by these types of cross flows leads to the separatrix configurations sketched in figure 11.5 where the shaded regions around the antisaddle in (0,0) indicates that the limit cycles in these regions are still to be studied.

Portrait (a) has straight line orbits along all three sides of the triangle; the antisaddle in (0,0) can be shown to be a center point and the triangle only contains closed orbits [62,T]. Portrait (b) has straight line orbits along two sides of the triangle; it corresponds to the Volterra–Volta equation for which it is known that no limit cycles exist [54,B]. The antisaddle is (un)stable if there is (out)inflow across the remaining side. The portrait(c) has a straight line orbit only along one side of the triangle with (out)inflow on the second side and (in)outflow on the third side or (in)outflow on both sides. A saddle loop occurs in (c_2^a) and a separatrix cycle with two saddles in (c_1^a). At most one (hyperbolic)limit cycle (or separatrix cycle) is possible as a result of the presence of the straight line orbit. Phase portrait(d) corresponds to (in)outflow across all sides of the triangle; for $b_{20}=0$ the antisaddle in (0,0) is a node and there exist no limit cycles. For $b_{20} \neq 0$ there also exist no limit cycles as can be shown using the theory of rotated vector fields [91,Z,Chapter1]. The phase portraits (e) correspond to (in)outflow across one side and (out)inflow across the remaining sides of the triangle. A saddle loop, in (1,0) attached to a saddle, is possible in (e^a) and a curved saddle connection between (α, β) and (1,0) in (e^e). The limit cycle problem for case (e) is the most difficult part in classifying class $e^{-1}e^{-1}e^{-1}E^1E^1E^1$. Obviously, limit cycles can only occur in the interior of the triangle since sides of the triangle, being isoclines, cannot be crossed by periodic solutions around (0,0).

A careful and extensive analysis of the possible bifurcation surfaces in parameter space was carried out by Zegeling [91,Z].

For the cases (a),(b) and (c), having at least one straight line orbit as a side of the triangle spanned by the saddle points, bifurcation diagrams are given. They yield the phase portraits for all values of the parameters $a_1, a_2, b_1, b_2, \alpha$ and β for these cases with the exception that the precise determination of the location in parameter space of the saddle loop in (c_2^a) and separatrix cycle in (c_1) should be further accomplished by numerical calculations.

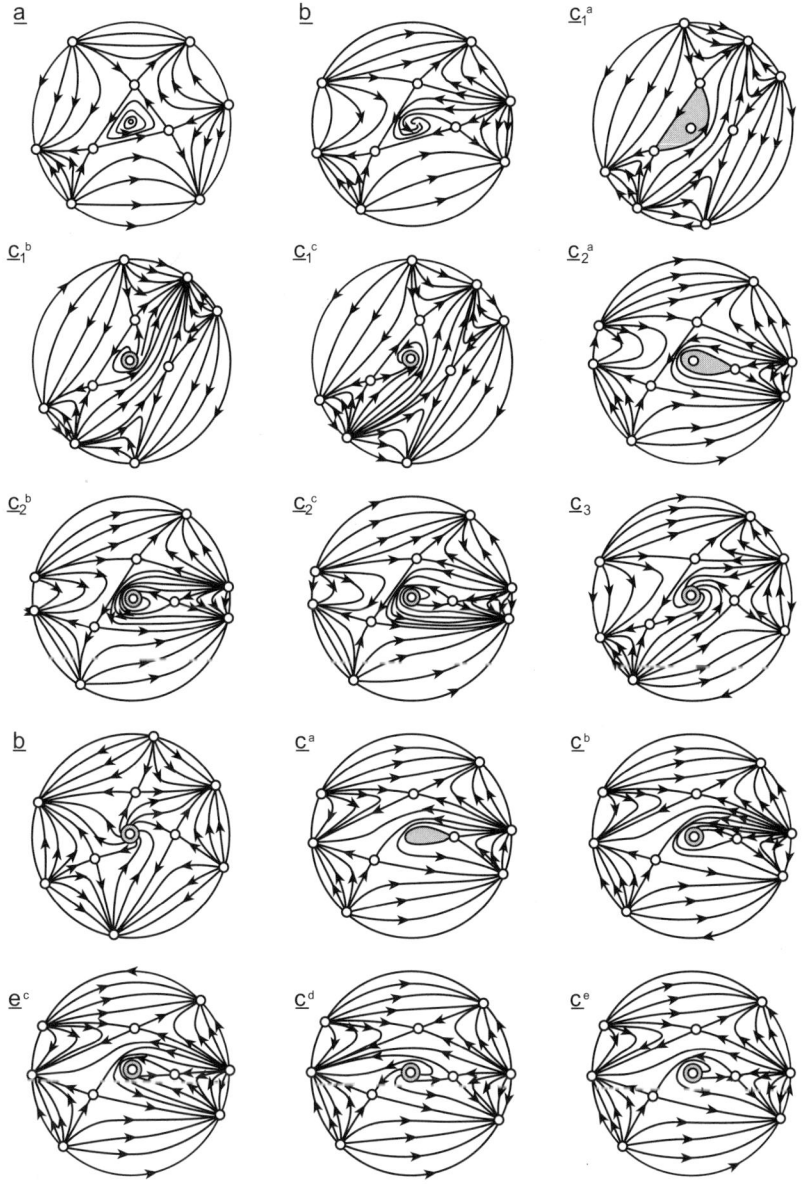

Figure 11.5 Possible separatrix structures in the class $e^{-1}e^{-1}e^{-1}e^1 E^1 E^1 E^1$

In case (d) no bifurcation occurs; (0,0) may be either a node or a focus.

In case (e) the exact location in parameter space of the first, second, third foci and center points in (0,0), the fine saddle in (1,0) and the surface representing the degenerate case of an infinite number of finite critical points, if $a_1 a_2 - b_1 b_2 = 0$, are given. Corresponding to the curved saddle connection in (e^e) between (α, β) and (1,0) and the saddle loop from (1,0) in (e^a), including those with a fine saddle, it is shown, using carefully selected families of rotated vector fields, that there exists a unique bifurcation surface in parameter space for these cases, such that for the cases $(e^a), (e^b), (e^c), (e^d)$ and (e^e) the corresponding regions in parameter space are indicated. A precise determination of these surfaces awaits numerical calculations. Finally, bifurcations of simple, semistable limit cycles and those of odd order are considered and possible bifurcation surfaces are constructed based on the assumption that around (0,0) there exist at most three (hyperbolic) limit cycles. It would be an interesting problem to support this assumption with numerical evidence.

In fact little work has been done so far to explore the class $e^{-1} e^{-1} e^{-1} e^1$ $E^1 E^1 E^1$ numerically. For specific values of the parameters the possibility of two or three limit cycles around (0,0) were demonstrated numerically in 1959 by Qin Yunan and Pu Fuquan [59,QP],[67,QP]; an example was further elaborated by Mieussens in 1978 [78,M] and Perko in 1984.

The limit cycle problem for the class $e^{-1} e^{-1} e^{-1} e^1 E^1 E^1 E^1$ can be studied using the classification of Ye Yanqian. Then it corresponds to class III with $A \equiv n(a^2 n + b^2 l - abm) < 0$ and $c_{45} + c_{56} = -am + bl - bn < 0$. However, for class III also more results need to be known. Phase portraits when (0,0) is a third order weak focus were given for a special case by Chen Ming Xiang [86,Ch2] and for all cases by Artes and Llibre [97,AL].

11.3   Systems with finite index $i_f = 0$

These systems are characterized by the condition $A \equiv c_{46}^2 - c_{45} c_{56} > 0$. Involved are the finite critical point combinations $e^{-1} e^1 e^1 e^1, e^{-1} c_1^0 c_1^0 e^1, c_1^0 c_1^0 c_1^0 c_1^0,$ $e^{-1} e^1 m_2^0, c_1^0 c_1^0 m_2^0, e^{-1} m_3^1, e^1 m_3^{-1}, m_2^0 m_2^0, c_2^0 c_2^0$ and the critical points indicated as m_4^0. For the infinite critical points there exist the combinations $E^{-1} E^1 E^1,$ $C_{0,1}^0 C_{0,1}^0 E^1, E^1 M_{0,2}^0$ and $M_{0,3}^1$.

11.3.1   Systems with a fourth order finite critical point

For systems with a critical point m_4^0 of the type present in a translated homogeneous system , the phase portraits are topologically characterized in Figure 3.3; they represent systems within the classes $m_4^0 E^{-1} E^1 E^1, m_4^0 C_{0,1}^0 C_{0,1}^0 E^1, m_4^0 E^1 M_{0,3}^1$.

An m_4^0 point can also be a semi-elementary or a nilpotent saddle node. The phase portraits for systems containing such points were classified by de Jager [89,J],[90,J] and, using invariant theory, by Vulpe and Nikolaev [93,VN].

If m_4^0 is a semi-elementary critical point we may work with

$$\dot{x} = a_{20}x^2 + a_{11}xy + a_{02}y^2, \tag{11.3.1}$$
$$\dot{y} = y + b_{20}x^2 + b_{11}xy + b_{02}y^2, \tag{11.3.2}$$

with $a_{20} = a_{11} = 0, a_{02}b_{20} \neq 0$ and scaling yields $a_{02} = b_{20} = 1$, so

$$\dot{x} = y^2, \tag{11.3.3}$$
$$\dot{y} = y + x^2 + \lambda_1 xy + \lambda_2 y^2, \tag{11.3.4}$$

with $\lambda_1, \lambda_2 \in R$.

Using a linear transformation such that one infinite critical point is located at the ends of the y axis, de Jager concludes at six possible phase portraits for the transformed system [89,J]. An alternative approach would be to retain (11.3.3),(11.3.4) bringing more symmetry in the parameter plane by letting $\lambda_1=3(a+b), \lambda_2=3(a-b)$; then this system becomes

$$\dot{x} = y^2, \tag{11.3.5}$$
$$\dot{y} = y + x^2 + 3(a+b)xy + 3(a-b)y^2, \tag{11.3.6}$$

of which the locations of the infinite critical points are determined by

$$f(u) \equiv u^3 - 3(a-b)u^2 - 3(a+b)u - 1 = 0. \tag{11.3.7}$$

For an $M_{0,q}^i$ point with $q \geq 2$, necessary conditions are $f(u)=f'(u)=0$ yielding $D=0$, where

$$D \equiv 1 + 6(a^2 - b^2) - 3(a^2 - b^2)^2 + 4(a-b)^3 - 4(a+b)^3. \tag{11.3.8}$$

The additional condition $f''(u)=0$ yields that an $M_{0,3}^1$ point exists for $a=0, b=-1$, located in $u=1$. The corresponding phase portrait is given in Figure 11.6.

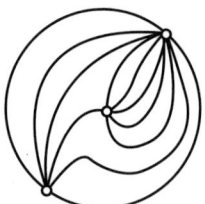

Figure 11.6 Phase portrait in class $m_4^0 M_{0,3}^1; m_4^0$ is an elementary saddle node

It is topologically equivalent to that for the class $m_4^0 C_{0,1}^0 C_{0,1}^0 E^1$ having a semi-elementary saddle node, since E^1 can be shown to be located in $u_1 > 0$, as $u_1 u_2 u_3 = u_1(\alpha^2 + \beta^2){=}1$, where $C_{0,1}^0$ is located in $u_{2,3} = \alpha \pm i\beta$. If m_4^0 is a semi-elementary saddle node the class $m_4^0 C_{0,1}^0 C_{0,1}^0 E^1$ is characterized by D$>$0, class $m_4^0 E^1 M_{0,2}^0$ by D$=$0, (a,b)$\neq$(0,$-$1), and the class $m_4^0 E^{-1} E^1 E^1$ by D$<$0.

A further analysis of (11.3.5),(11.3.6) is needed to be able to compare with the results in [89,J].

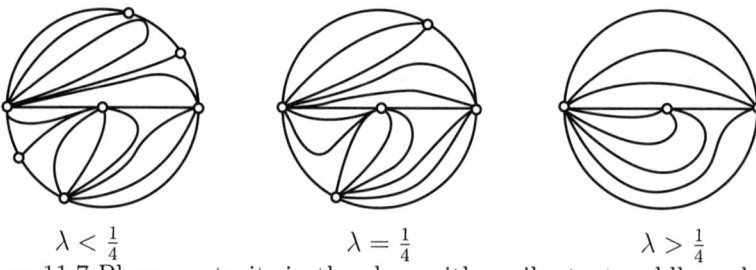

$$\lambda < \tfrac{1}{4} \qquad\qquad \lambda = \tfrac{1}{4} \qquad\qquad \lambda > \tfrac{1}{4}$$

Figure 11.7 Phase portraits in the class with a nilpotent saddle node m_4^0

If m_4^0 is a nilpotent saddle node we may work with

$$\dot{x} = y + a_{20}x^2 + a_{11}xy + a_{02}y^2, \qquad (11.3.9)$$
$$\dot{y} = b_{20}x^2 + b_{11}xy + b_{02}y^2, \qquad (11.3.10)$$

with $b_{20} = b_{11} = 0, a_{20}b_{02} \neq 0$, and scaling yields $a_{20} = b_{02}{=}1$, so

$$\dot{x} = y + x^2 + \lambda_1 xy + \lambda_2 y^2, \qquad (11.3.11)$$
$$\dot{y} = y^2, \qquad (11.3.12)$$

with $\lambda_1, \lambda_2 \in R$ or with $x \to x - \tfrac{1}{2}\lambda_1 y, y \to y$

$$\dot{x} = y + x^2 + \lambda y^2, \qquad (11.3.13)$$
$$\dot{y} = y^2, \qquad (11.3.14)$$

with $\lambda \in R$.

The phase portraits can now easily be determined and are given in Figure 11.7. They correspond to the classes $m_4^0 E^{-1} E^1 E^1, m_4^0 E^1 M_{0,2}^0$ and $m_4^0 C_{0,1}^0 C_{0,1}^0 E^1$.

Obviously, systems with an m_4^0 point do not contain limit cycles; this is different for systems with an m_3^i point in $m_f=4$, $i_f=0$, to be studied next.

11.3.2 Systems with a third order finite critical point

A classification of the phase portraits of quadratic systems with a third order point m_3^i was given by de Jager [89,J,90,J] and later by Gasull and Prohens [96,GP] who used the theory of rotated vector fields. For these systems Coppel gave a proof that there exists at most one limit cycle, which if it exists is hyperbolic. An m_3^i point is either semi-elementary or nilpotent.

If m_3^i is *semi-elementary* we may work with

$$\dot{x} = a_{20}x^2 + a_{11}xy + a_{02}y^2, \tag{11.3.15}$$
$$\dot{y} = y + b_{20}x^2 + b_{11}xy + b_{02}y^2, \tag{11.3.16}$$

with $a_{20}=0$; $a_{11}b_{20}>0$ if m_3^i is a third order saddle m_3^{-1} and $a_{11}b_{20}<0$ if a third order node m_3^1.

If m_3^i is a *saddle*, scaling yields $a_{11} = b_{20}=1$, whereas $A \equiv c_{46}^2 \quad c_{45}c_{56}= b_{02} - a_{02}b_{11} + a_{00}^2 >0$, so that $m_f=4,i_f=0$.

(i)If $a_{02}=0$, (11.3.15),(11.3.16) reads

$$\dot{x} = xy, \tag{11.3.17}$$
$$\dot{y} = y + x^2 + \lambda_1 xy + \lambda_2 y^2, \tag{11.3.18}$$

where $\lambda_1 \geq 0$, if needed by letting $x \to -x, \lambda_1 \to -\lambda_1$, and $\lambda_2 >0$ as $A>0$. Apart from the saddle in $(0,0)$ there is an elementary node in $(0,-\frac{1}{\lambda_2})$, as a result of which the system has no limit cycles.

The infinite critical points are given by

$$(1 - \lambda_2)u^2 - \lambda_1 u - 1 = 0. \tag{11.3.19}$$

For $\lambda_1 \geq 0$, $\Delta \equiv \lambda_1^2 - 4(\lambda_2 - 1) <0$ there exists one phase portrait in class $e^1 m_3^{-1} C_{0,1}^0 C_{0,1}^0 E^1$, topologically equivalent to that for $\lambda_1 = 0, \lambda_2=1$ in class $e^1 m_3^{-1} M_{0,3}^1$, where $M_{0,3}^1$ is an infinite node. For $\lambda_1 >0$, $\Delta=0$ as well as for $\lambda_2 \equiv 1$ there exists one phase portrait in class $e^1 m_3^{-1} E^1 M_{0,2}^0$. For $\lambda_1 \geq 0$, $\lambda_2 \neq 2, \Delta >0$ there exist two phase portraits in class $e^1 m_3^{-1} E^{-1} E^1 E^1$[89,J].

(ii)If $a_{02} \neq 0$, system (11.3.15),(11.3.16) is studied in [89,J] by transforming it first to a system such that one infinite critical point is at the ends of the x axis. An alternative approach would be to let $a_{02}=c$, $b_{11}=3(a+b)c^{\frac{1}{3}}$, $b_{02}=1+3(a-b)c^{\frac{2}{3}}$, then (11.3.15),(11.3.16) can be written as

$$\dot{x} = xy + cy^2, \qquad\qquad\qquad\qquad (11.3.20)$$
$$\dot{y} = y + x^2 + 3(a+b)c^{\frac{1}{3}}xy + [1 + 3(a-b)c^{\frac{2}{3}}]y^2, \qquad (11.3.21)$$

with $a,b \in \mathbb{R}$, $c \neq 0$ and $A=1+3(a-b)c^{\frac{2}{3}}-3(a+b)c^{\frac{4}{3}} + c^2 > 0$.

The location of the infinite critical point is then given by

$$f(u) = cu^3 - 3(a-b)c^{\frac{2}{3}}(a-b)u^2 - 3(a+b)c^{\frac{1}{3}}u - 1 = 0, \qquad (11.3.22)$$

so that $c^{\frac{1}{3}}u$ can be taken as variable.

For the existence of an $M^i_{0,q}$ point with $q \geq 2$, necessary conditions are $f(u)=f'(u)=0$ yielding $D=0$ where

$$D = 1 + 6(a^2 - b^2) - 3(a^2 - b^2)^2 + 4(a-b)^3 - 4(a+b)^3, \qquad (11.3.23)$$

coincides with (11.3.8) so that the analysis in section 11.3.1 can be used.

The additional condition $f''(u)=0$ yields that an $M^1_{1,3}$ point exists for $a=0$, $b=-1$ which is a node located in $c^{\frac{1}{3}}u=1$ and it may be shown that $A>0$. There appears to be four phase portraits in the class $e^1 m_3^{-1} M^1_3$; a simple limit cycle occurs as well as a Hopf bifurcation and a saddle loop (homoclinic orbit) bifurcation. They are topologically equivalent to the four phase portraits in the class $e^1 m_3^{-1} C^0_{0,1} C^0_{0,1} E^1$, where E^1 can be shown to be located in $c^{\frac{1}{3}}u>0$ as $cu_1 u_2 u_3 = c^{\frac{1}{3}}u_1(\alpha^2 + \beta^2)=1$ and $C^0_{0,1}$ in $c^{\frac{1}{3}}u_{2,3} = \alpha \pm i\beta$. Class $e^1 m_3^{-1} C^0_{0,1} C^0_{0,1} E^1$ is characterized by $D>0$, class $e^1 m_3^{-1} E^1 M^0_{0,2}$ by $D=0$, $(a,b) \neq (0,-1)$, class $e^1 m_3^{-1} E^{-1} E^1 E^1$ by $D<0$ and class $e^1 m_3^{-1} M^1_{0,3}$ by $D=0$, $(a,b)=(0,-1)$. In [89,J] nine phase portraits are given in class $e^1 m_3^{-1} E^1 M^0_{0,2}$ and five phase portraits in class $e^1 m_3^{-1} E^{-1} E^1 E^1$. In both cases a hyperbolic limit cycle occurs as well as a Hopf bifurcation and a saddle loop (homoclinic) bifurcation. Further analysis of (11.3.20),(11.3.21) is needed to compare with the results of de Jager in [89,J].

If m^i_3 is a *semi-elementary node* m^1_3, then $a_{20}=0$ and $a_{11}b_{20}<0$ in (11.3.15), (11.3.16) and scaling yields $a_{11} = -1, b_{20}=1$, whereas $A \equiv c^2_{46} - c_{45}c_{56} = b_{20} + b_{11}a_{02} + a^2_{02} > 0$, so that $m_f=4, i_f=0$.

(i) If $a_{20}=0$, system (11.3.15),(11.3.16) can be written as

$$\dot{x} = -xy, \tag{11.3.24}$$
$$\dot{y} = y + x^2 + \lambda_1 xy + \lambda_2 y^2, \tag{11.3.25}$$

where $\lambda_1 \geq 0$, if needed by letting $x \to -x$, $\lambda_1 \to -\lambda_1$, and where $\lambda_2 > 0, A > 0$. Apart from the node in $(0,0)$ there is a saddle in $(0,-\frac{1}{\lambda_2})$ as a result of which the system has no limit cycles.

The infinite critical points are given by

$$(1 + \lambda_2)u^2 + \lambda_1 u + 1 = 0, \lambda_2 > 0. \tag{11.3.26}$$

For $\lambda_1 \geq 0$, $\Delta \equiv \lambda_1^2 - 4(1 + \lambda_2) < 0$ there exists one phase portrait in class $e^{-1}m_3^1 C_{0,1}^0 C_{0,1}^0 E^1$. For $\lambda_1 \geq 2$, $\Delta = 0$ there exists one phase portrait in class $e^{-1}m_3^1 E^1 M_{0,2}^0$ and for $\lambda_1 > 2$, $\Delta > 0$ one phase portrait in the class $e^{-1}m_3^1 E^{-1} E^1 E^1$ [89,J].

(ii) If $a_{02} \neq 0$, system (11.3.15),(11.3.16) is studied in [89,J] using a linear transformation to locate one critical point at the ends of the x axis. As before, an alternative approach would be to let $a_{02}=c, b_{11}=3c^{\frac{1}{3}}(a+b)$, $b_{02} = -1+3c^{\frac{2}{3}}(a-b)$; then (11.3.15),(11.3.16) can be written as

$$\dot{x} = -xy + cy^2, \tag{11.3.27}$$
$$\dot{y} = y + x^2 + 3c^{\frac{1}{3}}(a + b)xy + [-1 + 3c^{\frac{1}{3}}(a - b)]y^2, \tag{11.3.28}$$

with a,b $\in \mathbb{R}$, $c \neq 0$ and $A=-1+3c^{\frac{2}{3}}(a-b)+3c^{\frac{4}{3}}(a+b)+c^2 > 0$. Apart from the m_3^1 node in $(0,0)$ there is a saddle e^{-1}; there exist no limit cycles.

The location of the infinite critical points is again given by

$$f(u) \equiv cu^3 - 3c^{\frac{2}{3}}(a - b)u^2 - 3c^{\frac{1}{3}}(a + b)u - 1 = 0, \tag{11.3.29}$$

and $c^{\frac{1}{3}}u$ can be taken as variable.

The analysis then can follow the line given for an m_3^{-1} point. Class $e^{-1}m_3^1 C_{0,1}^0 C_{0,1}^0 E^1$ is characterized by D>0, where D is given by (11.3.23), class $e^{-1}m_3^1 E^1 M_{0,2}^0$ by D=0, (a,b)$\neq$(0,1), class $e^{-1}m_3^1 E^{-1} E^1 E^1$ by D<0 and class $e^{-1}m_3^1 M_{0,3}^1$ by D=0, (a,b)=(0,−1). For these classes A>0 should be satisfied as well. In [89,J] one phase portrait is given for classes $e^{-1}m_3^1 C_{0,1}^0 C_{0,1}^0 E^1$, three for class $e^{-1}m_3^1 E^1 M_{0,2}^0$, two for class $e^{-1}m_3^1 E^{-1} E^1 E^1$ and one for class $e^{-1}m_3^1 M_{0,3}^1$.

If m_3^i is *nilpotent* we may work with the system

$$\dot{x} = y + a_{20}x^2 + a_{11}xy + a_{02}y^2, \qquad (11.3.30)$$
$$\dot{y} = b_{20}x^2 + b_{11}xy + b_{02}y^2, \qquad (11.3.31)$$

with $b_{20}=0$; the point is an m_3^{-1} saddle if $a_{20}b_{11} <0$ and a m_3^1 point with an elliptic and a hyperbolic sector if $a_{20}b_{11} >0$.

If m_3^i is a saddle m_3^{-1}, scaling and use of a linear transformation yields from this system

$$\dot{x} = y + \lambda_1 x^2 + \lambda_2 xy + \lambda_3 y^2, \qquad (11.3.32)$$
$$\dot{y} = xy, \qquad (11.3.33)$$

with $\lambda_1 <0$, $\lambda_2 \in \{0,1\}$ and $\lambda_3 <0$ since A$=\lambda_1\lambda_2 >0$.

Apart from the saddle in (0,0) there is an antisaddle in $(0,-\frac{1}{\lambda_3})$, which is a center point for $\lambda_2=0$ and a focus or node for $\lambda_2 \neq 0$. Using a Dulac function it can be shown that the system contains no limit cycles [89,J]. The critical points at infinity are given by

$$u(\lambda_3 u^2 + \lambda_2 u + \lambda_1 - 1) = 0. \qquad (11.3.34)$$

Class $e^1 m_3^{-1}C_{0,1}^0 C_{0,1}^0 E^1$ is characterized by $\Delta \equiv \lambda_2^2 + 4\lambda_3(1 - \lambda_1) <0$, class $e^1 m_3^{-1}E^1 M_{0,2}^0$ by $\Delta=0$ and class $e^1 m_3^{-1}E^{-1}E^1 E^1$ by $\Delta >0$. An $M_{0,3}^1$ point is not possible. In class $e^1 m_3^{-1}C_{0,1}^0 C_{0,1}^0 E^1$ two phase portraits are found, in class $e^1 m_3^{-1}E^1 M_{0,2}^0$ one and in class $e^1 m_3^{-1}E^{-1}E^1 E^1$ one [89,J],[96,GP2].

If m_3^i is a *point with an elliptic and a hyperbolic sector*, scaling and use of a linear transformation yields (11.3.32),(11.3.33) again with $\lambda_1 >0$, $\lambda_2 \in \{0,1\}$ and $\lambda_3 >0$ since A$=\lambda_1\lambda_3 >0$. Apart from the m_3^i point there is a saddle in $(0,-\frac{1}{\lambda_3})$, as a result of which the system contains no limit cycles.

As for m_3^{-1} the locations of the infinite critical points are given by (11.3.34). Class $e^{-1} m_3^1 C_{0,1}^0 C_{0,1}^0 E^1$ is characterized by $\Delta <0$, class $e^{-1} m_3^1 E^1 M_{0,2}^0$ by $\Delta=0$, $(\lambda_1,\lambda_2) \neq(1,0)$, class $e^{-1} m_3^1 E^{-1}E^1 E^1$ by $\Delta >0$ and class $e^{-1} m_3^1 M_{0,3}^1$ by $\Delta=0,(\lambda_1,\lambda_2)=(1,0)$.

In class $e^{-1} m_3^1 C_{0,1}^0 C_{0,1}^0 E^1$ two phase portraits were found, in class $e^{-1} m_3^1 E^1 M_{0,2}^0$ two, in class $e^{-1} m_3^1 E^{-1}E^1 E^1$ five and in class $e^{-1} m_3^1 M_{0,3}^1$ one, see [89,J] and [96,GP2].

Invariant theory was used by Vulpe and Nikolaev [92,VN1] to locate the class $e^i m_3^i$ in coefficient space. This work was later extended by Voldman [96,V].

11.3.3 Systems with a finite critical point of the second order m_2^0 or c_2^0

Involved are the finite critical point combinations $e^{-1}e^1m_2^0$, $c_1^0c_1^0m_2^0$, $m_2^0m_2^0$ and $c_2^0c_2$. An m_2^0 may be either a semi-elementary saddle node or a cusp.

The phase portraits containing a cusp were classified by de Jager [89,J] and[90,J], and to some extent by Gasull and Prohens [96,GP]. A systematic classification of all phase portraits containing a semi-elementary saddle node seems not yet to have been made so far, although phase portraits containing such a point occasionally appear in some classifications. For instance, the phase portraits in the class $c_1^0c_1^0m_2^0$ (both for $m_f=2$ and $m_f=4$) were determined by Vulpe and Nikolaev, using invariant theory [91,VN].

A system having a *semi-elementary* saddle node m_2^0 in $(0,0)$ may be written as

$$\dot{x} = a_{20}x^2 + a_{11}xy + a_{02}y^2, \qquad (11.3.35)$$
$$\dot{y} = y + b_{20}x^2 + b_{11}xy + b_{02}^2, \qquad (11.3.36)$$

where $a_{20} \neq 0$ and A>0 in order that $m_f=4$, $i_f=0$. Scaling to reduce the number of parameters leads to $a_{20}=1$ and if $b_{20} \neq 0$ to $b_{20}=1$. Then

$$\dot{x} = x^2 + \alpha xy + \beta y^2, \qquad (11.3.37)$$
$$\dot{y} = y + x^2 + \gamma xy + \delta y^2, \qquad (11.3.38)$$

with $\alpha, \beta, \gamma, \delta \in \mathbb{R}$ and A=$(\beta-\delta)^2 + (\alpha-\gamma)(\alpha\delta-\beta\gamma) > 0$. This is the largest class for which a classification of phase portraits with a saddle node m_2^0 is needed.

If $b_{20}=0$, $a_{02} \neq 0$ we may scale further to yield $a_{02}=1$ and the system then becomes

$$\dot{x} = x^2 + \alpha xy + y^2, \qquad (11.3.39)$$
$$\dot{y} = y + \gamma xy + \delta y^2, \qquad (11.3.40)$$

with $\alpha, \gamma, \delta \in \mathbb{R}$ and A=$\gamma^2 - \alpha\gamma\delta + \delta^2 > 0$.

For this system the x axis consists of straight line orbits. Possible limit cycles in this class are thus unique and hyperbolic. With $\Delta \equiv \alpha^2 - 4\beta$ there exists the combinations $e^{-1}e^1m_2^0$ for $\Delta > 0$, $m_2^0m_2^0$ for $\Delta=0$ and $c_1^0c_1^0m_2^0$ for $\Delta < 0$. An extensive investigation of the phase portraits of quadratic systems with a straight line orbit, including the class $e^{-1}e^1m_2^0$ was given by Rozet in his thesis of which, however, only a limited number of copies is available. Instead of considering the cases $b_{20} \neq 0$ and $b_{20}=0$, alternatively, may be considered the cases $a_{11} \neq 0$ and $a_{11}=0$ or $b_{02} \neq 0$ and $b_{02}=0$, diminishing the number of parameters without obtaining a straight line orbit.

Instead of doing so, consider the cases $b_{20} = a_{02}=0$, thus $b_{02} \neq 0$ since A>0 and scaling yields $b_{20}=1$ and the Volterra−Lotka system

$$\dot{x} = x(x+\alpha y), \qquad (11.3.41)$$
$$\dot{y} = y(1+\gamma x+y), \qquad (11.3.42)$$

with $\alpha, \gamma \in \mathbb{R}$ and $A=1-\alpha\gamma >0$, is obtained.

All phase portraits of the Volterra−Lotka system are known and can be found in [87,R]. In particular it is known that no limit cycles can exist; yet periodic solutions around center points may occur.

A system having a nilpotent *cusp* m_2^0 in (0,0) may be written as

$$\dot{x} = y + a_{20}x^2 + a_{11}xy + a_{02}y^2, \qquad (11.3.43)$$
$$\dot{y} = b_{20}x^2 + b_{11}xy + b_{02}y^2, \qquad (11.3.44)$$

where $b_{20} \neq 0$ and A>0 in order to be in class $m_f=4$, $i_f=0$.

For $\Delta \equiv b_{11}^2 - 4b_{20}b_{02} <0$ there exists the combinations $c_1^0 c_1^0 m_2^0$, for $\Delta=0$ the combination $m_2^0 m_2^0$ and for $\Delta >0$ the combination $e^{-1}e^1 m_2^0$.

The phase portraits containing the combination $c_1^0 c_1^0 m_2^0$ are classified by de Jager [89,J],[90,J]. Since there exists at least one real infinite critical point, say $u = \frac{1}{\alpha}$, letting $x \to x + \alpha y, y \to y$ yields $a_{02}=0$, whereas by scaling there may be put $b_{20}=1$, and, as $\Delta <0, b_{02} >0$, so $b_{02}=1$.

Then (11.3.43),(11.3.44) becomes

$$\dot{x} = y + \lambda_1 x + \lambda_2 xy, \qquad (11.3.45)$$
$$\dot{y} = x^2 + \lambda_3 xy + y^2, \qquad (11.3.46)$$

where $\Delta = \lambda_3^2 - 4 <0$, $A=\lambda_1^2 - \lambda_1\lambda_2\lambda_3 + \lambda_2^2 >0$.

Apart from an infinite critical point "at the ends" of the y axis, there are infinite critical points given by

$$(\lambda_2 - 1)u^2 + (\lambda_1 - \lambda_3)u - 1 = 0, \qquad (11.3.47)$$

thus class $c_1^0 c_1^0 m_2^0 C_{0,1}^0 C_{0,1}^0 E^1$ prevails for $D\equiv (\lambda_1 - \lambda_3)^2 + 4(\lambda_2 - 1) <0$, class $c_1^0 c_1^0 m_2^0 E^1 M_{0,2}^0$ for $D=0,(\lambda_1 - \lambda_3, \lambda_2) \neq (0,1)$, class $c_1^0 c_1^0 m_2^0 E^{-1} E^1 E^1$ for D>0 and class $c_1^0 c_1^0 m_2^0 M_{0,3}^0$ for $(\lambda_1 - \lambda_3), \lambda_2)=(0,1)$ thus D=0. In [89,J],[90,J] one phase portrait is given for $c_1^0 c_1^0 m_2^0 C_{0,1}^0 C_{0,1}^0 E^1$, three for $c_1^0 c_1^0 m_2^0 E^1 M_{0,2}^0$, two for $c_1^0 c_1^0 m_2^0 E^{-1} E^1 E^1$ and one for $c_1^0 c_1^0 m_2^0 M_{0,3}^1$. Obviously there are no limit cycles.

The phase portraits containing the combination $m_2^0 m_2^0$ with at least one cusp were investigated by de Jager [89,J],[90,J] and using the theory of rotated vector fields by Gasull and Prohens [96,GP2]. Obviously there are no limit cycles either in these phase portraits. Invariant theory is used by Vulpe and Nikolaev [92,VN1] to locate class $m_2^0 m_{0,2}^0$ in coefficient space; this was later extended by Voldman [96,V].

Scaling yields $b_{20}=1$ and by letting $x \to x - \frac{1}{2}b_{11}y, y \to y$, there is $b_{11} = b_{02}=0$, whereas since A=$a_{02}^2 > 0$, scaling yields $a_{02}=1$ so that (11.3.43), (11.3.44) may be written as

$$\dot{x} = y + \lambda_1 x^2 + \lambda_2 xy + y^2, \qquad (11.3.48)$$
$$\dot{y} = x^2, \qquad (11.3.49)$$

where $\lambda_1, \lambda_2 \in \mathbb{R}$ and A>1.

Apart from the cusp in $(0,0)$ there is an m_2^0 point in $(0,-1)$, being a saddle node if $\lambda_2 \neq 0$ and a cusp if $\lambda_2=0$.

The infinite critical points are given by

$$f(u) \equiv u^3 + \lambda_2 u^2 + \lambda_1 u - 1 = 0, \qquad (11.3.50)$$

and letting $\lambda_1 = -3(a + b), \lambda_2 = -3(a - b)$ symmetry in the parameter plane may again be obtained. From (11.3.23) then again follows the expression for D=D(a,b). So class $m_2^0 m_2^0 C_{0,1}^0 C_{0,1}^0 E^1$ is characterized by D>0, class $m_2^0 m_2^0 E^1 M_{0,2}^0$ by D=0, (a,b)$\neq$(0,−1), class $m_2^0 m_2^0 E^{-1} E^1 E^1$ by D<0 and class $m_2^0 m_2^0 M_{0,3}^1$ by D=0, (a,b)=(0,−1). For these classes de Jager arrives at the possibility of 24 phase portraits whereas the later work of Gasull and Prohens shows 22 possible phase portraits.

Finally, the phase portraits containing the combination $e^{-1} e^1 m_2^0$ where m_2^0 is a cusp were determined by de Jager [89,J],[90,J]. Some of these phase portraits have a unique limit cycle, which is hyperbolic as it occurs. Also phase portraits with a center point occur. As there exists at least one real infinite critical point, say at u=$\frac{1}{\alpha}$, letting $x \to x + \alpha y, y \to y$ yields $a_{20}=0$, whereas by scaling there may be put $b_{20}=1$. Since A=$(a_{20}^2 - a_{20}a_{11}b_{11} + a_{11}^2)b_{02} > 0$, there is $b_{02} \neq 0$ and scaling yields $b_{02} = \pm 1$. Then (11.3.43),(11.3.44) becomes

$$\dot{x} = y + \lambda_1 x^2 + \lambda_2 xy, \qquad (11.3.51)$$
$$\dot{y} - x^2 + \lambda_3 xy \pm y^2, \qquad (11.3.52)$$

with $\lambda_1, \lambda_2, \lambda_3 \in \mathbb{R}$.

Using a different system an extensive classification results in more than 30 phase portraits for the system with an $e^{-1}e^1m_2^0$ combination [89,J],[90,J].

If there are complex second order critical points there exist only the combination $c_2^0 c_2^0$. Without loss of generality the complex critical points may be thought to be located in P_1:(i,0) and P_2:$(-$i,0), which can be achieved by letting $a_{00} = a_{20}, b_{00} = b_{20}, a_{10} = b_{10}=0$ in (1.1),(1.2). Then

$$\dot{x} = a_{20} + a_{01}y + a_{20}x^2 + a_{11}xy + a_{02}y^2, \qquad (11.3.53)$$
$$\dot{y} = b_{20} + b_{01}y + b_{20}x^2 + b_{11}xy + b_{02}y^2, \qquad (11.3.54)$$

with the result that $c_{12} = c_{14}=0, c_{13} = -c_{34}, c_{15} = c_{45}, c_{16} = c_{46}, c_{23} = c_{24} = c_{25} = c_{26}=0$. (Following Theorem 2.3 the definition of these determinants are given.) In order for P_1 and P_2 to be of second order, (2.1.5) should be read as $A(x^2 + 1)^2=0$ and (2.1.6) as $Ay^4=0$, with A$>$0. As a result there follows $c_{34} = c_{35} = c_{45}=0, c_{46} \neq 0$. The cases $b_{20}=0$ and $b_{20} \neq 0$ will be distinguished.

(i)If $b_{20}=0$, then $a_{20} \neq 0$ as $c_{46} \neq 0$, $b_{01}=0$ as $c_{34}=0$, $b_{11}=0$ as $c_{45}=0$, whereas $b_{02} \neq 0$ as $c_{46} \neq 0$. Then (11.3.53),(11.3.54) reads, after scaling and by letting $x \to x + \frac{1}{2}a_{11}y, y \to y$,

$$\dot{x} = 1 + \lambda_1 y + x^2 + \lambda_2 y^2, \qquad (11.3.55)$$
$$\dot{y} = y^2, \qquad (11.3.56)$$

where $\lambda_1, \lambda_2 \in \mathbb{R}$ and A=1$>$0.

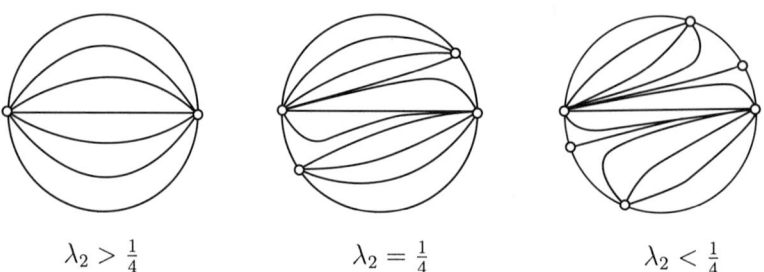

$\lambda_2 > \frac{1}{4}$ $\lambda_2 = \frac{1}{4}$ $\lambda_2 < \frac{1}{4}$

Figure 11.8 Phase portraits in the class with a $c_2^0 c_2^0$ combination; $b_{20}=0$

The location of the infinite critical points is given by

$$(\lambda_2 u^2 - u + 1)u = 0, \qquad (11.3.57)$$

so there is a $C_{0,1}^0 C_{0,1}^0 E^1$ combination for $\lambda_2 > \frac{1}{4}$, a $E^1 M_{0,2}^0$ combination for $\lambda_2 = \frac{1}{4}$ and an $E^{-1} E^1 E^1$ combination for $\lambda < \frac{1}{4}$, whereas an $M_{0,3}^0$ point cannot occur. In all classes $c_2^0 c_2^0 C_{0,1}^0 C_{0,1}^0 E^1$, $c_2^0 c_2^0 E^1 M_{0,2}^0$ and $c_2^0 c_2^0 E^{-1} E^1 E^1$ there exists one phase portrait for $b_{20}=0$. They are given in Figure 11.8.

(ii)If $b_{02} \neq 0$, then letting $x \to x + \frac{a_{20}}{b_{20}} y$, $y \to y$, using $c_{34} = c_{45}=0$ and scaling yields from (11.3.53),(11.3.54)

$$\dot{x} = y^2, \tag{11.3.58}$$
$$\dot{y} = 1 + \lambda_1 y + x^2 + \lambda_2 xy + \lambda_3 y^2, \tag{11.3.59}$$

with $\lambda_1, \lambda_2, \lambda_3 \in \mathbb{R}$ and $A=1>0$.

The location of the infinite critical points is given by

$$u^3 - \lambda_3 u^2 - \lambda_2 u - 1 = 0. \tag{11.3.60}$$

Let $\lambda_2=3(a+b)$, $\lambda_3 =3(a-b)$ and (11.3.29) is obtained. With D as in (11.3.29), it may be seen that the $C_{0,1}^0 C_{0,1}^0 E^1$ combination occurs for D>0, the $E^1 M_{0,2}^0$ for D=0, (a,b)$\neq$(0,−1), the $E^{-1} E^1 E^1$ for D<0 and a $M_{0,3}^0$ point for (a,b)= (0,−1), so D=0.

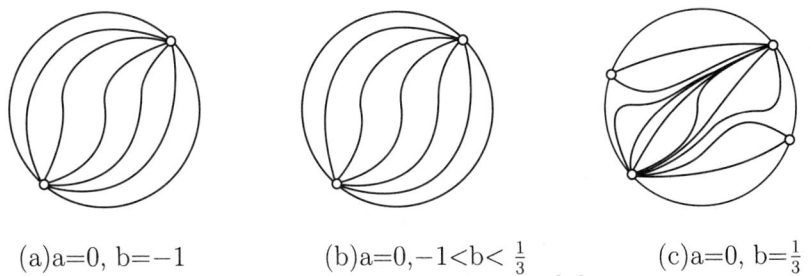

$\quad$(a)a=0, b=−1 $\qquad\qquad$ (b)a=0,−1<b< $\frac{1}{3}$ $\qquad\qquad$ (c)a=0, b=$\frac{1}{3}$

Figure 11.9 Phase portraits in the class with a $c_2^0 c_2^0$ combination: $b_{20} \neq 0$

The phase portraits for a=0, $-1 \leq b \leq \frac{1}{3}$ are readily determined and give a first idea of the phase portraits in the class with a $c_2^0 c_2^0$ combination and $b_{02} \neq 0$. The phase portrait for $c_2^0 c_2^0 M_{0,3}^1$ is given in Figure 11.9(a): it corresponds to a=0, b=−1 and $M_{0,3}^1$ is located in u=1. Increasing b into the interval $-1<b< \frac{1}{3}$ makes the $M_{0,3}^1$ point bifurcate into two complex points $C_{0,1}^0$ and an elementary node E^1, remaining located in u=1 for all b> −1. The phase portrait for a=0, $-1<b< \frac{1}{3}$ is in class $c_2^0 c_2^0 C_{0,1}^0 C_{0,1}^0 E^1$ and is given in Figure 11.9(b). For b$-\frac{1}{3}$, apart from the node E^1 in u=1, an $M_{0,2}^0$ saddle node appears in $u = -1$, as illustrated in Figure 11.9(c). Further increase of b leads to bifurcation of $M_{0,2}^0$ and the class $c_2^0 c_2^0 E^{-1} E^1 E^1$ is entered. Decrease of b below -1 also leads to this class, however, this takes place by bifurcation of the $M_{0,3}^1$ node.

The systems in the class with a $c_2^0 c_2^0$ combination belong to the chordal quadratic systems ; they also occur in the classes $m_f{=}0$ and $m_f{=}2$.

11.3.4   Systems with four finite critical points

In these classes occur the combinations $e^{-1} e^{-1} e^1 e^1$, $c_1^0 c_1^0 e^{-1} e^1$ and $c_1^0 c_1^0 c_1^0 c_1^0$, the latter also belonging to the class of chordal quadratic systems.

For the class $c_1^0 c_1^0 c_1^0 c_1^0$ we begin with system (11.3.53),(11.3.54), representing a system with a complex critical point in $(i,0)$ and $(-i,0)$,

$$\dot{x} = a_{20} + a_{00}y + a_{20}x^2 + a_{11}xy + a_{02}y^2, \qquad (11.3.61)$$
$$\dot{y} = b_{20} + b_{01}y + b_{20}x^2 + b_{11}xy + b_{02}y^2. \qquad (11.3.62)$$

The coordinates of the other critical points (x_-, y_-) and (x_+, y_+), where

$$x_\pm = \frac{1}{2A}[-2c_{34}c_{56} + c_{35}c_{46} \pm \sqrt{c_{35}^2 - 4c_{34}c_{36} - 4A}], \qquad (11.3.63)$$
$$y_\pm = \frac{1}{2A}[2c_{34}c_{46} - c_{35}c_{45} \pm \sqrt{c_{35}^2 - 4c_{34}c_{36} - 4A}], \qquad (11.3.64)$$

are obtained as the solutions of

$$(x^2 + 1)[Ax^2 + (2c_{34}c_{56} - c_{35}c_{46})x + c_{34}c_{36} + c_{46}^2] = 0, \quad (11.3.65)$$
$$y^2[Ay^2 + (c_{35}c_{45} - 2c_{34}c_{46})x + c_{34}^2 + c_{45}^2] = 0, \quad (11.3.66)$$

so that class $c_1^0 c_1^0 c_1^0 c_1^0$ occurs for $\Delta \equiv c_{35}^2 - 4c_{34}c_{36} - 4A <0$, A$>$0.

For the infinite critical points we follow the procedure mentioned before. The locations of these points are given by

$$f(u) \equiv a_{20}u^3 + (a_{11} - b_{02})u^2 + (a_{20} - b_{11})u - b_{20} = 0. \qquad (11.3.67)$$

Let $a_{02} \neq 0$, as $a_{02}{=}0$ may be treated as a limiting case. We subsequently take $b_{20}{=}0$ and $b_{20} \neq 0$.

(i)If $b_{20}{=}0$, apart from u=0, there exist the solutions of

$$f(u) \equiv a_{02}u^2 + (a_{11} - b_{02})u + a_{20} - b_{11} = 0, \qquad (11.3.68)$$

and there exists one phase portrait in class $c_1^0 c_1^0 c_1^0 c_1^0 C_{0,1}^0 C_{0,1}^0 E^1$ for D$\equiv (a_{11} - b_{02})^2 - 4a_{02}(a_{20} - b_{11}) <0$; it is topologically equivalent to Figure 11.8(a), one in class $c_1^0 c_1^0 c_1^0 c_1^0 E^1 M_{0,3}^1$ for D=0, $(a_{20} - b_{11})^2 + (a_{11} - b_{02})^2 \neq 0$, equivalent to Figure 11.8(b), one in class $c_1^0 c_1^0 c_1^0 c_1^0 E^{-1} E^1 E^1$ for D>0, equivalent to Figure 11.8(c) and one in class $c_1^0 c_1^0 c_1^0 c_1^0 M_{0,3}^1$ for $a_{20} - b_{11} = a_{11} - b_{02}=0$ so D=0, also equivalent to Figure 11.8(a).

(ii)If $b_{20} \neq 0$, let a$=-\frac{1}{6}a_{02}^{-\frac{1}{3}} b_{20}^{-\frac{2}{3}}(a_{20} + a_{11} - b_{11} - b_{02})$, b$=-\frac{1}{6}a_{02}^{-\frac{1}{3}} b_{20}^{-\frac{2}{3}}(a_{20} - a_{11} - b_{11} + b_{02})$, c$=a_{02}b_{20}^{-1}$; then with $\bar{u} = a_{02}^{\frac{1}{3}} b_{20}^{-\frac{1}{3}}$, (11.3.67) becomes

$$\bar{u}^3 - 3(a - b)\bar{u}^2 - 3(a + b)\bar{u} - 1 = 0. \qquad (11.3.69)$$

The analysis may then be continued as for the class $c_2^0 c_2^0, b_{20} \neq 0$ in section 11.3.3 leading for a=0, $-1 \le b \le \frac{1}{3}$ to the phase portraits in Figure 11.9.

For the class $c_1^0 c_1^0 e^{-1} e^1$ without loss of generality we may locate e^{-1} in $P_1:(-1,0)$ and e^1 in $P_2:(1,0)$, then (1.1),(1.2) reads

$$\dot{x} = -a_{20} + a_{01}y + a_{20}x^2 + a_{11}xy + a_{02}y^2, \qquad (11.3.70)$$
$$\dot{y} = -b_{20} + b_{01}y + b_{20}x^2 + b_{11}xy + b_{02}y^2, \qquad (11.3.71)$$

$\Lambda>0$ and since there is $\Omega(x,y) = -2c_{34}x - c_{35}y + 2c_{45}x^2 + 4c_{46}xy + 2c_{56}y^2$, there should be $c_{34} <0, c_{34} < c_{45} < -c_{34}$ to let e^{-1} be in P_1 and e^1 in P_2.

The coordinates of the critical points (x_+, y_-) and (x_-, y_+) are then given by the solutions of

$$(x^2 - 1)[Ax^2 + (2c_{34}c_{56} - c_{35}c_{46})x - c_{34}c_{36} + c_{46}^2] = 0, \quad (11.3.72)$$
$$y^2[Ay^2 + (c_{35}c_{45} - 2c_{34}c_{46})y + c_{34}^2 - c_{45}^2] = 0, \quad (11.3.73)$$

which are

$$x_\pm = \frac{1}{2A}[-2c_{34}c_{56} + c_{35}c_{46} \pm \sqrt{c_{35}^2 - 4c_{34}c_{36} + 4A}], \quad (11.3.74)$$
$$y_\pm = \frac{1}{2A}[-c_{35}c_{45} + 2c_{34}c_{56} \pm \sqrt{c_{35}^2 - 4c_{34}c_{36} + 4A}], \quad (11.3.75)$$

and class $c_1^0 c_1^0 e^{-1} e^1$ occurs for $\Delta \equiv c_{35}^2 - 4c_{34}c_{36} + 4A <0$, A>0, $c_{34} <0$, $c_{36} <0$, $c_{34} < c_{45} < -c_{34}$.

The critical points at infinity may be analyzed as for class $c_1^0 c_1^0 c_1^0 c_1^0$. About the phase portraits in class $c_1^0 c_1^0 e^{-1} e^1$, information in the literature seems to be scarce. The antisaddle in (1,0) can be a focus, weak up to the third order and at least three limit cycles can exist around the focus. Bifurcation of limit cycles in this class using the theory of rotated vector fields were studied by Cherkas and Gaiko [87,CG] and Ye Yanqian [92,Y1]. Also phase portraits with a center point exist, as was demonstrated in Chapter 6.

As was stated in section 3.4.1 of Chapter 3 the class $e^{-1}e^{-1}e^1e^1$ may be represented by the system

$$\dot{x} = a_1x(1-x) + b_1y(1-y) + (\frac{\alpha-1}{\beta}a_1 + \frac{\beta-1}{\alpha}b_1)xy, \quad (11.3.76)$$

$$\dot{y} = b_2x(1-x) + a_2y(1-y) + (\frac{\beta-1}{\alpha}a_2 + \frac{\alpha-1}{\beta}b_2)xy, \quad (11.3.77)$$

with A=$\frac{\alpha+\beta-1}{\alpha\beta}(a_1a_2 - b_1b_2)^2 > 0$, and we may take $\alpha > 0$, $\beta > 0$, $\alpha+\beta-1 > 0$.

No complete classification of the phase portraits of this system seems to exist in the literature, although partial results can be found.

Obviously, if $b_2{=}0$ or $b_1{=}0$, straight line orbits exist and uniqueness and hyperbolicity of possible limit cycles are ascertained. Phase portraits for (11.3.76),(11.3.77) in this case are investigated by Artes [90,A]. If both $b_1 = b_2{=}0$, a Volterra–Lotka equation is obtained of which all phase portraits are known [87,R]. No limit cycles occur in this case, yet center points and surrounding periodic solutions are possible.

If $b_1b_2 \neq 0$ and a focus exists, it may be weak up to the third order [97,AL] and three limit cycles may occur around one focus [67,QP]. In case two nests of limit cycles occur only precisely one (hyperbolic) limit cycle exists in each nest [93,Z2]. Also phase portraits with a center point exist.

Results on the limit cycle behaviour in class $e^{-1}e^{-1}e^1e^1$ are also available from the studies of class III of the classification of Ye Yanqian using the condition n($a^2n + b^2$l–abm)>0 and from those in class II, n$\neq$0. Many of these results are mentioned in the second edition of the book of Ye Yanqian on limit cycles [65,Y], wherein also many references are given. Most results are obtained in class II, n$\neq$0.

An extensive compilation of then known phase portraits, including not only those in the classes under discussion in this chapter, was given in 1986 by Liang Zhaojun [86,L2] (in Chinese). For subsequent progress on class II we refer to work of Zhang Pingguang [90,Z4],[91,Zh] and Zhang Xiang [99,Z4].

11.4 Systems with finite index $i_f=2$

These systems are characterized by the condition $A \equiv c_{46}^2 - c_{45}c_{56} < 0$, $c_{45} + c_{56} > 0$. Involved are the finite critical point combinations $e^{-1}e^1e^1e^1$, $c_1^0c_1^0e^1e^1$, $e^1e^1m_2^0$, $e^1m_3^1$ and the critical point m_4^2. For the infinite critical points there exist the combinations $E^{-1}E^{-1}E^1$, $E^{-1}C_{0,1}^0C_{0,1}^0$, $E^{-1}M_{0,2}^0$ and $M_{0,3}^{-1}$.

11.4.1 Systems with less than four finite critical points

In systems with a critical point m_4^2 there exists only the type present in a translated homogeneous system; the phase portraits are topologically characterized in Figure 3.3 of Chapter 3. They represent systems within the classes $m_4^2E^{-1}E^{-1}E^1$, $m_4^2E^{-1}C_{0,1}^0C_{0,1}^0$, $m_4^2E^{-1}M_{0,2}^0$ and $m_4^2M_{0,3}^{-1}$.

In the class with the $e^1m_3^1$ combination the third order point m_3^1 point may be either semi-elementary or nilpotent. Invariant theory is used by Vulpe and Nikolaev [92,VN1] to locate this class in coefficient space. This work was later extended by Voldman [96,V]. Phase portraits in this class were classified by de Jager [89,J],[90,J] and later by Gasull and Prohens [96,GP2] for the case m_3^1 is nilpotent.

If m_3^1 is semi-elementary it is a node and we may work with

$$\dot{x} = a_{20}x^2 + a_{11}xy + a_{02}y^2, \tag{11.4.1}$$
$$\dot{y} = y + b_{20}x^2 + b_{11}xy + b_{02}y^2, \tag{11.4.2}$$

with $a_{20}=0$, $a_{11}b_{20} < 0$. Scaling yields $a_{11}=-1$, $b_{20}=1$, whereas there should be $A = a_{02}^2 + a_{02}b_{11} + b_{02} < 0$, then also $c_{45} + c_{56} = 1 - a_{02}b_{11} - b_{02} > 0$, so that $m_f=4$, $i_f=2$.

(i)If $a_{02}=0$, (11.4.1),(11.4.2) can be written as

$$\dot{x} = -xy, \tag{11.4.3}$$
$$\dot{y} = y + x^2 + \lambda_1 xy + \lambda_2 y^2, \tag{11.4.4}$$

where $\lambda_1 \geq 0$,if needed by letting $x \rightarrow x, \lambda_1 \rightarrow -\lambda_1$, and where $\lambda_2 < 0$ as $A<0$. Apart from the node in $(0,0)$ there is an antisaddle in $(0,-\frac{1}{\lambda_2})$ which also appears to be a node. As a result there exists no limit cycle. Apart from the infinite critical point "at the ends of the y axis" there exist the infinite critical points given by

$$(1 + \lambda_2)u^2 + \lambda_1 u + 1 = 0, \lambda_1 \geq 0, \lambda_2 < 0. \tag{11.4.5}$$

For $\lambda_1 \geq 0$, $-1 < \lambda_2 < 0$, $\Delta \equiv \lambda_1^2 - 4(1+\lambda_2) < 0$, and for $\lambda_1 > 0$, $\lambda_2 = -1$ there exists one phase portrait in class $e^1 m_3^1 C_{0,1}^0 C_{0,1}^0 E^1$. For $\lambda_1 > 0$, $-1 < \lambda_2 < 0$, $\Delta = 0$ one phase portrait in class $e^1 m_3^1 E^1 M_{0,2}^0$ and for $\lambda_1 > 0$, $-1 < \lambda_2 < 0$, $\Delta > 0$ one phase portrait in class $e^1 m_3^1 E^{-1} E^{-1} E^1$. For $\lambda_1 > 0$, $\lambda_2 = -1$ one phase portrait in class $e^1 m_3^1 E^{-1} M_{0,2}^0$ and for $\lambda_1 \geq 0$, $\lambda_2 < -1$ one phase portrait in class $e^1 m_3^1 E^{-1} E^{-1} E^1$ [89,J].

(ii) If $a_{02} \neq 0$ the phase portraits of (11.4.1),(11.4.2) were also determined by de Jager[89,J]. Apart from the node in $(0,0)$ there exists an antisaddle, which may be either a node or a focus, possibly weak of the first order. A unique hyperbolic limit cycle bifurcates from the weak focus which also occurs at an unbounded separatrix cycle. As was remarked earlier, it was shown by Coppel [90,Co] that there exists at most one (hyperbolic) limit cycle in a system with a semi-elementary m_3^i point.

The critical points at infinity may be analyzed as in section 11.3.2. In [89,J] two possible phase portraits are given in class $e^1 m_3^1 C_{0,1}^0 C_{0,1}^0 E^{-1}$, six in $e^1 m_3^1 E^{-1} M_{0,2}^0$, five in $e^1 m_3^1 E^{-1} E^{-1} E^1$ and one in $e^1 m_3^1 M_{0,3}^{-1}$.

If m_3^1 is nilpotent, it is a critical point with an elliptic and a hyperbolic sector, and we can work again with

$$\dot{x} = y + a_{20}x^2 + a_{11}xy + a_{02}y^2, \qquad (11.4.6)$$
$$\dot{y} = b_{20}x^2 + b_{11}xy + b_{02}y^2, \qquad (11.4.7)$$

with $b_{20}=0$, $a_{20}b_{11} > 0$. Scaling and a linear transformation again yields

$$\dot{x} = y + \lambda_1 x^2 + \lambda_2 xy + \lambda_3 y^2, \qquad (11.4.8)$$
$$\dot{y} = xy, \qquad (11.4.9)$$

with $\lambda_1 > 0$, $\lambda_2 \in \{0,1\}$, $\lambda_3 < 0$ and $A=\lambda_1\lambda_3 < 0$ so then also $c_{45} + c_{56} = \lambda_1 - \lambda_3 > 0$.

Apart from the node in $(0,0)$ there exists an antisaddle in $(0,-\frac{1}{\lambda_3})$, which is a center point if $\lambda_2=0$ and a node or focus if $\lambda_2=1$. There exist no limit cycles.

For $\lambda_2=0$, $0 < \lambda_1 < 1$ there exist two phase portraits in the class $e^1 m_3^1 C_{0,1}^0 C_{0,1}^0 E^{-1}$, for $\lambda_1=1$ one in the class $e^1 m_3^1 M_{0,3}^{-1}$ and for $\lambda_1 > 1$ one in the class $e^1 m_3^1 E^{-1} E^{-1} E^1$. For $\lambda_2=1$ there exist two phase portraits in the class $e^1 m_3^1 C_{0,1}^0 C_{0,1}^0 E^{-1}$, three in the class $e^1 m_3^1 E^{-1} M_{0,2}^0$ and three in the class $e^1 m_3^1 E^{-1} E^{-1} E^1$ [89,J],[90,J].

For the class $e^1 e^1 m_2^0$ we work again with

$$\dot{x} = -a_{20} + a_{01}y + a_{20}x^2 + a_{11}xy + a_{02}y^2, \qquad (11.4.10)$$

$$\dot{y} = -b_{20} + b_{01}y + b_{20}x^2 + b_{11}xy + b_{02}y^2, \qquad (11.4.11)$$

which has the finite critical points P_1:(1,0) and P_2:(−1,0).

The degenerate isocline l connecting these critical points is given by l;$y(c_{34} - c_{45}x - c_{56}y)=0$, where we may put $c_{46}=0$, if needed by replacing $c_{45}x$ by $c_{45}x - c_{46}y$.

Since A$\equiv c_{46}^2 - c_{45}c_{56} <0$, $c_{45}+c_{56} >0$ there is $c_{45} >0$, $c_{56} >0$. Furthermore $\Omega(x,y) = -2c_{34}x - c_{35}y + 2c_{45}x^2 + 2c_{56}y^2$ so P_1 and P_2 are antisaddles if $-1< \frac{c_{34}}{c_{45}} <1$. The other critical points are on $c_{34} - c_{45}x=0$ with $y_{\pm} = \frac{1}{2c_{45}c_{56}}[c_{35}c_{45} \pm \sqrt{\Delta}]$ with $\Delta \equiv 4c_{34}^2 c_{45}c_{56} + c_{35}^2 c_{45}^2 - 4c_{45}^3 c_{56}$. If $\Delta=0$ class $e^1 e^1 m_2^0$ occurs with an m_2^0 point in $(\frac{c_{34}}{c_{45}}, \frac{c_{35}}{2c_{56}})$; it is a cusp if d $\equiv 2b_{01}c_{45}c_{56} + 2(2a_{20} + b_{11})c_{34}c_{56} + (a_{11} + 2b_{02})c_{35}c_{45}=0$ and a saddle node if d$\neq 0$.

The phase portraits when a cusp occurs were determined by de Jager [89,J],[90,J]. Two phase portraits are given in class $e^1 e^1 m_2^0 C_{0,1}^0 C_{0,1}^0 E^{-1}$, six in class $e^1 e^1 m_2^0 E^{-1} M_{0,2}^0$, five in class $e^1 e^1 m_2^0 E^{-1} E^{-1} E^1$ and one in class $e^1 e^1 m_2^0 M_{0,3}^{-1}$. Except in class $e^1 e^1 m_2^0 M_{0,3}^{-1}$, in these classes a unique (hyperbolic) limit cycle occurs [88,Co], as well as unbounded separatrix cycles.

The phase portraits with a saddle node seem not to have been determined systematically for class $e^1 e^1 m_2^0$. The limit cycles in this class are not completely studied yet; this applies in particular to the question whether two nests of limit cycles can exist. Bifurcation of an m_2^0 point into an $e^{-1}e^1$ combination, however, leads to at most a (1,1) configuration of limit cycles in an $e^{-1}e^1 e^1 e^1$ combination of finite critical points.

11.4.2 Systems with four finite critical points

Involved are the combinations $c_1^0 c_1^0 e^1 e^1$ and $e^{-1}e^1 e^1 e^1$.

As was indicated in Chapter 7, the class with the $c_1^0 c_1^0 e^1 e^1$ combination is the most interesting class in quadratic systems from the limit cycle point of view. Examples with (at least) four limit cycles seem, so far, to have been given only in this class. Also limit cycle distributions over two nests other than the (1,1) distribution appear in this class, since (1,2) and (1,3) distributions can be shown to exist. (With the unlikely exception for the class $e^1 e^1 m_2^0$, where m_2^0 is a saddle node, in all other classes it can be shown that only the (1,1) distribution is possible.)

Despite this interest a complete classification of the phase portraits for this class, possibly leaving aside a strict proof of the limit cycle behavior, seems not to have been given yet. Apart from (11.4.10),(11.4.11) with Δ <0, other forms may be used for its classification. Cherkas and Gaiko [87,CG]proposed a system, containing a focus (center) in (0,0) and an anti-saddle in (1,0), constructed in such a way that two parameters, rotating the vector field, are determined. It reads

$$\dot{x} = \alpha x - y - \alpha x^2 + (a + \alpha\gamma)xy + (b - \gamma + \alpha c)y^2, \quad (11.4.12)$$
$$\dot{y} = x + \alpha y - x^2 + (\gamma - \alpha a)xy + (\alpha\gamma + c - \alpha b)y^2, \quad (11.4.13)$$

where $b^2 - 4(a - 1)c$ <0,a>1.

If $\alpha \neq 0$, then (0,0) is a focus, and if $\alpha=0$, then it is a weak focus or a center point. The critical point in (1,0) is an antisaddle since $\Omega(x,y) = (1 + \alpha^2)^2(a-1)>0$ and (1,0) is real. From Theorem 2.2 there follows that $y^2(Ay^2 + B_2y + C_2)=0$, whereas $C_2=-(1+\alpha^2)^2 (a-1)<0$ and $B_2^2 - 4AC_2 = a^2(1+\alpha^2)^4(b^2-4(a-1)c)$ <0 so A<0 and there exist two finite critical points c_1^0 as well.

The possible phase portraits for (11.4.12),(11.4.13), without considering the limit cycles, were given by Artes and Llibre [94,AL1]. Three phase portraits were found in class $c_1^0c_1^0e^1e^1$ ($C_{0,1}^0, C_{0,1}^0, E^{-1}$ or $M_{0,3}^{-1}$), thirteen in class $c_1^0c_1^0e^1e^1E^{-1}M_{0,2}^0$ and thirteen in class $c_1^0c_1^0e^1e^1E^{-1}E^{-1}E^1$. Single and simultaneous unbounded separatrix cycles appear as well as center points.

Results on the limit cycles in the class $c_1^0c_1^0e^1e^1$ using the classification of Ye Yanqian may be found in those belonging to class 3, $n(a^2n + b^2l - abm)$ <0,$-am + bl - bn$ >0, $b+n \neq 0$, $B_1^2 - 4AC_1$ <0 (see Table 5). If a=0, obviously, uniqueness of a limit cycle is ascertained since a straight line orbit exists. For a discussion on the limit cycles in the class $c_1^0c_1^0e^1e^1$ we refer to section 7.4. Class $c_1^0c_1^0e^1e^1$ seems a fruitful subject for numerical exploration.

The classification of the phase portraits in the class with an $e^{-1}e^1e^1e^1$ combination seems also incomplete, although partial results may be found in the literature.

For instance, discussing separatrix cycles in quadratic systems, Gaiko uses the normal form

$$\dot{x} = -y(1 + x) + \alpha Q(x,y), \quad (11.4.14)$$
$$\dot{y} = x + \lambda y + ax^2 + \beta y(1 + x) + cy^2 \equiv Q(x,y), \quad (11.4.15)$$

which is of class $e^{-1}e^1e^1e^1$ if c<0, and 0<a<1($\lambda^2 - 4c(a - 1)$ >0) or a>0 whereas α, β are two parameters rotating the vector field. Over a hundred phase portraits are presented in [00,G].

As far as the limit cycle problem is concerned it is known that only a (1,1) distribution of hyperbolic limit cycles can occur [93,Z2], whereas the conjecture that at most three (hyperbolic) limit cycles in one nest are possible still has to be shown to be true.

Results on the limit cycles in class $e^{-1}e^1e^1e^1$ using the classification of Ye Yanqian may be found in those belonging to class 3 with n($a^2n + b^2l - abm$) < 0, $-am + bl - bn > 0, b + n \neq 0, B_1^2 - 4A_1C_1$ >0 (see Table 5). As mentioned above, in class $3_{a=0}$ at most one (hyperbolic) limit cycle can exist. The phase portraits for the class $e^{-1}e^1e^1e^1$ having a straight line orbit were investigated by Artes [90,A]. Further results were obtained by Ye Yanqian [92,Y2],[97,Y1],[98,Y] and Zhang Xiang [01,Z1].

11.5 Structurally stable quadratic systems

As was mentioned in Chapter 7 on limit cycles, a given quadratic system will only contain a finite number of limit cycles. However, it has not been ex-cluded rigorously yet that certain variations of the coefficients of a particular system may lead to a sequence of phase portraits with an unbounded increase of the number of limit cycles; a proof of the existence of H(2) has still to be given. On the other hand all experience so far leads to the assumption that a quadratic system can have at most four limit cycles; H(2)=4. Based on this assumption the number of possible phase portraits would be finite; an estimate would probably yield a number of 2000 qualitatively (topologically) different phase portraits. Most of them are structurally unstable, meaning in the present context that for these systems arbitrarily small changes in the coefficients exist such that a qualitatively different phase portrait, as defined on the Poincaré disk, results. In fact, then all systems with finite multiplic-ity m_f <4 are structurally unstable since they contain at least one multiple infinite critical point $M_{p,q}^i$(p,q≥1,q≥1). Also in m_f=4 finite multiple critical points, center points, weak foci, saddle connections, separatrix cycles and multiple limit cycles can be present as structurally unstable elements. This limits considerably the number of possible phase portraits of structurally sta-ble quadratic systems, which, however, does not diminish their significance; the coefficients of the structurally stable quadratic systems forming a dense subset in the $\mathbb{R}^{12}$ coefficient space, meaning that in a neighbourhood of a point in this space which represents a structurally unstable system, part of this neighbourhood represent structurally stable systems.

The possible portraits of the structurally stable quadratic systems are not yet all known, mainly because the limit cycle problem is not solved yet. A feasible problem, then, is to determine the possible phase portraits within the class of quadratic systems without limit cycles. Work on this problem was started in 1976 by Tavares dos Santos [76,T], whereas a contribution appeared two years later by Cai Suilin [79,C]. Further contributions were subsequently given in 1980 by Shi Songling [80,S2], in 1984 by Ye Yanqian [86,Y], in 1986 by Zhu Deming [86,Z1], in 1994 by Kooij and van Horssen [94,KH], whereas a conclusive account was finally given in 1995 by Artes, Kooij and Llibre, which resulted in an extensive report [98,AKL].

It appears that for structurally stable quadratic systems without limit cycles there exist in class $i_f=-2$ one phase portrait in $e^{-1}e^{-1}c_1^0c_1^0E^1E^1E^1$ and four portraits in class $e^{-1}e^{-1}e^{-1}e^1E^1E^1E^1$. In class $i_f=0$ there exists one phase portrait in $c_1^0c_1^0c_1^0c_1^0C_{0,1}^0C_{0,1}^0E^1$, one in $e^{-1}c_1^0c_1^0C_{0,1}^0C_{0,1}^0E^1$, five in $e^{-1}e^{-1}e^1e^1C_{0,1}^0C_{0,1}^0E^1$, one in $c_1^0c_1^0c_1^0c_1^0E^{-1}E^1E^1$, three in $e^{-1}c_1^0c_1^0e^1E^{-1}E^1E^1$, sixteen in $e^{-1}e^{-1}e^1e^1E^{-1}E^1E^1$. In class $i_f=2$ there exists one portrait in $c_1^0c_1^0e^1e^1C_{0,1}^0C_{0,1}^0E^{-1}$, one in $e^{-1}e^1e^1e^1C_{0,1}^0C_{0,1}^0E^{-1}$, three in $c_1^0c_1^0e^1e^1E^{-1}E^{-1}E^1$ and seven in $e^{-1}e^1e^1e^1E^{-1}E^{-1}E^1$, being 44 phase portraits all together. The same portraits are found if quadratic systems with limit cycles are represented such that a region with (a) limit cycle(s) the outmost limit cycle is identified with an antisaddle with the same stability as the outer stability of this limit cycle.

Finding the regions in the $\mathbb{R}^{12}$ coefficient space corresponding to the 44 phase portraits mentioned above is still not a simple problem, for one thing because the structurally stable quadratic systems without limit cycles do not constitute a dense subset in the coefficient space.

Bibliography

A short list of books is given to make reading this book possible, easier or more fruitful. Further references can be found in these books.

[A] V.I.Arnold,"Geometrical Methods in the Theory of Differential Equations, "Springer-Verlag,Berlin,1988.

[ALGM1] A.A.Andronov, E.A.Leontovich, I.I.Gordon and A.G.Maier, "Qualitative Theory of Second Order Dynamic Systems," Translated from the Russian, Israel Program for Scientific Translations, Jerusalem-London,1973.

[ALGM2] A.A.Andronov, E.A.Leontovich, I.I.Gordon and A.G.Maier, "Theory of Bifurcations of Dynamic Systems on a Plane," Translated from the Russian, Israel Program for Scientific Translations, Jerusalem-London,1973.

[BP] W.E.Boyce and R.C.DiPrima, "Elementary Differential Equations and Boundary Value Problems," John Wiley and Sons, New York,5th ed. 1992.

[CL] E.A.Coddington and N.Levinson, "Theory of Ordinary Differential Equations," McGraw-Hill, New York, 1955.

[G] R.Grimshaw, "Nonlinear Ordinary Differential Equations," Blackwell Scientific Publications, Oxford, 1990.

[GH] J.Guckenheimer and P.Holmes, "Nonlinear Oscillations, Dynamical Systems and Bifurcations of Vector fields,"Applied Mathematical Science, Vol. 42, 2nd Printing, Springer-Verlag,New York, 1986.

[JS] D.W.Jordan and P.Smith, "Nonlinear Ordinary Differential Equations," Clarendon Press, Oxford, 1992.

[V] F Verhulst, "Nonlinear Differential Equations and Dynamical Systems," Universitext, 2nd Edition, Springer, Berlin, 1996.

Classic articles that are basic to the content of this book are:

[P] H.Poincaré, Sur les courbes définies par une équation différentielle, Oeuvres, Vol.1, Paris, Gauthier-Villars, 1928.

306

[H] D.Hilbert, Mathematical problems, Bull. Amer. Math. Soc. Vol 8, 437-479, 1902.

Starting from the idea that all information on the qualitative theory of quadratic systems in the plane, including books, papers and preprints in any language, can be, in principle, useful for the classification of the phase portraits of these systems, the author collected all material he could find, as produced in the 20th century and the beginning of the 21th century. As a result a bibliography, usually giving also a summary, and ordered chronologically in order to be able to follow the historic development of the subject, was produced which can be found on the site http://ta.twi.tudelft.nl/DV/Staff/J.W. Reyn.html

A selection of this material was used for the present book and listed below; also in chronological order and with a notation such that the year of publication is obvious (so [ab,L] means that in the year 19ab a paper was published by an author with a family name beginning with an L, with the exception that, at the end of the list [ab,L] indicates 20ab.)

Obviously, the flow of papers on quadratic systems will continue to yield information that will enable the completion of the classification of the phase portraits of quadratic systems. Hopefully further editions of this book will eventually yield a complete collection.

[04,B] W.Büchel, *Zur Topologie der durch eine gewöhnliche Differentialgleichung erster Ordnung und ersten Grades definierten Kurvenschar.* (On the topology of the curves defined by an ordinary differential equation of the first order and the first degree), Mitteil. der Math. Gesellsch. in Hamburg, Band IV,4,33-68.(25figures)

[08,B] W.Büchel, *Die physikalischen Bedeutingen der durch die Differentialgleichung $\frac{dy}{dx} = \frac{Y(x,y)}{X(x,y)}$ definierten Kurvenschar,* Mitt. Math. Ges. Hamburg.

[08,D] H.Dulac, *Détermination et intégration d'une certaine classe d' é quations différentielles ayant pour point singulier un centre.* (Determination and integration of a certain class of differential equations having a center as singular point.) Bulletin des Sciences Mathématiques, (2), 32, 230-252.

[11,K] W.Kapteyn, *Over de middelpunten der integraalkrommen van dif ferentiaalvergelijkingen van de eerste orde en den eerste graad.* (On the center points of integral curves of differential equations of the first order and the first degree.) Koninkl. Nederl. Akad. Wet. Versl. 19, 1446-1457.

[12,K] W.Kapteyn, *Nieuw onderzoek omtrent de middelpunten der inte-gralen van differentiaal vergelijkingen van de eerste orde en den eerste graad.* (New investigation on the center points of integral curves of differential equa-tions of the first order and the first degree.) Koninkl. Nederl. Wet. Versl. 20, 1354-1365, 21, 27-33.

[20,L] A.J.Lotka, *Undamped oscillations derived from the law of mass action,* Journal of the American Chemical Society, 42, 8, 1595-1599.

[23,D] H.Dulac, *Sur les cycles limites,* Bulletin de la Société Mathématique de France, 55, 45-188. (There exists a Russian translation in the form of a book with the same title, translated by G.I.Shilov under the redaction of N.F.Otrokov and published by Nauka, Moscow, 1980, 156 pp.)

[29,S] A.Sommerfeld, "Atombau und Spektrallinien Wellenmechanischer Ergänzungsband," Vieweg, Braunschweig.

[30,T] G.I.Taylor, *Recent work on the flow of compressible fluids,* Journ. London Mathematical Society 5, 224-240.

[31,V] V.Volterra, *Lecons sur la théorie mathématique de la lutte pour la vie,* (Lessons on the mathematical theory of the struggle for life.) Gauthier-Villar, Paris, 14-16, 27-34.

[34,G] G.F.Gause, *The struggle for existence,* Williams and Wilkins, Bal-timore.

[34,F] M.Frommer, *Uber das Auftreten von Wirbeln und Strudeln (geschlossener und spiraliger Integralkurven) in der Umgebung rationaler Unbestimmtheitsstellen.* (On the occurrence of center points and foci (closed and spiralled integral curves) in the neighborhood of rational points of un-determinacy.) Math. Annalen 109, 395-424.

[35,M] A.Maijer, *On the theory of coupled vibrations of two self-excited generators,* Technical Physics of the U.S.S.R., Vol. II,no.5, 465-481.

[36,M] A.Maijer, *A contribution to the theory of forced oscillations in a generator with two degree of freedom,* Technical Physics of the U.S.S.R.,Vol. III, no.12, 1056-1071.

[39,B1] N.N.Bautin, *Du nombre de cycles limites naissant en cas de varia tion des coefficients d'un état d'équilibre du type foyer ou centre.* (On the num-ber of limit cycles which appear with the variation of coefficients from an equi-librium position of focus or center type.) C.R.(Doklady)Akad.Sci.U.S.S.R.(N.S.) 24, 1939, 669-672. See the Russian paper in 1952, which was translated into English by the Amer. Math. Soc. in 1954 as Translation Number 100.

[39,B2] N.N.Bautin, *On a case of nonharmonic oscillations,* (Russian), Uch. Zap. Gork. Gos-Univ., no. XII, 231-237.

[39,B3] N.N.Bautin, *On a differential equation with a limit cycle,* (Russian), Zhurn. Tekhn. Fiz. 9, no.7, 601-611.

[39,C] S.Chandrasekhar, *An introduction to the study of stellar structure,* Chapter 4, Chicago Univ. Press (reprint by Dover, New York, 1957.)

[39,R] L.F.Richardson, *Generalized foreign politics,* Brit. Journ. Psychol. Monog. Suppl.23, Cambridge University Press, 23-26.

[42,B] N.V.Butenin, *Mechanical systems with gyroscopic forces showing self excited oscillations,* (Russian), Prikl. Mat. Mech. 6, 327-346.

[42,L] H.Lemke, *Ueber die Differentialgleichungen, welche den Gleichgewichtszustand eines gasförmigen Himmelskörpers bestimmen, dessen Teile gegeneinander nach dem Newtonschen Gesetze gravitieren,* Journ. Reine und Angew. Math. p.1912.

[42,W] H.Weyl, *On the differential equations of the simplest boundary layer problems,* Ann. of Math. 43, 381-407.

[43,FS] D.A.Frank-Kamenetskii and J.E.Sal'nikov, *On the possibility of self excited oscillations in a homogeneous chemical system with quadratic auto-katalyse,* (Russian), Zhurn. Phys. Chem., Vol XVII, no 2, 79-86.

[45,B] N.V.Butenin, *Effect of external sinusoidal forces with gyroscopic forces,* (Russian), Trud. AKVA, no 7, 138-148.

[48,S] N.A.Sakharnikov, *On Frommer's conditions for the existence of a center* (Russian), Prikl.Mat.Mech.,12, 669-670.

[49,B] N.V.Butenin, *On the theory of forced oscillations in a nonlinear mechanical system with two degrees of freedom,* (Russian), Prikl. Mat. i Mech., Vol 13, 337-349.

[51,GY] G.Guderley and Yoshihara, *An axial-symmetric transonic flow pattern,* Quarterly of Appl.Math.,8,333-339.

[51,L] L.S.Lyagina, *The integral curves of the equation* $y' = \frac{ax^2 + bxy + cy^2}{dx^2 + exy + fy^2}$, (Russian), Usp. Mat. Nauk.,6,no 2(42),171-183.

[52,B] N.N.Bautin, *On the number of limit cycles which appear with the variation of coefficients from an equilibrium position of center type,* Mathem. Sbornik (N.S)30(72),181-196,1952. Translated into English by F.V.Atkinson and published by the AMS in 1954 as Translation Number 100, 396-413; reprint as AMS Transl.(1)5(1962), 396-413.

[53,J] C.W.Jones, *On reducible nonlinear differential equations occurring in mechanics,* Proc. of Roy. Soc. of London, Series A, 217, 327-343.

[54,B] N.N.Bautin, *On the periodic solutions of a system of differential equations,* (Russian), Prikl.Mat.i Mech. 18, 128.

[54,KZ] K.Kestin and S.K.Zaremba, *Adiabatic one dimensional flow of a perfect gas through a rotating tube of uniform cross section*, The Aeronautical Quarterly, Volume IV, february, 373-399.

[54,S] K.S.Sibirskii, *On the conditions for the presence of a center and a focus*, (Russian), Kishinev Gos. Univ. Uc. Zap., II, 115-117.

[54,W] A.B.Wasileva, *On the mathematical theory of catalysis*, (Russian), Vestnik Moscow University, no. 6, 39-46.

[55,PL] I.G.Petrovskii and E.M.Landis, *On the number of limit cycles of the equation $\frac{dy}{dx} = \frac{P(x,y)}{Q(x,y)}$,where P and Q are polynomials of the 2nd degree*, Mat. Sbornik (N.S) 37(79), 209-250, 1955. (There is an early announcement in D.A.N. SSSR (NS)102, 29-35, 1955.) Translated into English by Edwin Hewitt and published as AMS Translations, Series 2, Vol. 10, 177-221, 1958.

[55,S] K.S.Sibirskii, *On the principle of symmetry and the problem of a center*, (Russian), Kishinev Gos. Univ. Uc. Zap., 17, 27-33.

[57,C] S.Chandrasekhar, "An introduction to the study of stellar structures", Dover, New York.

[57,Q] Qin Yuanshun (Chin Yuan Shun), *On algebraic limit cycles of degree two of quadratic systems*, (Chinese), Science Record, N.S. 1(2),1-3. (Summary of [58,Q])

[57,Y] Ye Yanqian (Yeh Yen-Chian), *Periodic solutions and limit cycles of certain nonlinear differential systems*, (Chinese), Science Record, N.S. 1(6), 359-361.

[58,Q] Qin Yuanshun (Chin Yuan Shun), *On the algebraic limit cycles of degree two of quadratic systems*, Acta Math. Sinica, 8, 23-35. (Chinese with German summary). Translated into English in Scientia Sinica, 7, 934-945, reprinted in Chinese Mat. Acta 8 (1966), 608-619.

[58,Y1] Ye Yanqian, *Periodic solutions and limit cycles of systems of nonlinear differential equations*, (Chinese), Nanjing Daxue Xuebao (Nanking Univ. Journ.), no.1, 7-17.

[58,Y2] Ye Yanqian, *Limit cycles of certain nonlinear differential systems*, II, Science Record (N.S.)(2), 276-279.

[59,B] A.N.Berlinskii, *Some questions of the qualitative investigation of the differential equation $\frac{P(x,y)}{Q(x,y)}$, where P and Q are polynomials not higher than of the second degree*, Kand. Dissertation, Tashkent, 7 pp.

[59,PL] I.G.Petrovskii and E.M.Landis, *Corrections to the article "On the number of limit cycles of the equation $\frac{dy}{dx} = \frac{P(x,y)}{Q(x,y)}$, where P and Q are polynomials not higher than of the second degree"* and *"On the number of limit *cycles of the equation $\frac{dy}{dx} = \frac{P(x,y)}{Q(x,y)}$, where P and Q are polynomials"*, (Russian), Mat. Sborn. 48(90), 253-255.

[59,QP] Qin Yuanxun and Pu Fuquan, *Concrete examples of existence of three limit cycles for the system* $\frac{dx}{dt} = \frac{X_2(x,y)}{Y_2(x,y)}$, (Chinese, English summary), Acta Math. Sinica, Vol.9, no.2, 213-225.

[59,T] Tung Chin-Chu (Dong Jinzhu), *Position of limit cycles of a quadratic system,* Sci. Record (N.S)2, 1958, 421-425; Acta Math. Sinica, 9,156-169, (Chinese,Russian summary); Sci. Sinica 8, 151-171, the latter in English, revising three previous papers.

[60,B] A.N.Berlinskii, *On the behaviour of integral curves of a certain differential equation,* (Russian), Izv. Vys. Uch. Zav. Mat. 2(15), 3-18. Translated into English by National Lending Library for Science and Technology, Russian Translating Programme RTS 5158, June 1969, Boston Spa, Yorkshire, U.K.

[60,C] W.A.Coppel, *On a differential equation of boundary layer theory,* Phil. Trans. of the Roy. Soc., Series A, Vol.253, 101-136.

[62,T] Tung Chin-Chu (Dong Jinzhu), *The structure of the separatrix cycles in quadratic systems,* Acta Math. Sinica, 12,251-257, 1962. (Chinese). Translated as Chinese Mathematics, 3,277-284, 1963.

[62,Y,H,W,X,L] Ye Yanqian, He Chongyou, Wang Mingschu, Xu Mingwei, Luo Dingjun, *A qualitative study of the integral curves of the equation* $\frac{dy}{dx} = \frac{q_{00}+q_{10}x+q_{01}y+q_{20}x^2+q_{11}xy+q_{02}y^2}{p_{00}+p_{10}x+p_{01}y+p_{20}x^2+p_{11}xy+p_{02}y^2}$, I, Acta Math.Sinica, 12,1-15, 1962. Translated as Chinese Mathematics, Vol. 3, 1-18, 1963.

[62,Y] Ye Yanqian, *A qualitative study of the integral curves of the differ ential equation* $\frac{dy}{dx} = \frac{q_{00}+q_{10}x+q_{01}y+q_{20}x^2+q_{11}xy+q_{02}y^2}{p_{00}+p_{10}x+p_{01}y+p_{20}x^2+p_{11}xy+p_{02}y^2}$, II, *Uniqueness of limit cycles,* Acta Math.Sinica, 12, 60-67, 1962. Translated as Chinese Mathematics, Vol.3, 62-70, 1963.

[63,C] L.A.Cherkas, *Algebraic solutions of the equation* $\frac{dy}{dx} = \frac{P(x,y)}{Q(x,y)}$ *where P and Q are second degree polynomials,* (Russian), Dokl.Akad.Nauk B.S.S.R., Vol. 7, no. 11, 732-735.

[63,LS] H.R.Latipov and I.I.Sirov, *On the behaviour of characteristics in the large of a differential equation,* (Russian), Studies in Differential Equations (Russian), Izd. Akad. Nauk Uzbek.S.S.R, Tashkent, 117-131.

[63,S1] K.S.Sibirskii, *Solution of the problem of the center for the differen tial equation* $\frac{dy}{dx} = \frac{c_{10}x+c_{01}y+c_{20}x^2+c_{11}xy+c_{02}y^2}{b_{10}x+b_{01}y+b_{20}x^2+b_{11}xy+b_{02}y^2}$ *for the Sakharnikov and Sibirskii cases of a center,* (Russian), Investigations of Differential Equations, Izd. Akad. Nauk Uzb. S.S.R.,Tashkent, 200-203.

[63,S2] K.S.Sibirskii, *Invariants of linear representations of the rotation group of the plane and the problem of the center,* (Russian), Dokl.Akad. Nauk S.S.S.R., 497-500.

[64,KK] I.S.Kukles and M.Khasanova, *On the behaviour of characteristics of a differential equation in the Poincaré circle,* (Russian), Dokl. A.N.T.S.S.R., Vol. VII, no.12, 3-6.

[64,L] W.S.Loud, *Behaviour of the period of solutions of certain plane autonomous systems near centers,* Contr. to Diff. Eq. 3, 21-36.

[65,L] N.A.Lukashevitch *Integral curves of a certain differential equation,* Diff. Uravn., Vol.1, no.1, 82-95, (English translation 60-70).

[65,Y] Ye Yanqian, "Theory of limit cycles," (Chinese), Shanghai Science Technology Press, Shanghai,1965,iii+369pp.(Second edition in 1984, English translation by AMS in 1986.)

[66,C] W.A.Coppel, *A survey of quadratic systems,* Journal of Differential Equations, 2, 293-304.

[67,QP] Qin Yuanxun and Pu Fuchian, *A concrete example of the existence of three limit cycles near the equilibrium of the system $\frac{dy}{dt}=P$, $\frac{dy}{dt}=Q$, where P and Q are polynomials of the second degree,* Acta Math. Sinica 9, no.2, 213-226, 1967. (Chinese Mathematics 9, no.4, 521-534,(1967).

[69,LM] M.Lefranc and J.Mawhin, *Étude qualitative des solutions de l' equation differentielle d'Emden-Fowler,* (English summary), Acad. Roy. Belg. Bull. Cl. Sci., 5, 55, 763-770.

[70,C] L.A.Cherkas, *The absence of limit cycles for a certain differential equation that has a structurally unstable focus,* (Russian), Diff. Uravn. 6, no.5, 779-783. (Engl. Transl. 589-592)

[70,CZ] L.A.Cherkas and L.I.Zhilevich, *Some criteria for the absence of limit cycles and for the existence of a single limit cycle,* (Russian), Different. Uravn., 6,no.7, 1170-1178. (Eng.Transl. 891-897)

[70,DP] R.J.Dickson and L.M.Perko, *Bounded quadratic systems in the plane,* Journ. of Diff. Equations, 7, 251-273.

[70,E] R.M.Evdokimenko, *Construction of algebraic paths and the qualitative investigation in the large of the properties of integral curves of a system of differential equations,* (Russian), Different. Uravn. 6. no 10, 1780-1791. (Engl.Transl. 1349-1358)

[70,J] A.I.Jablonskii, *Algebraic integrals of a system of differential equations,* (Russian), Diff. Uravn. 6, no.11, 1752-1760, (Engl.Transl. 1326-1333).

[71,E] R.M.Evdokimenko, *Integral curves of a dynamical system,* (Russian), Diff. Uravn. 7, no.8, 1522-1524. (Engl.Transl. 1156-1158)

[71,J] A.I.Jablonskii, *The qualitative behaviour of the tracjectories of a certain class of differential systems,* Diff. Uravn. 7, 279-285, (Engl.Transl. 219-223)

[71,PS] I.I.Pleskan and K.S.Sibirskii, *Isochronism of a system with quadratic nonlinearities,* (Russian), Mat. Issled. 6, no.4, (22), 140-154.

[71,R1] I.G.Rozet, *Nonlocal bifurcation of limit cycles and quadratic differential equations in the plane,* (Russian), Dissertation kand. phys. mat., Samarkand, 179 pp.

[71,Z] L.J.Zhilevich, *Separatrices and limit cycles of a certain differential equation,* (Russian), Diff. Uravn. 7, 782-790. (Engl.Transl. 597-602)

[72,CZ1] L.A.Cerkas and L.I.Zhilevich, *The absence of limit cycles for a differential equation,* (Russian), Diff.Uravn.8,2271-2273.(Engl.Transl.1760-1762)

[72,CZ2] L.A.Cerkas and L.I.Zhilevich, *The limit cycles of certain differential equations,* (Russian), Diff. Uravn. 8,no.7,1207-1213. (Engl. Transl. 924-929)

[72,R] G.S.Ryckov, *The limit cycles of the equation $u(x+1)du=(-x+ax^2+bxy+cu+du^2)dx$,* (Russian), Diff. Uravn. 8, 2257-2259. (Engl.Transl.1748-1750)

[73,A] B.P.Averin, *On the limit cycles of a system, close to a linear conservative system,* (Russian), III All Union Conference on the Qualitative Theory of Differential Equations, Samarkand, October 1973, Abstracts 5-6.

[73,F1] V.F.Filipcov, *On the question of the algebraic integrals of certain systems of differential equations,* (Russian), Diff. Uravn. 9, 469-476, 588. (Engl. Transl. 359-363)

[73,F2] V.F.Filipcov, *Algebraic limit cycles* (Russian), Diff. Uravn. 9, 1281-1288, 1358. (Engl.Transl. 983-988)

[74,AJ] C.Atkinson and C.W.Jones, *Similarity solution in some nonlinear diffusion problems and in boundary layer flow of a pseudo plastic fluid,* Quarterly Journal of Mechanics and Applied Mathematics, Vol. XXVII, Part 2, 193-211.

[74,A] B.P.Averin, *On the limit cycles of a system, close to a conservative system,* (Russian), Diff. Eq. Sbornik Trud. Mat. Kat. Ped. R.S.F.S.R., no. 3, Rjazan, 3-6.

[74,B] F.Brauer, *On the populations of competing species,* Mathematical Bioscience 19, 299-306.

[74,CZ] L.A.Cherkas and L.I.Zhilevich, *The limit cycles of a quadratic differential equation,* (Russian), Diff.Uravn.10,947-949.(Engl.Transl.732-734)

[74,T] F.Takens, *Forced oscillations and bifurcations,* Comm.Math. Inst. R.U.Utrecht, no.3, 1-59.

[74,VS] N.I.Vulpe and K.S.Sibirskii,*Affine classification of a quadratic system*, (Russian), Diff. Uravn. 10, 2111-2124. (Engl. Transl. 1633-1643)

[75,BDD] E.Bernard-Weil, M.Duvelleroy and J.Droulez, *Analogical study of a model for the regulation of ago-antagonistic couples, Application to adrenal-postpituitary interrelationships*, Mathematical Biosciences 27, 333-348.

[75,D1] T.A.Druzhkova, *A certain differential equation with algebraic integrals*, (Russian), Diff. Uravn. 11, no.2, 262-267. (Engl.Transl. 199-203)

[75,D2] T.A.Druzhkova, *The order of an algebraic integral curve of a differential equation*, (Russian), Diff. Uravn. 11, no.7, 1338. (Engl. Transl. 1006)

[75,KPS] E.F.Kirnitskaja, M.N.Popa and K.S.Sibirskii, *New applications of affine invariants of quadratic differential systems*, (Russian), Akad. Nauk Moldavian S.S.R., Mat. Issl. Kishineb, II(36), 137-149.

[75,P] L.M.Perko, *Rotated vector fields and the global behaviour of limit cycles for a class of quadratic systems in the plane*,Journal of Differential Equations, 18, 63-86.

[75,YC] Ye Yanqian and Chen Lansun, *Uniqueness of the limit cycle of the system of equations $dx/dt=y+\delta x+lx^2+xy+ny^2$, $dy/dt=x$*, (Chinese), Acta Math. Sinica 18, 219-222.

[76,C1] L.A.Cherkas, *Absence of cycles for equations $y'=\frac{Q_2(x,y)}{P_2(x,y)}$ having foci of the third degree of coarsness*, (Russian), Diff.Uravn.12, no.12, 2281-2282. (Engl.Transl. 1594-1595)

[76,C2] L.A.Cherkas, *Number of limit cycles of an autonomous second order system*, (Russian), Diff. Uravn. 12, no.5, 944-946. (Engl.Transl.666-668)

[76,E] R.M.Evdokimenko, *Behaviour of integral curves of a dynamic system*, (Russian), Diff. Uravn. 12, no.9, 1557-1567. (Engl.Transl. 1095-1103)

[76,S] K.S.Sibirskii, "Algebraic invariants of differential equations and matrices," (Russian), Izdatv. "Stinca", Kishinev, 268 pp. Translated into English as; "Introduction to the algebraic theory of invariants of differential equations", Series of Nonlinear Science, Theory and applications, Manchester University Press.

[76,T] G.Tavares dos Santos, *Classification of generic quadratic vector fields with no limit cycles*, Lecture Notes in Math. LN 597, Geometry and Topology, Springer Verlag 1977, 605-940.

[77,Ce] L.A.Cherkas, Estimation of the number of limit cycles of autonomous systems, (Russian), Diff.Uravn. 13, no.5, 779-802. (Engl.Transl. 529-547,1978)

[77,C] Chen Lansun, *Uniqueness of the limit cycle of a quadratic system in the plane,* (Chinese), Acta Mat. Sinica 20, no.1, 11-13.

[77,M] R.F.Matlak, *A new proof of the absence of limit cycles in a quadratic autonomous system,* Journ.Austral.Mat.Soc,Ser.A.23,no.3,284-291.

[77,V.S] N.I.Vulpe and K.S.Sibirskii, *Geometric classification of quadratic systems,* (Russian), Diff.Uravn.13,no.5,803-814.(Engl.Transl.548-556)

[78,A] E.A.Andranova,*On the limit cycles of quadratic systems of differential equations,* (Russian), Methods in the qualitative theory of differential equations, University of Gorki, 186-193.

[78,KS] E.F.Kirnitskaya and K.S.Sibirskii,*Conditions for a quadratic differential system to have two centers,* (Russian) Diff. Uravn. ,14,no.9, 1589-1593. (Engl.Transl. 1129-1133)

[78,M] M.Mieussens, *Sur les cycles limites de* $y' = -\frac{x+(1+\varepsilon)x^2+2xy-y^2}{y-2xy+y^2}$ *et de* $y' = \frac{x+\lambda y+x^2-(5+\varepsilon)xy-y^2}{\lambda x-y+(2-\delta)xy-y^2}$, Université de Bordeaux I, U.E.R. de Mathématiques et Informatique, no.7803, 19 pp.

[78,N] T.A.Newton, *Two dimensional homogeneous quadratic differential systems,* SIAM Review, 29, no.1, 120-138.

[78,OI] N.F.Otrov and A.I.Ilanamov, *On differential equations not having multiple limit cycles,* (Russian), Differential and Integral Equations, Gorki, no.2, 3-10.

[78,S] N.N.Serebriakova, *On the existence of a limit cycle of multiplicity three in a certain differential equation; bifurcation from a curve in a conservative system,* Methods in the qualitative theory of differential equations, University of Gorki, 181-185. (Russian)

[78,V] H.R. van der Vaart, *Conditions for periodic solutions of Volterra differential systems,* Bulletin of Mathematical Biology, 40,133-160.

[79,C] Cai Suilin, *A note to the thesis "Classification of generic quadratic vector fields without limit cycles",* (Chinese), Zhejiang University Journal, no. 4, 105-113.

[79,CW] Chen Lansun and Wang Mingshu, *Relative position and number of limit cycles of a quadratic differential system,* (Chinese), Acta Math.Sinica 22, 751-758.

[79,D] Tsutomu Date, *Classification and analysis of two-dimensional real homogeneous quadratic differential equation systems,* Journal of Differential Equations 32, 311-334.

[79,E1] C.Escher, *Models of chemical reaction systems with exactly evaluable limit cycle oscillations,* Zeits. fur Physik B 35, 351-361.

[79,E2] C.Escher, *On chemical reaction systems with exactly evaluable limit cycle oscillations*, J.Chem.Phys. 70(9).

[79,E] R.M.Evdokimenko, *Investigation in the large of a dynamical system with a given integral curve*, (Russian), Diff. Uravn. 15, no.2, 215-221. (Engl.Transl. 145-149)

[79,R] J.W.Reyn, *Generation of limit cycles from separatrix polygons in the phase plane*, Geometrical Approaches to Differential Equations, Proceedings, Scheveningen, The Netherlands, Lecture Notes in Mathematics, LN 810, 264-289.

[80,C] P.Curz, *Stabilité locale des systèmes quadratiques*, Ann.Scient.Ec. Norm. Sup. Serie 4, Tome 13, 293-302.

[80,E1] C.Escher, *Global stability analysis of open two variable quadratic mass action systems with elliptical limit cycles*, Z.Phys B, Condensed Matter 40, 137-141.

[80,E2] C.Escher, *Models of chemical reaction systems with exactly evaluable limit cycle oscillations and their bifurcation behaviour*, Ber. Bunsenges. Phys. Chem. 84, 387-391.

[80,K] J.N.Kapur, *The effect of harvesting on competing populations*, Institute of Technology, Kanpur,India,Mathematical Biosciences 51,175-185.

[80,M] M.Mieussens, *Sur les cycles limites des systémes quadratiques*, C.R.Acad.Sc., Paris, Tom. 291, Serie A, 337-340.

[80,S1] Shi Songling, *A concrete example of the existence of four limit cycles for plane quadratic systems*, Scientia Sinica, Vol. 23, no.2,February, 1980, 153-158. Appeared in Chinese in Sc. Sin. Vol. 11, 1051-1056.

[80,S2] Shi Songling, *Topological classification of generic quadratic vector fields with no limit cycles*, Report at the Third National Conference on Diff. Eq., Kunming, 5 pp.

[81,B1] R.I.Bogdanov, *Versal deformation of a singularity of a vector field on the plane in the case of zero eigenvalues*, Selecta Math.Sov.1,no.4, 389-421. (Russian original in Trudy Sem. Petrovsk. 2, 37-65, 1976)

[81,B2] R.I.Bogdanov, *Bifurcation of the limit cycle of a family of plane vector fields*, Selecta Math. Sov. 1, no.4, 373-387. (Russian original in Trudy Sem. Petrovsk. 2, 23-35, 1976)

[81,C] Cai Suilin, *A quadratic system with a third order fine focus*, (Chinese, English summary), Chinese Ann. of Math.,Vol.2, no.4, 475-478.

[81,Ch] L.A.Cherkas, *Bifurcation of limit cycles of a quadratic system with variation of the parameter rotating the field*, (Russian), Diff. Uravn. 17, no 11, 2002 2016. (Engl. Transl. 1265-1270)

[81,E] C.Escher, *Bifurcation and coexistence of several limit cycles in models of open two variable quadratic mass action systems*, Chemical Physics 63, 337-348.

[81,K] J.P.Keener, *Infinite period bifurcation and global bifurcation branches*, SIAM Journal of Applied Mathematics, 41, no.1, 127-144.

[81,S] Shi Songling, *On limit cycles of plane quadratic systems*, Scientia Sinica, Vol. XXIV, no.2, 153-159.

[82,BV] W.J.van den Broek and F.Verhulst, *A generalized Lane-Emden-Fowler equation*, Math. Meth. in the Appl. Sci., 259-271.

[82,CT] C.Chicone and Tian Jinhuang, *On general properties of quadratic systems*, Amer. Math. Monthly 89, no. 3, 161-178.

[82,D] Du Xingfu, *Quadratic systems with a weak focus of order three*, Science Bull. 16, 1020.

[82,E] C.Escher, *Double Hopf bifurcation in plane quadratic mass action systems*, Chemical Physics 67, 239-244.

[82,I] Ju.S.Il'jasenko, *The multiplicity of limit cycles arising from perturbations of the form $w' = \frac{P_2}{Q_1}$ of a Hamiltonian equation in the real and complex domain*, Trudy Sem. Petrovsk. 3, 49-60. (Translated by the AMS Transl.(2), Vol. 118,191-202,1982)

[82,LOT] A.A.Lacey, J.R.Ockedon and A.B.Taylor, *"Waiting time" solutions of a nonlinear diffusion equation*, Siam Journ. of Appl. Math., Vol.42, no.6, 1252-1264.

[82,LS] V.A.Lunkevich and K.S.Sibirskii, *Integrals of a general quadratic system in case of a center*, (Russian), Diff. Uravn. 18, no.15, 786-792.(Engl. Transl. 563-568)

[82,QSC] Qin Yuanxun, Shi Songling and Cai Suilin, *On the limit cycles of planar quadratic systems*, Scientia Sinica (Series A), 25, no.1,41-50.

[82,S] Shi Songling, *An example of a quadratic system with two separate limit cycles not including each other*, (Chinese), Acta Math. Sinica 25, no.6, 657-659.

[82,SR] D.W.Storti and R.H.Rand, *Dynamics of two strongly coupled van der Pol oscillations*, Int. Journ. of Nonlinear Mechanics, Vol. 17, no.3, 143-152.

[82,Y] Yang Xinan, *The number of limit cycles of a bounded quadratic system with two finite singular points*, I, (Chinese), Acta Math. Sinica 25, no. 3, 297-301.

[82,Ye] Ye Yanqian, *Some problems in the qualitative theory of ordinary differential equations,* Journ. of Diff. Eq. 46, no.2, 153-164.

[82,WL] Wang Mingshu and Lin Yingju, *The non existence of limit cycles of some quadratic differential systems,* (Chinese, Engl. summary), Chinese Ann. Math. 3,(6),721-724.

[83,A] E.A.Andronova, *Quadratic systems that are close to conservative with four limit cycles,* (Russian), Methods of the qualitative theory of differential equations, Gorkov. Gos. Univ. Gorki, 118-126, 166-167.

[83,B] R.Bamon, *A class of planar quadratic vector fields with a limit cycle surrounded by a saddle loop,* Proc. Amer. Math. Soc.,Vol.88, no.4,719-724.

[83,CS] C.Chicone and D.S.Schafer, *Separatrix and limit cycles of quadratic systems and Dulac's theorem,* Transactions of the Amer. Math. Soc., Vol. 278, no.2, 585-612.

[83,L] Li Chengzhi, *Two problems of plane quadratic systems,*Scientia Sinica (Series A),Vol.26, no.5, 471-481.

[83,QSC] Qin Yuanxun Suo, Guangjian and Du Xingfu, *On limit cycles of planar quadratic quadratic systems,* Sci. Sinica Ser. A 26, no. 10, 1025-1033.

[83,V] N.I.Vulpe, *Affine invariant conditions for the topological discrimination of quadratic centers with a center,* (Russian), Diff. Uravn. 19, no.3, 371-379. (Engl.Transl. 273-280)

[83,W] G.Wanner, *On Shi's counterexample for Hilbert's 16th problem,* (In German),Yearbook: Surveys of Mathematics 16 (1983), 9-24, Mannheim.

[83,Y] Yang Xian, *Number of limit cycles of the bounded quadratic system with two finite singular points,* Chin.Ann. of Math. 4B(2), 217-223.

[83,YW] Ye Yanqian and Wang Mingshu, *An important property of the weak focus of a quadratic differential system,* (Chinese), Chinese Ann. Math., Ser.A 4,no.1,65-69.(Engl.summ.in Chinese Ann. Math. Ser.B 4, no.1, 130.)

[84,BL] T.R.Blows and N.G.Lloyd, *The number of limit cycles of certain polynomial differential equations,* Proc. of the Royal Society of Edinburgh, 98A, 215-239.

[84,CW] Cai Suilin and Wang Zhongwei, *A quadratic system with second order fine focus,* (Chinese, English summary), Chinese Ann. of Math., Vol. 5A, no.6, 765-770.

[84,PS] L.M.Perko and Shu Shihlung, *Existence, uniqueness and non existence of limit cycles for a class of quadratic systems in the plane,* Journ. of Diff. Eq.,53, no.2, 146-171.

[84,P] L.M.Perko, *Limit cycles of quadratic systems in the plane,* Rocky Mountain Journ. of Math., Vol. 14, no.3, 619-645.

[85,C] Chen Weifeng, *The uniqueness of the limit cycle for a class of quadratic systems with a fine focus of order two*, (Chinese, English summary), Zhejiang Daxue Xuebao 19, no.3, 107-113.

[85,Ch] Chen Shueping, *Limit cycles of a quadratic system with a parabola as a particular integral*, (Chinese), Kexue Tongbao 30, no.6, 401-405.

[85,H] Han Maoan, *Uniqueness of limit cycles of a quadratic system of type $(III)_{n=0}$ around a focus of order two*, (Chinese, English summary), Chinese Ann. Math. Ser. A 6, no.6, 1, 124-125.

[85,HL1] Huang Qiming and Li Jibin, *Study by computer of a planar quadratically distributed Hamiltonian system with two centers*, (Chinese, English summary), Journal of Yunnan University, Vol.7, no.4, 392-399.

[85,HL2] Hu Qinxun and Lu Yugi, *On the singular points at infinity of second order planar systems*, (Chinese, English summary), Journal of Beijing College of Technology, no.2, 16-28.

[85,LY] Liang Zhaojun and Ye Yanqian, *The global structure and bifurcation curves of systems of type $II_{\delta=n=0}$*, Journal of Eng. Math., Vol.2, no.1, 32-37. (Chinese, English summary)

[85,L1] Li Chengzhi, *The quadratic systems possessing two weak foci*, (English), Annals of Differential Equations, Vol.1, no.2, 161-169.

[85,L2] Li Chengzhi, *Planar quadratic systems possessing two centers*, (Chinese), Acta Math. Sinica 28, no.5, 644-648.

[85,Q] Qin Yuanxun, *On surfaces defined by ordinary differential equations: a new approach to the Hilbert's 16th problem.* Conference on Ordinary and Partial Differential Equations, Dundee 1984, Lecture Notes in Mathematics 1151, Springer Verlag, 115-131.

[85,R] A.E.Rudenok, *Limit cycles of a two dimensional autonomous system with quadratic non linearities*, (Russian), Diff. Uravn. 21, no.12, 2072-2082, 2203-2204. (English Translation 1390-1398)

[85,WC] Wang Do Da and Chen Lansun, *The (3,1) distribution of the limit cycles of a quadratic system*, Acta Mathem. Sinica 28, no.3,407-413. (Chinese)

[85,Y] Ye Yanqian, *On the impossibility of (2,2) distribution of limit cycles of any real quadratic differential system*, Nanjing Daxue Xuebao Shuxue Bannian Kan 2, no.2, 161-182.

[86,A] E.A.Andronova, *On the topology of quadratic systems with four (or more) limit cycles*, (Russian), Uspekhi Mat. Nauk 41, no.2, 183-184. (Engl. Transl. Russian Math. Surveys 41, no.2, 191-192)

[86,B] R.Bamon, *Quadratic vector fields in the plane have a finite number of limit cycles,* Inst. Hautes Etudes Sci. Publ. Math.,no. 64, 111-142.

[86,Cg] Chen Guangqing, *On the limit cycles of the quadratic system $E_2(p)$,* Proc. of the 1983 Being Symposium on Differential Geometry and Differential Equations, 411-412.

[86,Ch1] Chen Mingxiang, *Global topological structures of the phase portraits of the equations $\dot{x}=-y+5axy+y^2,\dot{y}=x+ax^2+5xy$,* (Chinese), Journ. Central China Normal University, no.2, (summ.4), 96-103.

[86,Ch] Chen Weifeng, *A class of quadratic systems with a second order weak focus,* (Chinese, English summary),Chin.Ann.Math.Ser.B 7, no.2, 257-258. Also in Chin.Ann.Math.Ser.A 7,no.2, 201-211.

[86,C] L.A.Cherkas, *Absence of limit cycles around a triple focus in a quadratic system in a plane,* (Russian), Diff.Uravn.,22,no.11, 2015-2017, 2024.

[86,Co1] W.A.Coppel, *A simple class of quadratic systems,* Journal of Differential Equations, 64, no. 3, 275-282.

[86,Co2] W.A.Coppel, *The limit cycle configuration of quadratic systems,* The Australian National University, Research Report no. 32, Dept. of Math. IAS., Ninth Conference on Ordinary and Partial Differential Equations, University of Dundee, 30 june-4 july 1986, Pitman Research Notes in Mathematics 157, 52-65, Longman Scientific and Technical.

[86,GSL] A.Gasull, Sheng Liren and J.Llibre, *Chordal quadratic systems,* Rocky Mountain Journal of Mathematics, 16, no.4, 751-782.

[86,JL] Jiang Jifa and Liu Zhengrong, *Quadratic systems with Poincaré bifurcations,* (Chinese), Journal of the Anhui University, no.3, 15-22.

[86,LX] Li Cunfu and Xu Yumin, *Poincaré bifurcations of a quadratic system with a pair of isochronous centers,* Journal Kunming Institute of Technology, no.4, 80-87.

[86,L] Li Chengzhi, *Non existence of limit cycles around a weak focus of order three for any quadratic system,* Chin. Ann. Math. Ser. B 7, no.2, 174-190. (Chinese summary in Chin. Ann. MAth. Ser. A 7,no.2, 239)

[86,LC] Li Jibin and Chen Xiaoqiu,*Poincaré branching of a class of second degree planar systems,* (Chinese), Kexue Tongbao 31 (1986), no.16, 1213-1217. (English edition in Kcxuc Tongbao 32,(1987), no.10, 655-660.)

[86,L1] Liang Zhaojun, *Introduction to a global analysis of the polynomial differential systems,* (Chinese), Journal of the Central China Normal University, no.2, (Sum.4), 1-71.

[86,SC] Suo Guangjian and Chen Yongshao, *The real quadratic system with two imaginary straight line solutions,* Ann. Diff. Eq. 2, no.2, 197-207.

[86,Ya] Yan Zhong, *The study of limit cycles for the system of type* $(II)_{n=0}$, Ann. Diff. Eqs, 2, no.4, 461-474.

[86,Y] Ye Yanqian, *Recent contributions of Chinese mathematicians to the theory of quadratic systems,* Proceedings of the 1983 Bejing Symposium on differential geometry and differential equations, Science Press, Bejing, 291-302.

[86,Z1] Zhu Deming, *The topological classification of structurally stable quadratic systems without limit cycles,* (Chinese, English summary), Nanjing Daxue Xuebao Ziran Kexue Ban 22, no.2, 263-273.

[86,Z2] Zhu Deming,*Planar quadratic differential equations with a weak saddle,* Ann. Diff. Eq. 2, no.4, 497-508.

[87,C] Cai Suilin, *The weak saddle and separatrix cycle of a quadratic system,* (Chinese), Acta Math. Sinica 30, no.4, 553-559.

[87,CZ] Cai Suilin and Zhang Pingguang, *A quadratic system with a weak saddle,* (Chinese), Journ. Math. Res. Exp., no 1, 63-68.

[87,CP] M.I.T. Camacho and C.F.B. Palmeira,*Non singular quadratic differential equations in the plane,* Trans.Amer.Math.Soc., 301, no.2, 845-859.

[87,Ch] Chen Shuping,*Limit cycles of second order systems having parabolic integral curves,* (Chinese), Journ. of Math. Res. Exp., no.1, 153-155.

[87,CG] L.A.Cherkas and V.A.Gaiko, *Bifurcation of limit cycles of a quadratic system with two critical points and two field rotating parameters,* (Russian), Diff. Uravn. Vol. 23, no.9, 1544-1553. (English Translation Diff. Eqs. 23, no.9, 1062-1069)

[87,CGL] B.Coll, A.Gasull and J.Llibre, *Some theorems on the existence, uniqueness and nonexistence of limit cycles for quadratic systems,* Journ. of Diff. Eqs. 67, no.3, 372-399.

[87,Col] B.Coll, *A qualitative study of some classes of planar vector fields,* (English, Title and Introduction in Catalan), Thesis at the University Autònoma de Barcelona, 167 pp.

[87,Co] R.Conti, *Centers of quadratic systems,* Ricerche di Matematica, Suppl. Vol. XXXVI, 117-126.

[87,DGZ] B.S.A.Drachman, S.A.van Gils and Zhang Zhifen, *Abelian integrals for quadratic vector fields,* Journ. fur Reine und Angewandte Math. 382, 165-180.

[87,DRS] F.Dumortier, R.Roussarie and J.Sotomayor, *Generic 3-parameter families of vector fields in the plane, unfolding a singularity with nilpotent linear part. The cusp case,* Ergod. Theory Dynam. Systems ,375-413.

[87,J] P.de Jager, *An example of a quadratic system with a bounded separatrix cycle having at least two limit cycles in its interior,* Delft Progress Report 11 (1986/87), no. 3-4, 141-150.

[87,LC] Li Jibin and Chen Xiaoqiu,*Poincaré bifurcations in the quadratic planar differential system,* Kexeue Tongbao, Vol. 32, no. 10, 655-660.

[87,R] J.W.Reyn, *Phase portraits of a quadratic system of differential equations occurring frequently in applications,* Nieuw Archief voor Wiskunde, (4)5, no.2, 107-155.

[87,ZC] Zhang Pingguan and Cai Suilin, *Quadratic systems with second and third order weak saddle points,* (Chinese), Acta Math. Sinica 30, no.4, 560-565.

[87,Z] Zhu Deming, *Saddle values and integrability of quadratic differential systems,* Chin. Ann. Math. Ser.B 8, no.4, 466-478. (Chinese summary in Chin.Ann. Math. Ser. A, no.5, 645)

[88,B] D.Bularas, *Existence d'un centre isochrone dans les systemes differentiels quadratiques plans,* Acts Edo 1987, Algers Cahiers Mathematiques, Fasc. 1,(1988), 43-47.

[88,CP] M.I.T. Camacho and C.F.B. Palmeira, *Errata to "Nonsingular quadratic differential equations in the plane",* Transactions of the American Mathematical Society, Vol. 307, no.1, 431.

[88,Ch1] Chen Weifeng, *A quadratic system with a second order weak focus,* (Chinese, English summary), Journal of Zhejiang University, Vol.22, no.1, 141-155.

[88,Ch2] Chen Weifeng, *A quadratic system with a second order weak focus III* (Chinese, English summary), Journal of Zhejiang University, Vol. 22, no. 5, 1-10.

[88,CGL] B.Coll, A.Gasull and J.Llibre,*Quadratic systems with a unique finite rest point,* Publicaciones Matematiques, Vol. 32, 199-259.

[88,Co] W.A.Coppel, *Quadratic systems with a degenerate critical point,* Bull. Austr. Math. Soc., Vol. 38, 1-10.

[88,GL] A.Gasull and J.Llibre,*On the nonsingular quadratic differential equations in the plane,* Proc. of the Amer. Math. Soc. 104, no.3, 793 794.

[88,H] C.Holmes, *Some quadratic systems with a separatrix cycle surrounding a limit cycle,* Journ. London Math. Soc. (2), 37, 545-551.

[88,LC] Li Jibin and Chi Yuehua, *Global bifurcation and chaotic behaviour in a distributed quadratic system with two centers,* Acta Math. Applicata Sinica 11, 312-323. (Chinese)

[88,L] N.G.Lloyd,*Limit cycles of polynomial systems; some recent developments, New directions in dynamical systems,* London Math. Soc. Lecture Notes Serie 127, 192-234.

[88,RS] C.Rousseau and D.Schlomiuk, *Generalized Hopf bifurcations and applications to planar quadratic systems,* Annales Polonici Mathematici LXIX, 1-16.

[88,Ya] Yang Xinan, *A bounded quadratic system with a weak focus,* Ann. of Diff. Eq., Vol.4, no.2, 231-242.

[88,Y] Ye Weiyin, *On real quadratic systems possessing two conjugate complex straight line solutions,* Ann. of Diff. Eq. 4(4), 491-501.

[89,BVS] D.Bularas, N.J.Vulpe and K.S.Sibirskii, *Solution of the problem "in the large" for a general quadratic differential system,* (Russian), Diff. Uravn. Vol. 25, no. 11, 1856-1862.

[89,Ch] C.Christopher, *Quadratic systems having a parabola as an integral curve,* Proceedings of the Royal Society of Edinburgh, 112A, 113-114.

[89,Co] W.A.Coppel, *Some quadratic systems with at most one limit cycle,* Dynamics Reported, Vol.2, 61-88, Eds. U.Kirchkraber and H.O.Walther, John Wiley and Sons Ltd. and B.G.Teubner.

[89,GRS] J.Guckenheimer, R.Rand and D.Schlomiuk, *Degenerate homoclinic cycles in perturbations of quadratic Hamiltonian systems,* Nonlinearity, 2, 405-418.

[89,H] J.Hulshof, *Similarity solutions of the porous medium equation with sign changes,* Math. Inst., University Leiden, The Netherlands,W 89-12,1-31.

[89,J] P. de Jager, *Phase portraits of quadratic systems. Higher order singularities and separatrix cycles,* Thesis,Technological University Delft,1-139.

[89,JR] P.Joyal and C.Rousseau, *Saddle quantities and applications,* Journ. of Diff. Eqs. 78, 374-399.

[89,ZC] Zhang Pingguang and Cai Suilin, *Quadratic systems possessing a weak focus of order two and a singular point with a zero characteristic root,* Kexue Tongbao, Vol.7, 486-489.

[89,Z1] Zhang Pingguang, *Research on the nonexistence of a limit cycle around a second or third order fine focus in a quadratic system,* (Chinese), Kexue Tongbao (Science Record), Vol.34. no. 18, 1365-1368.

[89,Z2] Zhu Deming, *A general property of quadratic differential systems,* Chin. Ann. Math. 10B(1), 26-32.

[90,A] J.C.Artes, *Quadratic differential systems,* (Catalan), Facultat des Ciencies, Universitat Autonoma de Barcelona, Departement de Matematques, Doctor Thesis, 268 pp.

[90,BVS] D.Bularas,N.J.Vulpe and K.S.Sibirskii, *The problem of a center "in the large" for a general quadratic system,* Soviet Math. Dokl. Vol.41, no.2, 287-290. (Translation by A.M.S.(1991) of Dokl. Akad. Nauk S.S.S.R., Vol. 311 (1990), no. 4, 777-780.)

[90,CZ] Chen Wenchen and Zhang Pingguang, *Uniqueness of limit cycles for a quadratic system with a second order weak focus,* (Chinese), Collected works on dynamical systems, Department of Mathematics, Zhejiang University, 49-67.

[90,Co] W.A.Coppel, *Quadratic systems with a critical point of higher multiplicity,* Differential and Integral Equations, Vol.3, no. 4, 709-720.

[90,GH] L.Gavrilov and E.Horozov, *Limit cycles and zeroes of Abelian integrals satisfying third order Picard-Fuchs equations,* Laboratoires de Mathématiques Appliquées, C.N.R.S. Unité associée 1204, Université de Pau et des Pays de l'Adour, 31pp..

[90,GM] J.Gratton and F.Minotti, *Self similar viscous gravity currents: phase plane formalism,* Journal of Fluid Mechanics, Vol. 26, 345-348.

[90,J] P.de Jager, *Phase portraits for quadratic systems with a higher order singularity with two zero eigenvalues,* Journal of Differential Equations 87, 169-204.

[90,KS] Ju.F.Kalin and K.S.Sibirskii, *Conditions for the different qualitative portraits for general quadratic systems with centers and $I_9 \neq 0$,* (Russian), Izv. Akad. Nauk Moldav. S.S.R., Mathematics, no.1, 17-26,77.

[90,K] Ju.F.Kalin, *Conditions for the qualitative portraits for general quadratic systems with centers and $I_9=0, K_1 \neq 0$,* (Russian), Izv. Akad. Nauk Moldav. S.S.S.R. Mathematics, no.3, 68-71.

[90,SGR] D.Schlomiuk, J.Guckenheimer and R.Rand, *Integrability of plane quadratic vector fields,* Expositiones Mathematicae 8, 3-25.

[90,S] D.Schlomiuk, *Une charactérisation géometrique générique des champs de vecteurs quadratique avec un centre,* C.R.Acad. Sci., Paris 310, Sectie I, 723-726.

[90,Z1] Zhang Pingguang, *Study of non existence of limit cycles around a weak focus of order two or three for quadratic systems,* Chinese Science Bulletin, Vol. 35, no.14, 1156-1161.

[90,Z2] Zhang Pingguang, *On the uniqueness of limit cycle and stability of separatrix cycles for a quadratic system with a weak saddle,* Northeastern Mathematical Journal, 6(2), 243-252.

[90,Z3] Zhang Pingguang, *Uniqueness of limit cycles for quadratic systems of class $(II)_{l-0}$,* (Chinese), Collected works on dynamical systems, Department of Mathematics, Zhejiang University, 38-48.

[90,Z4] Zhang Pingguang, *The uniqueness of limit cycles of quadratic system* $(II)_{m=0}$, Chinese Science Bulletin, Vol.35, no. 5, 360-365.

[90,ZC] Zhang Pingguang and Cai Suilin, *Quadratic systems possessing a weak focus of order two and a singular point with a zero characteristic root*, Chinese Science Bulletin, Vol. 35, no.6, 459-463.

[91,CZ] Chen Wencheng and Zhang Pingguang, *On the uniqueness of limit cycles of a quadratic system with a weak focus of order two*, (Chinese, English summary), Journal of System Science and Mathematical Sciences, 11(3), 217-226.

[91,CJ] C.Chicone and M.Jacobs, *Bifurcation of limit cycles from quadratic isochrones*, Journal of Differential Equations, 91, no.2, 268-326.

[91,Co] W.A.Coppel, *A new class of quadratic systems*, Journal of Differential Equations, Vol.92, 360-372.

[91,DF] F.Dumortier and P.Fiddelaers, *Quadratic models for generic local 3-parameter bifurcations on the plane*, Trans. of the Amer. Math. Soc., Vol.326, no.1, 101-126.

[91,DRS] F.Dumortier, R.Roussarie and J.Sotomayor, *Generic 3-parameter families of planar vector fields; unfoldings of saddle, focus and elliptic singularities with nilpotent linear part*, Springer Lecture Notes in Mathematics, 1480, 1-164.

[91,I] Yu. S. Il'yashenko, *Finiteness theorems for limit cycles*, Algebra and Analysis, (Kemerovo 1988), 55-64. Translated by A.M.S. Series 2, 148, Providence, RI.(1991).

[91,KR] R.E.Kooij and J.W.Reyn, *On the phase plane analysis of self similar gravity currents*, Delft Progress Report 15, 21-31.

[91,MNP] D.S.Malkus, J.A.Nohel and B.J.Plohr, *Analysis of new phenomena in shear flow of non- Newtonian fluids*, SIAM Journal of Appl. Math. 51, no.4, 899-929.

[91,SGZ] F.M.A.Salam, S.A.van Gils and Zhang Zhifen, *Global bifurcation analysis of an adaptive control system*, Differential and Integral Equations, Vol.4, no.6, 1353-1374.

[91,S1] Shen Boqian, *The problem of the existence of limit cycles and separatrix cycles of cubic curves in quadratic systems*, (Chinese), Chinese Ann. of Math. Ser. A 12, no.3, 382-389.

[91,S2] Shen Boqian, *A necessary and sufficient condition for the existence of parabolic separatrix cycles in a quadratic system*, (Chinese, Engl. summary), Journal Math. Res. Expositions 11, no.3, 425-431.

[91,V] N.I,Vulpe,*Conditions on the coefficients for the number and multiplicity of singular points of quadratic systems,* Diff. Uravn. tom 27, no.4, 572-578. (Engl. transl. 397-400)

[91,VN] N.I.Vulpe and I,V.Nikolaev, *Topological classification of quadratic systems with a unique second order singular point; the case with a non zero eigenvalue),* Qualitative theory of complex systems,RGPU,Leningrad,25-43.

[91,Z] A.Zegeling, *Limit cycles in quadratic systems,* Doctoral thesis, University of Technology, Delft, The Netherlands, 1-160.

[91,ZC] Zhang Pingguang and Cai Suilin, *Quadratic systems with a weak focus,* Bull. Austral. Mathem. Soc., vol.44, 511-526.

[91,Zh] Zhang Pingguang, *The uniqueness problem of limit cycles of quadratic systems $(II)_{l=0}$,* Ann. of Diff. Eqs. 7(2), 243-249.

[91,Zy] Yu Zhuang, *Limit cycles of quadratic systems with parabolic solutions,* (Chinese, Engl. summ.), Shandong Kuangye Xueyuan Xue Bao, 10, no.2, 203-207.

[92,BV] V.A.Baltag and N.I.Vulpe, *The number and multiplicity of the singular points of the quadratic system,* (Russian), Dokl. Akad. Nauk 323, no.1, 9-12.

[92,F] P.Fiddelaers, *Local bifurcations of quadratic vector fields,* Thesis, Limburgs Universitair Centrum, Faculteit Wetenschappen, Diepenbeek, Belgie, 116 pp..

[92,K] Yu.F.Kalin, *Conditions for the topological differences in quadratic systems with a center for $K_1 \equiv 0$,* Bulletin Academici de Stince a Republici Moldova, Matematika, 3(9), 43-45.

[92,P] L.M.Perko, *A global analysis of the Bogdanov-Takens system,* SIAM Journal of Applied Mathematics, Vol. 52, no.4, 1172-1192.

[92,SS] Shen Boqian and Song Yan, *Topological structure of quadratic systems which have a parabolic solution,* (Chinese), Journal Liaoning Normal University (Natural Science), Vol.15, no. 3, 177-182.

[92,VN1] N.I.Vulpe and I.V.Nikolaev,*Conditions for the coefficients for quadratic systems having two finite singular points, the sum of their multiplicities being four,* Buletinal Academici de Stinte a Republici Moldova, Matematica, 1(7), 51-59.

[92,VN2] N.I.Vulpe and I.V.Nikolaev, *Topological classification of quadratic systems with a unique third order singular point,* Buletinal Academici de Stinte a Republici Moldova, Matematica, 2(8), 37-44.

[92,X] Xie Xiangdong, *On the uniqueness of limit cycles of bounded quadratic systems with a weak focus,* Ann.of Diff.Eqs., Vol.8, no.3, 1-5 (379-383).

[92,Xu] Xu Changgjiang, *Quadratic systems that have curves of the third degree as particular integrals*, (Chinese),Acta Mathematica Scientia, Vol. 12, no.1, 9-18.

[92,Y1] Ye Yanqian, *Bifurcation diagrams of quadratic differential systems having one focus and one saddle*, Centre de Recerca Matematica Institut d'Estudis Catalans, no.153, 11 pp..

[92,Y2] Ye Yanqian, *Limit cycles and bifurcation curves for the quadratic differential system $III_{m=0}$ having three antisaddles*, Research Report of CRM, Inst. d'Estudis Catalans, no. 159.

[92,ZL] Zhang Pingguang, and Li Wenhua,*A quadratic system with a weak focus and a strong focus*, Ann. of Diff. Eqs.,8(1), 122-128.

[93,BV] V.A.Baltag and N.I.Vulpe,*Affine invariant conditions for determining the number and multiplicity of singular points of quadratic differential systems*, Buletinal Academici de Stinte a Republicii Moldova, Matematica, (English), 1(11), 39-48.

[93,G] A.Gasull, *On polynomial systems with enough invariant algebraic curves*, EQUADIFF-91 International Conference on Differential Equations, 26-31 August 1991, Barcelona, Vol. 2, 531-537.

[93,GH] L.Gavrilov and E.Horozov, *Limit cycles of perturbations of quadratic Hamiltonian vector fields*, Journ. Math. Pure and Appl. 72, 213-238.

[93,NV] I.V.Nikolaev and N.I.Vulpe, *Topological classification of quadratic systems with a unique finite second order singularity with two zero eigenvalues*, Buletinul Academiei de Stinte a Republici Moldova, Matematica, (English), 1(111), 3-8.

[93,R] J.W.Reyn, *Classes of quadratic systems of differential equations in the plane*, Proc. Special Program at the Nankai Institute of Mathematics, Tianjin, P.R.China,(September 1990- June 1991), Nankai Series in Pure and Applied Mathematical and Theoretical Physics, Vol.4, Dynamical Systems, 146-180.

[93,S1] D.Schlomiuk, *Algebraic particular integrals, integrability and the problem of the center*, Trans. Amer. Math. Soc. 338, no.2, 799-841.

[93,S2] D.Schlomiuk, *Algebraic and geometric aspects of the theory of polynomial vector fields*, Proc. of the NATO Asi 1992, Kluwer, 1993.

[93,VN] N.I.Vulpe and I.V.Nikolaev, *Topological classification of quadratic systems with a fourth order singular point*, Diff. Uravn. 29, no.10, 1669-1673. Translated as Diff. Eqs., Vol.29, no.10, 1449-1453 (1994).

[93,XC] Xie Xiangdong and Cai Suilin, *Bifurcations of limit cycles for quadratic systems with an invariant parabola*, Center for Mathematical Sciences, Zhejiang University, 9307, 15 pp..

[93,Z1] Zhang Pingguang, *Quadratic systems with a fine homoclinic loop*, Nanjing Daxue Xuebao Shuxue Banmin Kan, Nanjing, 91-96.

[93,Z2] Zhang Pingguang, *Quadratic systems with a fine infinite homoclinic loop*, Nanjing Daxue Xuebao Shuexue Banmin Kan, Nanjing, 91-96.

[93,Zy] Zhuang Yu, *A global phase portrait of another class of quadratic differential systems having a parabola as a particular integral curve*, (Chinese, Engl. summ.), Shandong Kuangye Xueyuan Xuebao 12, no.4, 408-414.

[94,AL1] J.C.Artes and J.Llibre, *Quadratic Hamiltonian vector fields*, Journal of Differential Equations , Vol. 107, no.1, 80-95.

[94,AL2] J.C.Artes and J.Llibre, *Phase portraits for quadratic systems having a focus and one antisaddle*, Rocky Mountain Journal of Mathematics, 24, no.3, 875-889.

[94,DRR1] F.Dumortier, R.Roussarie and C.Rousseau, *Elementary graphics of cyclicity 1 and 2*, Nonlinearity 7, 1001-1043.

[94,DRR2] F.Dumortier, R.Roussarie and C.Rousseau, *Hilbert 16th problem for quadratic vector fields*, Journal of Differential Equations, Vol. 110, no.1, 86-133.

[94,GW] Gao Suzhi and Wang Xuejin, *A class of quadratic systems with a fine focus of order two*, (Chinese, English and Chinese summaries), Beijing Shifan Daxue Xuebao, 30, no.1, 12-15.

[94,HI1] E.Horozov and I.D.Iliev, *On saddle loop bifurcations of limit cycles in perturbations of quadratic Hamiltonian systems*, Journal of Differential Equations, 113, 84-105.

[94,HI2] E.Horozov and I.D.Iliev, *On the number of limit cycles in perturbations of quadratic Hamiltonian systems*, (English summary), Proc. of the London Mathem. Soc. (3) 69, no.1, 198-224.

[94,HK] W.T.van Horssen and R.E.Kooij, *Bifurcations of limit cycles in a particular class of quadratic systems with two centers*, Journal of Differential Equations, Vol.114, no.2, 538-569.

[94,KH] R.E.Kooij and W.T.van Horssen, *Structurally stable quadratic systems without limit cycles*, Ann. of Diff. Eqs. 10(3), 259-274.

[94,P] L.M.Perko, *Coppel's problem for bounded quadratic systems*, Department of Mathematics, Northern Arizona University, Flagstaff, U.S.A., 218 pp.

[94,T] L.A.Timochouk, *Focal values for quadratic systems with four real singular points*, Reports of the Faculty of Technical Mathematics and Informatics, Report 94-64, Delft University of Technology, 23pp..

[94,WSE] A.J.T.M.Weeren, J.M.Schumacher and J.C.Engwerda, *Asymptotic analysis of Nash equilibria in nonzero-sum linear-quadratic differential games, The two-player case,* Research Memorandum FEW 634, Report of the Katholieke Universiteit Brabant, The Netherlands, 22 pp..

[94,Z] A.Zegeling, *Separatrix cycles and multiple limit cycles in a class of quadratic systems,* Journal of Differential Equations, Vol. 113, 355-380.

[94,ZK] A.Zegeling and R.E.Kooij, *Uniqueness of limit cycles in polynomial systems with algebraic invariants,* Bull. Austral. Math. Soc., Vol.49, no.1, 7-20.

[94,Z] H.Zoladek, *Quadratic systems with a center and their perturbations,* Journal of Differential Equations, Vol.109, 223-273.

[95,CX] Cai Suilin and Xie Xiangdong, *A note of "Quadratic systems having a parabola as a integral curve,* Proc. Roy. Soc. Edinburgh, Sect. A112 (1989), no. 1-2, 113-134, by C.Christopher", (English summary), Ann. Diff. Eqs. 11 (1995), no.3, 260-263.

[95,HR1] W.T.van Horssen and J.W.Reyn, *Bifurcation of limit cycles in a particular class of quadratic systems,* Diff. Int. Eqs. 8, no.4, 907-920.

[95,HR2] Huang Xianhua and J.W.Reyn, *Separatrix configurations of quadratic systems with finite multiplicity three and a $M_{1,1}^0$ type of critical point at infinity,* Report 95-115, Faculty of Technical Mathematics and Informatics, Delft University of Technology, 1-38.

[95,HR3] Huang Xianhua and J.W.Reyn, *On the limit cycle distribution over two nests in quadratic systems,* Bulletin Australian Mathematical Society, Vol.52, 461-474.

[95,LZ] Li Baoyi and Zhang Zhifen, *A note on a G.S.Petrov's result about the weakened 16th Hilbert problem,* Journal of Mathematical Analysis and Applications, 190, 489-516.

[95,LLZ1] Li Chengzhi, J.Llibre and Zhang Zhifen, *Weak focus, limit cycles and bifurcations for bounded quadratic systems,* Journal of Differential Equations, Vol.115, no. 1, 193-223.

[95,LLZ2] Li Chengzhi, J.Llibre and Zhang Zhifen, *Abelian integrals of quadratic Hamiltonian vector fields with an invariant straight line,*Publ.Mat., Vol.39, 355-366.

[95,RK] J.W.Reyn and R.E.Kooij, *Infinite singular points of quadratic systems in the plane,* Journal of Nonlinear Analysis, Theory, Methods and Applications, Vol.24, no.6, 895-927.

[95,SZ] D.S.Shafer and A.Zegeling, *Bifurcations of limit cycles from quadratic centers,* Journal of Differential Equations, Vol.122, no.1, 48-70.

[95,VV] A.Voldman and N.I.Vulpe, *Affine invariant conditions for the topologically distinguishing quadratic systems without finite critical points,* Buletinel A.S. a R.M. Matematika, no.2,(18)- 3(19), 100-112.

[95,ZL] Zhang Zhifen and Li Baoyi, *On the number of limit cycles of quadratic Hamiltonian systems under quadratic perturbations,* Proc.of the International Conference on Dynamical Systems and Chaos, Vol.1, Tokyo, Japan, 23-27 May 1994, (1995), 306-315.

[95,Z] H.Zoladek, *The cyclicity of triangles and segments in quadratic systems,* Journal of Differential Equations, 122, no.1, 137-159.

[96,AL] J.C.Artes and J.Llibre, *Corrections on [94,AL],* Journal of Differential Equations, 129, no.2, 559-560.

[96,C] W.A.Coppel, *Non-coprime quadratic systems,* Bulletin Australian Mathematical Society, Vol.53, no.83-90.

[96,DER] F.Dumortier, M.El.Morsalini and C.Rousseau, *Hilbert's 16th problem for quadratic systems and cyclicity of elementary graphics,* Nonlinearity 9, no.5, 1209-1261.

[96,GP1] A.Gasull and R.Prohens, *Quadratic and cubic systems with degenerate infinity,* Journal of Mathematical Analysis and Applications, 198, 25-34.

[96,GP2] A.Gasull and R.Prohens, *On quadratic systems with a degenerate critical point,* Rocky Mountain Journal of Mathematics, 26, no.1, 135-164.

[96,HI] E.I.Horozov and I.D.Iliev, *Perturbations of quadratic Hamiltonian systems with symmetry,* (English and French summaries), Ann. Inst. H.Poincaré Analyse Nonlinéaire 13, no.1, 17-56.

[96,H] Huang Xianhua, *Qualitative analysis of certain nonlinear differential equations; Quadratic Systems and Delay Equations,* Thesis, Delft University of Technology, The Netherlands, 1-162.

[96,I] I.D.Iliev, *The cyclicity of the period annulus of the quadratic Hamiltonian triangle,* (English summary), Journal of Differential Equations, 128, no.1, 309-326.

[96,R] J.W.Reyn, *Phase portraits of quadratic systems without finite critical points,* Nonlinear Analysis, Theory, Methods and Applications, Vol. 27, no.2, 207-222.

[96,SWZ] D.S.Shafer, Wu Xiaonan and A.Zegeling, *A numerical approach to bifurcation from quadratic centers,* Journal of Mathematical Analysis and Applications, 202, 90-107.

[96,V] A.Voldman, *On the topological character of the singular points of quadratic systems with one triple and one simple real critical point,* Symposium Septimum Tiraspolense Generalis Topologiae et Suae Applicationem, 5-11 August 1996, Chisinau, 277-230.

[97,AL] J.C.Artes and J.Llibre, *Quadratic vector fields with a weak focus of third order,* (English summary), Proceedings of the Symposium on Planar Vector Fields (Lleida, 1996), Publ. Mat. 41, no.1, 7-39.

[97,DLZ] F.Dumortier, Li Chengzhi and Zhang Zhifen, *Unfolding of a quadratic integrable system with two centers and two unbounded heteroclinic loops,* Journal of Differential Equations, 139, 146-193.

[97,G] V.Gaiko, *Qualitative theory of two-dimensional polynomial dynamical systems: problems, approaches, conjectures,* Nonlinear Analysis, Theory, Methods and Applications, Vol.30, no.3, 1385-1394.(1997), Proceedings of the Second World Congress of Nonlinear Analysis.

[97,HL] He Yue and Li Chengzhi, *On the number of limit cycles arising from perturbations of homoclinc loops of quadratic integrable systems,* (Engl. summary), Symposium on Planar Nonlinear Dynamical Systems (Delft,1995), Differential Equations and Dynamical Systems, 5, (1997), no. 3-4, 303-316.

[97,HR] Huang Xianhua and J.W.Reyn, *Weak critical points and limit cycles in quadratic systems with finite multiplicity three,* Symposium on Planar Nonlinear Dynamical Systems (Delft,1995), Differential Equations and Dynamical Systems, 5,(1997), no.3-4, 243-266.

[97,LZ] Li Baoyi and Zhang Zhifen, *Bifurcation phenomenon of a class of planar codimension 3 polycycle $S^{(3)}$ with two saddles resonating,* (English summary), Sci. China Ser. A 40, no.12, 1259-1271.

[97,NV] I.V.Nikolaev and N.I.Vulpe, *Topological classification of quadratic systems at infinity,* (English summary), Journal of the London Mathematical Society (2), 55, no.3, 473-488.

[97,R1] J.W.Reyn, *Phase portraits of quadratic systems with finite multiplicity one,* Nonlinear Analysis, Theory, Methods and Applications, Vol. 28, no.4, 755-778.

[97,R2] C.Rousseau, *Hilbert's 16th problem for quadratic vector fields and cyclicity of graphics,* (English summary), Proceedings of the Second World Congress of Nonlinear Analysis, Part 1 (Athens, 1966), Nonlinear Analysis, 30, (1997), no.1, 437-445.

[97,RH] J.W.Reyn and Huang Xianhua, *Phase portraits of quadratic systems with finite multiplicity three and a degenerate critical point at infinity,* Rocky Mountain Journal of Mathematics, 27, no.3, 929-978.

[97,RK] J.W.Reyn and R.E.Kooij, *Phase portraits of non-degenerate quad-*

ratic systems with finite multiplicity two, Symposium on Planar Nonlinear Dynamical Systems (Delft 1995), Differential Equations and Dynamical Systems, Vol.5,(1997), 3-4, 355-414.

[97,T] L.A.Timochouk, *Constructive algebraic methods for some problems in non-linear analysis,* Thesis, Delft University of Technology, 1997, 1-181.

[97,VV] M.Voldman and N.I.Vulpe, *Real quadratic systems with two imaginary invariant straight lines,* Buletinul Acad. de Stintje a Rep. Moldova, 1(29),31-36.

[97,Y1] Ye Yanqian, *Qualitative theory of the quadratic differential systems, I,* (English summary), Ann. of Diff. Eqs. 13, no.4, 395-407.

[97,Y2] Ye Yanqian, *Limit cycles and bifurcation curves for the quadratic differential system $III_{m=0}$ having three antisaddles, II,* (English summary),

[98,A] E.A.Andronova, *Partition of the parameter space of a quadratic system with a triple focus or a center at the origin,* (Russian, Russian summary), Diff. Uravn. 34, no.4, 441-450. (Translation in Differential Equations 34, no.4, 437-446.)

[98,AKL] J.C.Artes, R.E.Kooij and J.Llibre,*Structurally stable quadratic vector fields,* Memoirs of the American Mathematical Society, Vol. 134, no.639, viii+108 pp.

[98,CL] J.Chavarriga and J.Llibre, *On the algebraic limit cycles of quadratic systems,* Proceedings of the IV Catalan Days of Applied Mathematics (Tarragona, 1998), 17-24, Univ. Rovira Virgili, Tarragona, 1998.

[98,KV] Yu.F.Kalin and N.I.Vulpe, *Affine-invariant conditions for the topologically discrimination of quadratic Hamiltonian differential systems,* (Russian, Russian summary), Diff. Uravn. 34, 298-302, 428.

[98,KZ] R.E.Kooij and A.Zegeling,*Limit cycles in quadratic systems with a weak focus and a strong focus,* Kyungpook Math.Journ.38,no.2, 323-340.

[98,VV] A.Voldman and N.Vulpe, *Affine invariant conditions for topologically distinguishing quadratic systems with $m_f = 1$,* Nonlinear Analysis, Theory, Methods and Applications, Vol. 31, no. 1/2, 171-179.

[98,Y] Ye Yanqian, *Qualitative theory of the quadratic differential systems, II, Ergodicity of limit cycles,* (English summary), Ann. Differential Equations 14, no.2, 392-401.

[98,W] Wang Xuejin,*A class of quadratic differential systems with a fine focus of order two,* (Chinese, English and Chinese summaries), Acta Math. Sinica 41, no.2, 399-404.

[98,Z] Zhang Xiang, *Limit cycles and bifurcation phenomena for quadratic differential systems of type $III_{n=0}$,* (Chinese, Chinese summary), Chinese Ann.

332

Math. Ser. A 19, no.2, 211-220.

[99,Z1] Zhang Pingguang, *Uniqueness of the limit cycles of quadratic systems with a second order weak focus,*(Chinese),Acta Math.Sinica 41,no.2,289-304.

[99,Z2] Zhang Pingguang, *Quadratic systems with a weak focus and a strong focus,* (English summary), (A Chinese summary appears in Gaoxiao Yingyong Shuxue Xuebao Ser. A 14 (1999), no.1, 7-14) Appl. Math. J. Chinese Univ. Ser. B 14, no.1, 7-14.

[99,Z3] Zhang Pingguang, *Quadratic systems with weakly paired infinite hyperbolic saddles,* (Chinese, English and Chinese summaries), Acta Math. Sinica 42, no.1, 175-180.

[99,Z4] Zhang Xiang, *Bifurcation and distribution of limit cycles for quadratic differential systems of type II,*(English summary), J.Math. Anal. Appl. 233, no.2, 508-534.

[99,Z5] Zhang Pingguang, *Uniqueness of the limit cycles of quadratic systems with a second order weak focus,* (Chinese, Chinese and English summary), Acta Math. Sinica 42, no.2, 289-304.

[99,ZK] A.Zegeling and R.E.Kooij, *The distribution of the limit cycles in quadratic systems with four finite singularities,* Journal of Differential Equations, 151, no.2, 373-385.

[00,DHP] F.Dumortier, C.Herssens and L.Perko, *Local bifurcations and a survey of bounded quadratic systems,* (English, English summary), Journal of Differential Equations,(2000),Vol. 165, no.2, 430-467.

[00,G] V.A.Gaiko, *"Global bifurcations of limit cycles and the sixteenth problem of Hilbert",* (Russian), At the University Minsk, 2000, 1-167.

[01,Z1] Zhang Xiang, *Ergodicity of limit cycles in quadratic systems,* Nonlinear Analysis, 44, (2001), 1-19.

[01,Z2] Zhang Pingguang, *On the distribution and number of limit cycles for quadratic systems with two foci,* (Chinese), Acta Math. Sinica 44 (2001), no.1, 37-44.

[01,Z3] Zhang Pingguang and Zhao Shenqi, *On the number of limit cycles for quadratic systems with a weak focus and a strong focus,* Appl. Math. J. Chin. Univ. Ser. B 16, (2001), no.2, 127-132.

[02,R] J.W.Reyn, *Non-existence of limit cycles for a quadratic system in class II,* Nonlinear Analysis, Theory, Methods and Applications, 50, (2002), 323-331.

Index